Food Gardens for a Changing World

Food Gardens for a Changing World

Daniela Soleri, David A. Cleveland and
Steven E. Smith

CABI is a trading name of CAB International

CABI	CABI
Nosworthy Way	745 Atlantic Avenue
Wallingford	8th Floor
Oxfordshire OX10 8DE	Boston, MA 02111
UK	USA

Tel: +44 (0)1491 832111
Fax: +44 (0)1491 833508
E-mail: info@cabi.org
Website: www.cabi.org

Tel: +1 (617)682-9015
E-mail: cabi-nao@cabi.org

A catalogue record for this book is available from the British Library, London, UK.

Library of Congress Cataloging-in-Publication Data

Names: Soleri, Daniela, author.
Title: Food gardens for a changing world / by Daniela Soleri, David A.
    Cleveland, Steven E. Smith.
Description: Boston, MA : CABI, [2019] | Includes bibliographical references
    and index.
Identifiers: LCCN 2018049624 (print) | LCCN 2018057522 (ebook) | ISBN
    9781789241006 (ePDF) | ISBN 9781789241013 (ePub) | ISBN 9781789240986 (hbk
    : alk. paper) | ISBN 9781789240993 (pbk)
Subjects: LCSH: Food crops. | Kitchen gardens. | Gardening.
Classification: LCC SB175 (ebook) | LCC SB175 .S628 2019 (print) | DDC
    630--dc23
LC record available at https://lccn.loc.gov/2018049624

ISBN-13: 9781789240986 (hardback)
        9781789240993 (paperback)
        9781789241006 (ePDF)
        9781789241013 (ePub)

Commissioning editor: Rebecca Stubbs
Editorial assistants: Tabitha Jay and Emma McCann
Production editor: James Bishop
Illustrations by Daniela Soleri

Typeset by SPi, Pondicherry, India
Printed and bound in the UK by Bell and Bain Ltd, Glasgow

Contents

Acknowledgments

We are grateful to the gardeners and farmers who, with great patience and good humor have spoken with us, explaining their experiences, observations, knowledge and hopes.

Our ability to access the peer reviewed scientific literature—either through open access publications, or as members of public university communities—was essential for educating ourselves and writing this book. Expanding public access to these resources, and encouraging scientists to communicate in ways that are more public-friendly, are essential for an engaged, informed, and more equitable global community within and beyond food gardens.

For in-depth comments we thank Elinor Halström (RISE, Research Institute of Sweden, Chapter 1), Oliver Chadwick (UCSB, Chapter 7), Deborah Letourneau (UC Santa Cruz, Chapter 9), and Norman Ellstrand (UC Riverside, Chapter 10), along with Gretchen LeBuhn (San Francisco State U, an early version of all chapters). We thank our colleague and friend Tom Orum for working with us as lead author on Chapter 9, thus improving it significantly.

For comments on various parts of the book we also thank Jordan Clark (UCSB), Michael Crimmins (U of Arizona), Gerri French (Sansum Clinic), Elisabeth Garcia (UC Davis), Mike Ottman (U of Arizona), David Pellow (UCSB), and Debra Perrone (UCSB); and for help with research, Emilie Wood (UCSB).

We thank the Environmental Studies Program at UCSB for logistical support.

At CABI, James Bishop, Tabitha Jay, Emma McCann, Rachael Russell, and Rebecca Stubbs have been a pleasure to work with—their efficiency and support made a daunting process possible.

Introduction

D. SOLERI, D.A. CLEVELAND, S.E. SMITH

In mid September 2017, the fall equinox was approaching, and things were different in our garden. The heat-loving basil plants that should have been slowing down as the days shorten and cool, were showing no sign of changing. The fruit on both varieties of our persimmon trees were turning deep orange, with some even dropping off, at least one month earlier than in previous years. The unusually warm, dry summer of 2017 in much of the western US contributed to similar experiences for many gardeners. That fall was an example of how the timing and duration of plant life cycles, and our garden activities, are changing from what we are familiar with, and like other gardeners and farmers, we need to figure out how to respond.

Farmers in many parts of the world whose livelihoods depend on their harvests have been noticing changes like these for quite a while. When DS was interviewing maize (corn) farmers in Oaxaca, Mexico, a few years ago, many commented on changes in the *canícula*, the short drought that typically comes during the summer rainy season. Instead of lasting a couple of weeks as it had in the past, the *canícula* is continuing into the late summer and early fall, when maize kernels typically fill out. Some years the *canícula* becomes the end of the summer rains, not just a temporary dry spell, and when this happens the result is reduced or failed harvests. Farmers are responding by looking for seed of short-cycle maize varieties from other farmers in their own or neighboring communities. These are varieties ready to harvest in about 90 days instead of more than 120, producing an earlier, although smaller, harvest, which is still better than no harvest. Sorghum farmers in southern Mali have been doing the same for years—seeking out shorter-cycle varieties in response to drier weather (Lacy et al., 2006). In the mountains of Nepal, farmers are observing significant changes, including warmer temperatures, dried up water resources, earlier flowering crops, and new crop pests, all of which

have also been seen by scientists (Chaudhary and Bawa, 2011).

These changes are different than the year-to-year variation that is so familiar to gardeners and farmers. Changes are a natural part of all life, and many are the cyclical kind we are accustomed to, like the changes in rainfall, sunlight, and temperature through the seasons, and the rise and fall of insect populations. Other changes gardeners face are more local, or personal, like elderly parents joining the household, or a new building next door blocking the morning sun. But the changes many of us are experiencing today are different, moving us into new territory. These changes are part of larger trends that are bringing new conditions not encountered before, with challenges that are making it more difficult for many people to live good lives, and harder to grow food gardens—*global trends with local consequences*. These trends include increases in: warming, variability of precipitation and other climate changes, resource scarcity, environmental degradation, loss of biodiversity, economic inequity, proportion of older people in our communities, cultural diversity, social conflict, psychological distress, and diseases related to lifestyle and diet.

Gardeners and farmers are also changing. During our work with farmers in southern Mexico, and Zuni and Hopi farmers in the US, we've seen that many young and middle-aged farmers are having a tough time. Because of the long hours they spend on nine-to-five jobs away from their farms and gardens to earn wage income, they don't have the time their parents did for close observation of the dynamic conditions in their gardens and fields. As a result, some today lack the depth of understanding necessary for successfully growing food in their difficult environments, especially with the trends that are bringing new challenges. They may know about the methods used by their grandparents, but without experience and understanding these become recipes

---

to follow, rather than adaptable concepts, and their ability to successfully respond to change declines.

Most of us are part-time gardeners and city dwellers, and many of us are new to gardening. Even more than the farmers just described, we often don't have sufficient understanding to be able to respond successfully to changes, or know how to acquire this understanding. And like those farmers, many of us want to understand how our individual lives affect other people and the Earth, and want to manage our gardens in ways that are positive for people and other forms of life, communities, and ecosystems, now and in the future. That is, we want our gardening to achieve our goals and be consistent with our values. This is important because while food gardens may seem very small in the scheme of things, many practices and processes that are a part of gardening, from how we compost to how we treat each other and work together, are relevant at much larger scales, far beyond the boundaries of our gardens—*local actions have global consequences*. Food gardens can be a place where household, community and school gardeners work together to learn and practice direct, positive responses to the changes we are all experiencing. And the cumulative impact of many gardeners and gardens can be large.

There are lots of recipes for gardening available—lists of botanical pest repellants, crops to grow in different seasons, basics of soil management, techniques for saving water, how to organize a community garden. Most of these are just a few seconds away on the internet, in local public libraries, or available from local government offices. Recipes can be great; they are quick and handy, and we all use them. But recipes don't provide enough understanding for adjusting gardening practices to many of the specific situations we encounter, or to conditions mentioned above that we haven't experienced before. Recipes usually emphasize mechanical or technical approaches to gardening, and while these are often useful, they can get in the way of more holistic responses that include the institutional and behavioral approaches needed to respond adequately. For example, making efficient use of water in your garden is important, but advocating for changes in municipal water policy that reflect residents' concern for water conservation and equity is also needed.

We wrote *Food Gardens for a Changing World* (*FGCW*) for students and instructors looking for a resource that approaches food gardens as a subject of natural and social science. We also intend it to be a resource for gardeners who want to grow food gardens for reasons including enjoyment, health, flavor, cultural identity, or savings, and want to do so in ways that support healthy people, communities, and ecosystems. We hope to help readers ask questions and find answers for understanding food gardens' roles in a changing world, and for reaching their learning and food gardening goals—for themselves, and for the world around them. It will also be a good introduction for researchers to important food garden issues needing further study. Even though the audience for *FGCW* is diverse, for clarity and simplicity, our "voice" throughout addresses gardeners directly.

The details of scientific research methods and results we provide for many topics will not only be useful for students, instructors, and researchers, but we believe can strengthen gardeners' ability to understand and solve problems. We give many practical examples of how to make the best use of available information and apply the five key ideas outlined in this Introduction in order to reach goals for gardening. Larger school and community gardens, as well as supporting organizations and researchers, will find the information on scientific methods and results useful in their search for alternative management strategies, and in documenting their progress.

While direct experience and understanding may be lacking, many gardeners, especially in the *global north* (the so-called "industrial countries") are familiar with scientific explanations of concepts and problem solving. And so, while "book" knowledge can never replace the versatility of experiential knowledge, especially when your livelihood depends on it, familiarity with basic concepts is the foundation for creatively and effectively solving problems, and enjoying successful gardens.

FGCW focuses on the increasingly hot, dry regions of the global north, and though most of our examples are from the western US, this book is useful for students, teachers, food gardeners, activists, and researchers everywhere, especially in other dry, warm areas, because we emphasize basic concepts. We have also included some examples from the *global south* (the "third world"), where gardeners and small-scale farmers typically can't afford to use lots of external inputs. As a result, to address the same challenges they have developed ingenious practices using minimal inputs. These practices are invaluable resources for gardeners everywhere, and their ingenuity is often an example of locally appropriate problem solving, and a demonstration of the value of different forms of knowledge.

Our approach in *FGCW* is similar to that of our 1991 book, *Food from Dryland Gardens* (Cleveland and Soleri, 1991), which was written for extension and project people in the global south. Many of the topics covered in *FGCW* are new, while others are similar to those of the previous book, and sometimes we refer readers to *Food from Dryland Gardens*, which is freely available online.

I.1. Our Framework

Figuring out how to garden starts with what you want to grow and eat, what you enjoy, and what makes sense for your location, schedule, and budget. But many of us hope our gardening will also strengthen our interconnection with each other, our environment, and the future, and support food, environmental, and social justice. For this to happen, we need to understand gardens as systems that are a part of their ecological and social contexts—not existing in isolation. This is the direction that work in food and agriculture has been going in, particularly in agroecology (Altieri, 1995; De Schutter, 2010; Vandermeer, 2011: 20–24; Gliessman, 2013, 2015; Mendez et al., 2016), and we build on and add to those efforts. We use "garden" in this book to mean small-scale food production by households, community groups, schools, or workplaces, and that includes vegetables, herbs, fruit trees, and sometimes flowers, growing in defined plots, as well as in less well defined locations, such as a tomato and basil plant in a pot on a balcony, a row of olive trees along the street, or a bitter melon plant in the corner of a courtyard.

To help understand food gardens in their ecological and social contexts, we use a framework of five key ideas from current research in the natural and social sciences. These ideas are essential for applying the food gardening concepts presented in *FGCW* to achieve positive, sustained outcomes. As we have already said, conceptual understanding is a powerful tool for adapting gardens to current conditions and future change, however, the best way to use scientific concepts is in combination with the *art of gardening*. Gardening is an art as well as a science because it requires integrating our understanding of the complexity of food gardens based on reading, empirical observations, and experiments, with our understanding from the intuitive interpretation of our experiences. It is a process of creative improvisation that pays off richly with discovery, and joy.

We illustrate the five key ideas in Fig. I.1, and describe them in more detail in the next two sections.

I.2. Key Ideas for Understanding Food Gardens in an Ecological Context

1) *Ecological thinking. Thinking of the entire lifecycle of biological and physical garden resources (e.g., water, seeds, and organic material) makes it easier to manage them in response to environmental, climate, and social change, while producing other benefits.* All the components of the garden—plants, insects, microorganisms, water, air, soil, compost—are connected, and they have a lifecycle—a history before they reach the garden, and a future after they leave it. For example, the fruits and vegetables we grow, harvest, and eat are made from the mineral nutrients in the soil and the compost we apply, water from irrigation and rain, carbon dioxide from the air, and energy from the sun. After we harvest them, they become part of our bodies, our body and food waste, and the *greenhouse gases* we emit—gases in the Earth's atmosphere that contribute to global warming. In other words, nothing disappears, it only changes its form and location. Every garden component has impacts throughout its lifecycle, which can be either positive or negative in terms of our values and gardening goals. Ecological thinking makes it easier to see how to manage gardens to maximize their positive impacts, both within and beyond their immediate boundaries. See Box I.1 for a fun way to stimulate ecological thinking. Ecological thinking also reminds us that, because everything is connected, a strategy that optimizes progress toward one goal may also reduce progress toward another—there will often be *tradeoffs* that require comparison of alternative strategies.

In formal, scientific studies, benefit:cost ratios are compared in order to understand tradeoffs and decide which strategy is best. For some community or home garden programs, formal *lifecycle assessments* that trace and quantify benefits and costs throughout the lifetime of a process, like composting, or a resource like water, can be very useful (Kulak et al., 2013). For most gardeners, formal studies won't be necessary, but ecological thinking in terms of recognizing entire lifecycles can help us meet our gardening goals. For example, your goals may include maximizing the nutrients returned to

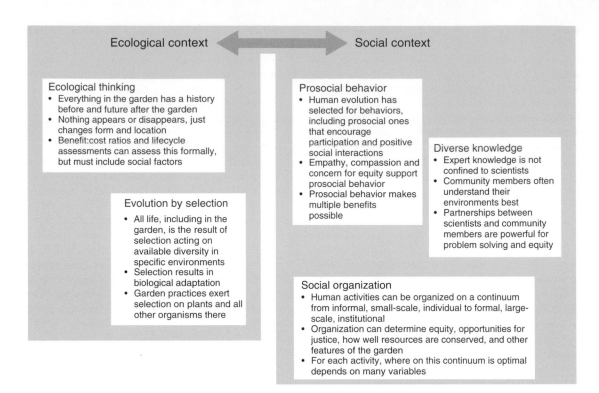

Figure I.1. Five key ideas for understanding food gardens in context

your garden from your food scraps and yard trimmings while minimizing greenhouse gas emissions. In this case, it's really helpful to realize that the carbon in organic material doesn't just disappear as your compost pile shrinks to a small percentage of its original size. Through a series of biochemical reactions it becomes part of microorganisms that use it for food, of non-living organic compounds in the soil that provide food for microorganisms and plants, and that sequester carbon, and of the greenhouse gases carbon dioxide and methane that are released into the air. The way you manage the

Box I.1. The Big Here Quiz

Our friend, the late anthropologist and naturalist Peter Warshall, created a Watershed Awareness Exercise in the 1970s (*CoEvolution Quarterly*, Winter 76–77), that was built on by others and republished in 1981 (Charles et al., 1981), and is now referred to as the "Big Here Quiz." It started as a short list of questions to stimulate thinking about your watershed (these examples are all taken from a revised, online edition posted by Kevin Kelly, 2005): Trace the water you drink from rainfall to your tap. How far do you have to travel before you reach a different watershed? Can you draw the boundaries of yours? When you flush, where do the solids go? What happens to the waste water?

The quiz was expanded over time to include other lifecycles that run through the place where you are. For example: How many days is the growing season here (from last frost to first frost)? Name three wild species that were not found here 500 years ago. Name one exotic species that has appeared here in the last five years. Who uses the paper/plastic you recycle from your neighborhood? Where does your electric power come from and how is it generated?

There are over 30 questions in the most recent version of this quiz. They can be surprisingly hard for many of us to answer, and are a great way to encourage discussion and ecological thinking.

compost pile, or where you send organic material for processing, will determine where that carbon ends up, and its net benefit and cost in terms of your goals—both in your garden, and beyond (Cleveland et al., 2017) (Chapter 7). While the benefit:cost approach is often applied only to economic and biophysical benefits and costs for an individual, limited group, or commercial business over the short term, we take a broader view in *FGCW*. We encourage an ecological, evolutionary, and prosocial framework that includes both benefit and cost in terms of the community and the environment, from the local to the global, and into the future.

2) *Evolution by selection. All living organisms are the result of evolution by selection, and how we garden continues this process by favoring some crop plants and other living organisms in the garden over others.* All life on Earth, including in gardens, developed through biological evolution. Life has been formed over time, primarily by selection acting on available diversity, and is often adapted to specific environments. The theory of evolution by natural selection was most famously outlined in the *Origin of Species* by Charles Darwin in 1859 (Darwin, [1859] 1967), and its core ideas have been tested and confirmed by many, many biologists since then. It is the foundation of our understanding of all life on Earth. When humans do the selecting it's called artificial selection, and a major inspiration for Darwin's ideas about natural selection was his observation of the changes farmers and gardeners made in plant varieties and animal breeds by selecting desired characteristics from one generation to the next (Darwin, [1883] 1868a, [1883] 1868b).

Because crop plants have largely evolved under selection by humans, they are uniquely adapted to environments altered by us, and to our selection. That means that most crops need to be cared for, including removing weeds that compete with them for resources, protecting them from pests and diseases, and supplying water and nutrients. This is one reason why "natural" gardening in a strict sense is an oxymoron. Evolution by selection also means that an important strategy for efficient food gardening—requiring fewer inputs, and with fewer negative impacts—is matching the adaptations of garden crops with the garden environment. For example, some varieties of crops like common bean, that evolved in the subtropics, may not receive enough hours of darkness each day to produce flowers and seeds in temperate zone summers (Section 5.8.1),

and so are not worth trying to grow in those areas. New selection pressures can also affect how well garden crops are adapted to a location. For example, as new pests like the bagrada bug (*Bagrada hilaris*) have arrived in southern California where they attack crops in the cabbage family, growing these crops may require more and more time and other inputs—the benefit:cost ratio may get smaller because those crops are poorly adapted to the new pest. As this pest population rises, growing more resistant leafy vegetable crops like spinach or leaf amaranth may have a larger benefit:cost ratio (Chapter 9).

Evolution by selection also means that the way you manage the garden creates selective pressures that act on the plants and other organisms in the garden. For example, if you minimize the amount of water you give open-pollinated crop varieties, and save seed for the next season from the most productive plants, you are likely exerting selection pressure. Over time, this may result in varieties evolving to be more resistant to water stress (Chapters 5 and 10). Humans can also drive evolution unintentionally—for example, using powerful synthetic herbicides or pesticides has resulted in the evolution of weed and insect pests resistant to these agrochemicals (Chapter 9).

I.3. Key Ideas for Understanding Food Gardens in a Social Context

3) *Prosocial behavior. Among humans' wide range of potential behaviors, prosocial ones are necessary for successful gardens that include management of resources and access to benefits in ways that support social justice.* Evolution has selected for a spectrum of human behaviors, from prosocial ones based on empathy, compassion, and a sense of equity, to antisocial ones based on selfishness, greed, and a sense of superiority (Sapolsky, 2017). The key to prosocial behavior is *empathy*, the capacity to understand another's thoughts and feelings, and even experience them. *Compassion*, the desire to relieve suffering, builds on empathy, enhancing our ability to feel caring and warm toward others, and ourselves. One reason empathy and compassion are so important for prosocial interactions is because they are necessary for our assessment of the *equity*, or lack of it, experienced by others (Grohn et al., 2014). Like empathy, concern for equity, or fairness, and especially justice, has been documented in many social animals,

including humans (De Waal, 2013). Recent studies have shown that individuals not only demand equity for themselves, but also for others, even if it costs them (Dawes et al., 2007, 2012).

There is evidence that inequity within societies harms everyone's health and happiness, and exacerbates our divisions (Chapter 1). For the poor and discriminated against, this is in part due to lack of basic resources such as shelter, healthy food, education, employment opportunities, and social networks. But the harm is also due to the subjective experience of being in lower social ranks (Marmot and Allen, 2014); the mental and physical health of poor people suffers as a result of being "made to feel poor" (Sapolsky, 2005). In the US, the harm of feeling poor exacerbates the profound burden of centuries of violence, abuse, and inequity experienced especially by Native and African American communities, in what is now called "historical trauma" (Brave Heart, 2003). But as members of society and inhabitants of Earth, the privileged also suffer because of society's inability to take action that benefits everyone, like mitigating climate change, reducing air pollution, or creating community gardens.

Empathy, compassion, and a sense of equity can be extended in space to people beyond those immediately around us. These feelings, and the behaviors they give rise to, are important for society, including the success of gardens, especially community gardens where individual actions can have an immediate effect on other gardeners. For example, not participating in group work days lays an extra burden on others, and creates feelings of mistrust, resentment, and exploitation, and a reluctance to cooperate in the future. Empathy, compassion, and equity can also be extended in time to future generations. For example, trying to use water more efficiently in your garden plot, even though you are not directly paying for it, can be an expression of compassion for future generations who will also need water. Consumption now that ignores our impact on the future for humans and all life results in what some call "intergenerational inequity" (Scott, 2013). Ultimately, and at larger scales, negative feelings—a lack of trust, empathy, and equity—corrode civic engagement and undermine the possibility of a just society.

4) *Social organization. How we organize ourselves, including for food gardening, affects our ability to reach the prosocial goals we have for our behavior, our communities, and our environment and resources, including the climate. The best scale of organization will depend on our specific goals and the methods we are using.* How humans and other social animals behave is strongly influenced by our social, institutional, and physical environments. This means that the way a garden is situated in a community, and how relations are organized among gardeners, and between gardeners and their neighbors, will affect what behaviors are favored or even possible. For example, when resources like water in a community garden are limited, access on a first-come, first-served basis puts gardeners in competition with one another, and encourages selfishness and overuse, even when people have better intentions. Organized this way, there is no mechanism for water use that would benefit more gardeners for a longer time, and would cause less social friction. In contrast, gardeners might say, "Let's figure out and agree on water use rules that minimize conflict, and would be fair for everyone in the garden, and also fair to other people in this watershed because we all need water now and in the future." That's a more cooperative, equitable way of organizing in response to the same problem (Section 3.3).

Human activities are organized on a continuous scale. Reaching our garden goals depends on finding the most effective way to organize activities, from the informal, small-scale and individual, to formal organization using large-scale institutions. Where on the continuum is optimal will depend on the goal, the activity, the context, the people involved, and the values being supported. With the same number of people and the same resources, how people organize themselves can make the difference between inequity and rapid resource depletion, or equity and long-term resource availability. Of course, different social organizations often involve tradeoffs, and the same benefit:cost analysis we described for ecological thinking may be necessary.

For example, if a goal is reducing greenhouse gas emissions, then you could compare the net emissions from home or community garden composting with emissions from sending organic waste to a municipal landfill or composting operation (Cleveland et al., 2017). Or, if a goal of your community garden is improving access to vegetables in the community, you could compare the net benefit of dividing the whole garden into individual plots for community members, with gardening a portion of the area in common and donating the harvest to a neighborhood center for distribution.

One way of organizing the use of resources that can encourage prosocial behavior and greater equity is common property management, an alternative to the dominant forms of management: private property or central control—for example by governments—and open access, where there are no rules governing access (Ostrom, 2010). Common property institutions have been developed by indigenous and local communities around the world, and are increasingly popular in industrialized countries (Section 3.3).

5) *Diverse knowledge. Local gardeners often understand many aspects of their physical and social environments better than scientists, and scientists can collaborate with them in research that supports gardeners' own goals for their gardens, including their values such as climate and food justice.* Many people, not only scientists, are competent observers of the world, and some have expertise as a result. While formal scientific knowledge is a powerful and effective tool for understanding some aspects of the world, it is often community members who know many characteristics of their local environments best. Curious, observant gardeners are community scientists, called "citizen" scientists by some, and their experience-based knowledge is central to asking questions and developing responses that best serve their goals (Ramirez-Andreotta et al., 2015). For example, farmers and gardeners know the local varieties of their crops and how those respond to their field and garden environments better than anyone; and neighborhood residents are often the only ones who know about past and present conditions, like toxic waste dumping, recurrent noxious emissions, neighborhood conflicts and priorities, cultural identity, and dietary preferences, all of which may have a huge effect on people's interest in food gardens. Gardeners can also make very accurate observations in their gardens, for example about the effects of changes in the frequency of irrigation or the amount of compost applied, or cycles of insect pest populations.

Research has shown that without genuine respect for different people's knowledge there is no hope for meaningful partnerships and lasting positive social change. While both local and scientific knowledge can include inaccurate or biased assumptions, collaboration between gardeners and formal scientists, built on mutual respect, can provide the motivation and capacity to test assumptions and discard those that aren't supported (Soleri and Cleveland, 2017). Combining local knowledge and formal scientific knowledge is a powerful approach for reaching gardeners' own goals in ways that make sense for them and the local environment (Box 3.1).

I.4. How this Book is Organized, and how to Use it

In the rest of *FGCW* we apply the five key ideas outlined above (Fig I.1) in discussing the basic concepts of food gardening. Part I, "Starting at the beginning: gardens and the big picture," takes a broad perspective in answering some central questions about the possible benefits of food gardens (Chapter 1), and the significant challenges facing gardeners as a result of environmental, climate, and social changes (Chapter 2). Chapter 3 outlines approaches for gardeners to respond to those changes in ways consistent with their goals and values.

Although we recommend reading Part I first, you can skip directly to Parts II and III if you want to start applying concepts in your garden right away. Part II, "Starting the garden," has information and methods for the siting and layout of food gardens (Chapter 4), basic plant biology (Chapter 5), and plant management and propagation (Chapter 6), all in light of the changes we are experiencing. Part III, "Garden management," focuses on managing soils (Chapter 7), water (Chapter 8), other organisms including pests and pathogens (Chapter 9), and saving and sharing seeds (Chapter 10).

At the start of each chapter we present the main ideas "in a nutshell." At the end of each chapter we provide some resources where more detailed information can be found online, in libraries, from other gardeners, Cooperative Extension workers, and educators. The appendices have worked examples (Appendix 3A) and other details (Appendix 1A) for readers who want to go deeper into a topic. At the end of every chapter is a list of references cited in the text. Hand-drawn graphs illustrate general concepts supported by data, while computer-generated graphs are used for specific data sets.

We hope *FGCW* will be useful for understanding the role of food gardens in a changing world, and help you enjoy your food garden, and engage with ideas and each other in exploring and developing your own effective strategies for gardening in ways that are better for you, your community, and for communities around the world.

I.5. References

Altieri, M. A. (1995) *Agroecology: The science of sustainable agriculture,* revised edn. Westview Press and IT Publications, Boulder, CO and London.

Brave Heart, M. Y. H. (2003) The historical trauma response among natives and its relationship with substance abuse: A Lakota illustration. *Journal of Psychoactive Drugs*, 35, 7–13, DOI: 10.1080/02791072.2003.10399988.

Charles, L., Dodge, J., Milliman, L. & Stockley, V. (1981) Where you at? A bioregional quiz. *CoEvolution Quarterly*, 32, 1.

Chaudhary, P. & Bawa, K. S. (2011) Local perceptions of climate change validated by scientific evidence in the Himalayas. *Biology Letters*, available at: http://rsbl.royalsocietypublishing.org/content/early/2011/04/16/rsbl.2011.0269, DOI:10.1098/rsbl.2011.0269.

Cleveland, D. A. & Soleri, D. (1991) *Food from dryland gardens: An ecological, nutritional, and social approach to small-scale household food production*. Center for People, Food and Environment (with UNICEF), Tucson, AZ. https://tinyurl.com/FFDG-1991

Cleveland, D. A., Phares, N., Nightingale, K. D., Weatherby, R. L., Radis, W., Ballard, J., Campagna, M., Kurtz, D., Livingston, K., Riechers, G., et al. (2017) The potential for urban household vegetable gardens to reduce greenhouse gas emissions. *Landscape and Urban Planning*, 157, 365–374, DOI: http://dx.doi.org/10.1016/j.landurbplan.2016.07.008.

Darwin, C. ([1883] 1868a) *The variation of animals and plants under domestication*, Vol. 2, 2nd revised edn. Johns Hopkins University Press, Baltimore, MA.

Darwin, C. ([1883] 1868b) *The variation of animals and plants under domestication*, Vol. 1, 2nd, revised edn. Johns Hopkins University Press, Baltimore, MA.

Darwin, C. ([1859] 1967) *On the origin of species by means of natural selection*, 1st facsimile edition. John Murray, London, Athenum, New York, NY.

Dawes, C. T., Fowler, J. H., Johnson, T., McElreath, R. & Smirnov, O. (2007) Egalitarian motives in humans. *Nature*, 446, 794–796.

Dawes, C.T., Loewen, P. J., Schreiber, D., Simmons, A. N., Flagan, T., McElreath, R., Bokemper, S. E., Fowler, J. H. & Paulus, M. P. (2012) Neural basis of egalitarian behavior. *Proceedings of the National Academy of Sciences*, 109, 6479–6483, DOI: 10.1073/pnas.1118653109.

De Schutter, O. (2010) *Agroecology and the right to food. Report submitted by the Special Rapporteur on the Right to Food*, Human Rights Council, United Nations General Assembly, available at: http://www.srfood.org/images/stories/pdf/officialreports/20110308_a-hrc-16-49_agroecology_en.pdf (accessed Oct. 4, 2018)

De Waal, F. (2013) *The bonobo and the atheist: In search of humanism among the primates*. W. W. Norton & Company, New York, NY.

Gliessman, S. (2013) Agroecology: Growing the roots of resistance. *Agroecology and Sustainable Food Systems*, 37, 19–31, DOI: 10.1080/10440046.2012.736927.

Gliessman, S. R. (2015) *Agroecology: The ecology of sustainable food systems*, 3rd edn. CRC Press, Taylor & Francis Group, Boca Raton, FL.

Grohn, J., Huck, S. & Valasek, J. M. (2014) A note on empathy in games. *Journal of Economic Behavior & Organization*, DOI: http://dx.doi.org/10.1016/j.jebo.2014.01.008.

Kelly, K. (2005) Cool tools: The big here quiz, available at: http://kk.org/cooltools/the-big-here-qu/ (accessed Oct. 11, 2017).

Kulak, M., Graves, A. & Chatterton, J. (2013) Reducing greenhouse gas emissions with urban agriculture: A life cycle assessment perspective. *Landscape and Urban Planning*, 111, 68–78, DOI: 10.1016/j.landurbplan.2012.11.007.

Lacy, S., Cleveland, D. A. & Soleri, D. (2006) Farmer choice of sorghum varieties in southern Mali. *Human Ecology*, 34, 331–353, DOI: 10.1007/s10745-006-9021-5.

Marmot, M. & Allen, J. J. (2014) Social determinants of health equity. *American Journal of Public Health*, 104, S517–S519. DOI: 10.2105/AJPH.2014.302200.

Mendez, V. E., Bacon, C. M. & Cohen, R. (2016) Introduction: Agroecology as a transdisciplinary, participatory and action-oriented approach. In Mendez, V. E., Bacon, C. M., Cohen, R. & Gliessman, S. R. (eds) *Agroecology: A transdisciplinary, participatory and action-oriented approach*, 1–21. CRC Press, Taylor & Francis Group, Boca Raton, FL.

Ostrom, E. (2010) Beyond markets and states: Polycentric governance of complex economic systems. *American Economic Review*, 100, 641–672, DOI: 10.1257/aer.100.3.641.

Ramirez-Andreotta, M. D., Brusseau, M. L., Artiola, J., Maier, R. M. & Gandolfi, A. J. (2015) Building a co-created citizen science program with gardeners neighboring a superfund site: The Gardenroots case study. *International Public Health Journal*, 7.

Sapolsky, R. M. (2005) The influence of social hierarchy on primate health. *Science*, 308, 648–652.

Sapolsky, R. M. (2017) *Behave: The biology of humans at our best and worst*. Penguin Press, New York.

Scott, C. A. (2013) Electricity for groundwater use: Constraints and opportunities for adaptive response to climate change. *Environmental Research Letters*, 8, 035005.

Soleri, D. & Cleveland, D. A. (2017) Investigating farmers' knowledge and practice regarding crop seeds: beware your assumptions! In Sillitoe, P. (ed.) *Indigenous knowledge: Enhancing its contribution to natural resources management*, 158–173. CAB International, Wallingford, UK.

Vandermeer, J. (2011) *The ecology of agroecosystems*. Jones & Bartlett Publishers, Sudbury, MA.

PART I STARTING AT THE BEGINNING: GARDENS AND THE BIG PICTURE

1

What can Food Gardens Contribute? Gardens and Wellbeing

D. SOLERI, D.A. CLEVELAND, S.E. SMITH

Chapter 1 in a nutshell.

- Eating fruits and vegetables from the garden can improve nutrition and health by providing compounds often lacking in the diet, replacing unhealthy, empty-calorie foods, and making foods more flavorful and meaningful.
- Community gardeners eat more fruits and vegetables than their neighbors who do not garden.
- Gardening can provide regular, enjoyable physical activity that can improve health.
- Feeling productive and interacting with plants, nature, and other gardeners supports a positive attitude about yourself and others.
- Growing food in a garden can be more environmentally beneficial than growing the same food in conventional agriculture.
- Food gardens can contribute to ecologically healthy and beautiful environments—for example, by increasing water infiltration, soil quality, carbon sequestration, and shade, and by providing homes for pollinators, and culturally and personally meaningful plants.
- Many of the individual and environmental benefits of gardens also have a positive impact on communities and society.
- Economic benefits from gardens can include fewer food purchases, and income from the sale or trade of garden produce.
- Still, gardens are not always or automatically beneficial—they can be used to exploit people and communities.
- Transparent and participatory garden organizations and projects help to avoid situations that aren't in the best interest of gardeners, their communities, or society.

A freshly picked ear of corn, a bowl of sautéed greens and herbs, a ripe peach, a handful of sweet jujube fruits, a bunch of bright flowers—what food gardens can contribute seems obvious when we enjoy the harvest. But in addition to these pleasures, there is more and more evidence of other benefits that food gardens and gardening can provide. In fact, in most cases, gardens provide multiple benefits, and in so doing they contribute to our personal *wellbeing*—that is, our physical, material, social, and emotional health and happiness. Gardens can contribute to better diets, increased physical activity, and healthy weight, that along with not smoking and moderate alcohol consumption, are estimated to have the potential to increase lifespans by about 20 years compared with people whose lifestyles do not include those practices (Li et al., 2018). Gardens can also contribute to environmental health that benefits individuals, communities, and the Earth. Food gardens can do this because, like many small-scale farms, especially in the global south, they can produce not only food, but herbs, flowers and medicines, protect biodiversity, support positive ecological processes, and beautify the environment (IAASTD, 2009).

Sometimes changing our habits is the most effective, but most challenging way to obtain the benefits gardens can offer. We are often adrift in our thoughts, which distract us from the present, and are frequently unhelpful (Killingsworth and Gilbert, 2010) (Fig. 1.1), so practicing focusing our attention on what we're doing and where we are, and understanding the possible benefits of doing this, can be very effective for changing some behaviors (Box 1.1). Attention and understanding can support positive behavior changes in diets, amount of physical activity, and social interactions that decrease *non-communicable diseases* (NCDs—non-infectious diseases that are often chronic, and progress over time). And gardening can contribute to increasing those positive behaviors. For example, it can be motivating to know that spending 15 minutes to walk to your plot in the community garden

Figure 1.1. Adrift in our thoughts

and working there for 30 minutes burns over 150 calories and can provide physical, psychological, and emotional benefits, whereas perusing the internet for the same amount of time burns only half the calories and has no health benefit. However, changing habits is not a simple matter of awareness and willpower. Many factors including poverty, racism, historical trauma, and environmental contamination

---

Box 1.1. Mindfulness and behavior change

Mindfulness is one of several approaches that people can use to help make positives changes in their behavior by changing the way they think (Hayes et al., 2011). The idea is to bring your mind to focus attention on the present and be able to observe your experiences without judging them. Judgment tends to shift the mind away from the present, tangling it up in explanatory stories that are not helpful, and are disconnected from reality. Careful observation of what is happening right now makes it easier to decide how we want to behave, rather than just following habitual patterns. Mindfulness has been shown to help people change behaviors such as unhealthy eating (Mason et al., 2016). It can also be calming, and interrupt negative thought patterns, helping people manage pain, anxiety, and depression, improving overall wellbeing (Hayes et al., 2011), and the way we interact with each other.

There are many practices for learning mindfulness. Here are two common ones. Focus on the breath: for five minutes sit quietly and attentively and focus on your breath coming in and out through your nose, notice the feeling and sound, count five complete breaths before starting the counting again—it can be very challenging to stay focused even for five breaths but this improves quickly with practice. Eat a raisin: sitting quietly and attentively, slowly eat one raisin, taking as long as possible, exploring it in your mouth, focusing on the taste, texture, smell, and any other sensations.

Working in the garden can also be an excellent way to practice focused attention, as documented in Wendy Johnson's memoir about the garden at Green Gulch Zen Center (Johnson, 2008): "Working in the garden is also meditation, though not in the conventional sense of calming down, moving slowly and deliberately, and dwelling in stillness. On the contrary, I am often most alert and settled in the garden when I am working hard, hip-deep in a succulent snarl of spring weeds."

---

undermine efforts to change, and are why gardens alone may not be adequate or even appropriate, when structural, social and policy changes are required (Section 1.6).

Every benefit is accompanied by costs, so we can think in terms of the benefit:cost ratio and the *net benefit* that occurs when that ratio is greater than one. There are many kinds of benefits and costs including financial, psychological, environmental, nutritional, and social. Whether something is a benefit or cost depends on how it is defined, based on subjective values and objective measurements, including how you define the boundaries of benefits and costs in time and space. Benefits can be multifaceted, and can also be intertwined with costs, and completely separating them may not be possible, or worthwhile. For example, as we discuss below in Section 1.5, time spent gardening can be assessed as a cost of gardening, or as a beneficial physical activity, which is very difficult to separate from your personal psychological benefits, and even the social benefits for you and your community. This does not mean benefit:cost assessments should not be made, but the multifunctional nature of gardens and gardening needs to be taken into account, including the importance to different people of different benefits and costs. In this chapter we outline some of the potential benefits of food gardens, particularly those that have been documented by scientific studies. However, lack of research evidence of benefits does not necessarily mean they don't exist. Because gardens have often been thought of as relatively unimportant, they have been overlooked for a long time and so there are not many systematic studies of gardens and their impacts, although this is changing.

Identifying and documenting benefits and costs is especially important in making the case for funding and public support of community and home gardens when they are competing with alternative uses for land, water, and time, as has been true, for example, since at least the late nineteenth century in the US (Lawson, 2005). Every situation is different, and we can't assume that benefits will or will not occur. Despite their many potential benefits, food gardens are not magical and should never be thought of or promoted as a cure-all. Sometimes benefits found in formal, controlled studies may not occur on the ground in real gardens, or vice versa, and sometimes food gardens are not the best response to a need or problem. As we emphasize throughout this book, keeping the framework described in the Introduction in mind, and being a good observer, will help when trying to figure out what contributions gardens can, and cannot, make in your situation. This chapter covers concepts important for assessing the value of existing or planned gardens, and for increasing their benefits.

1.1. Diet and Nutrition

The full effect of individual foods on nutrition and health is hard to detect, because it's difficult to separate the effects of foods from the meals and diets they are part of, from the variation in nutrients they contain, and from other factors that affect nutrition and health. For example, in the US, many vegetables are eaten in combination with foods containing lots of calories and sodium, and in forms with dietary fiber removed (Guthrie and Lin, 2014). But at the same time, an increasing number of studies are showing that eating diets high in fruits, vegetables, and whole grains, and low in meat, refined grains, added sugars, and salt—especially when all of these are done together—is associated with a lower risk of acquiring NCDs and a greater likelihood of living longer (Scarborough et al., 2012; Aune et al., 2013, 2017; Esselstyn et al., 2014; Hu et al., 2014; Tilman and Clark, 2014). "Fruit" and "vegetable" can be confusing terms because botanically, "fruit" is a seed-bearing structure and "vegetable" is not used, while in culinary usage, "vegetable" often refers to all edible plant parts, and in popular use the two terms frequently reflect what is thought of as sweet vs. savory, which often don't align with botanical definitions (IARC, 2003, 1–21). In *FGCW* we use these terms in the popular sense, for example when discussing diets, but use the more precise botanical definitions in Chapters 5, 6 and 10.

NCDs include many types of cancers, heart disease, type 2 diabetes (previously referred to as adult onset diabetes mellitus), and hypertension (high blood pressure). The *metabolic syndrome* is the name given to the combination of risk factors for heart disease, stroke, and type 2 diabetes, including a large waistline due to fat storage in the abdomen, high fasting blood sugar levels, and abnormal levels of fat in the blood (Bremer et al., 2012). The metabolic syndrome and NCDs are increasingly common among people living in societies that have experienced a nutrition transition to more unhealthy diets (Box 1.4).

While the details are often controversial, there now appears to be overwhelming evidence that a

What Can Food Gardens Contribute? Gardens and Wellbeing

13

diet containing lots of minimally processed plant foods is important for good health at all ages: "The case that we should, indeed, eat true food, mostly plants, is all but incontrovertible" (Katz and Meller, 2014, 94). Yet, while plant rich, vegetarian, or completely plant-based (vegan) diets can be nutritionally adequate and protect against NCDs, they can also contain excess salt, sugar, and fat, and need some planning to avoid deficiencies, especially of vitamins B_{12}, D and K2, omega-3 fatty acids, calcium, iron, and zinc (Craig, 2010; Li, 2014).

In this section we focus on what food gardens can contribute to nutritious and delicious diets

that support healthy, active, long lives through: a) the nutrient and other content of garden foods; b) the replacement of less healthy foods with garden foods; and c) the flavors added by garden foods.

Many factors affect how nutritious our diets are, from the chemical composition of the food we eat, to the way the body processes it, and from the foods available in our garden and grocery store and how they are grown and processed, to the domestic and international economic, trade, and development policies that determine food prices and availability (Box 1.2).

Box 1.2. Food environments

The reason many people don't eat enough fruits and vegetables is that a diet is not simply the result of individual choices and preferences. More than any other factor, it is our food environment that can make eating a healthy diet difficult.

The *food environment* includes the type, abundance, and quality of foods we are exposed to, how they are presented and advertised, their availability in terms of price and convenience, and how these are in turn influenced by socioeconomic, political, and cultural factors (Nestle, 2002; Drewnowski, 2009; Nestle and Nesheim, 2012). The influence of socioeconomic and political agendas on our food environment and diets is not new. There is evidence that one of the earliest tasks of nutrition science in the US was a political one: identifying the lowest cost, nutritionally adequate diets for the working class that could maintain their capacity for work without raising their wages (summarized in Drewnowski, 2010; see also Levenstein, 1996).

Nutritional problems resulting from persuasive marketing of unhealthy foods were noticed decades ago and given the name *commerciogenic malnutrition* (Jelliffe, 1972). Although originally describing the consequence of promoting infant formulas to replace breast milk in Africa, today commerciogenic malnutrition aptly describes the epidemic of overweight and obesity in the world that is a direct result of eating so much highly processed, aggressively advertised fast food and sugary drinks. For example, in 2009, corporations spent $149 million USD on marketing foods in US schools, 90% of that for sweetened beverages (Harris and Fox, 2014), despite the ample evidence that those drinks are making young people overweight and sick (Malik et al., 2010).

At the center of the typical US food environment today are cheap, abundant processed foods manu-

factured from commodities such as maize (corn, as sweetener), wheat, soybean (as starch, fat), meat, and dairy. Current US policies give subsidies or tax relief to producers and manufacturers of many commodity crops or foods. The problem is prominent in the US but is not limited to that country. In 2012, industrialized countries subsidized beef and milk production with approximately $18.0 and $15.3 billion USD respectively (Heinrich Böll Foundation, 2014). In the US in 2015, soybean, livestock (meat), and maize producers received $1.6 billion, $1.3 billion, and $5.9 billion USD in direct subsidies, respectively (EWG, 2017). Commodity foods are cheap, and low income and minority populations in the US eat a disproportionately large amount of these foods, and have a higher level of cardiometabolic risk (Siegel et al., 2016).

There are powerful psychological and social reasons why we are eating so much unhealthy food as well, including "mindless" eating in response to loneliness, stress, boredom, and exposure to relentless advertising. Children and the poor are often targeted most aggressively in the media, including ads in schools, often with celebrities as spokespeople (Harris et al., 2015). One US study estimated that eliminating food advertising targeting children 6–12 years old could reduce the number that are obese by 14–33% (Veerman et al., 2009).

Low income communities and households are encouraged to choose unhealthy, empty-calorie, processed foods because these are cheaper and more available than healthier alternatives, and made tantalizingly palatable with lots of salt, sugar, and fat. As the evidence accumulates for the health benefits from whole-food diets based mostly on plants, this information is becoming more widely known through public

Continued

Box 1.2. Continued.

discussion and health education programs. However, in the current US food environment this may simply widen the income-based inequality in health if poorer households cannot access or afford fruits and vegetables (Capewell and Graham, 2010). For all of these reasons the UN's Special Rapporteur on the Human Right to Food declared unhealthy diets a greater threat to global health than tobacco, urging support for both regulation of unhealthy foods and for equitable access to healthy, nutritious foods (De Schutter, 2014).

Gardens and nutrition programs alone cannot effectively counter the powerful influence of the agrifood industry whose goal is corporate profits (Section 10.4). Creating a food environment that supports healthy, active lives for all will require policy changes that restrict this negative corporate influence (Hastings, 2012). However, food gardens can play a role in creating positive, community controlled food environments and increasing access to healthy food.

1.1.1. Garden foods can provide compounds important for health

Isolating the health effect of vegetables and fruits is challenging. Experiments are difficult because people don't like keeping to specified diets for very long, so results are often ambiguous, as Nestle and Nesheim show in their review of diet and weight loss studies (2012, 165–173). It's much easier to set up controlled feeding experiments in animals, but results cannot be directly applied to humans. Therefore, many human studies are *observational*, for example where respondents answer survey questions, including about what they eat, and undergo physical examinations, such as the US National Health and Nutrition Examination Survey (NHANES). These answers plus other health data can be used to track the relationship of diet and health through time, but have the obvious problem of inaccurate reporting (Archer et al., 2013), as well as many uncontrolled variables that can influence results. Clearly there is a lot we do not understand, however, as stated above, the cumulative evidence is consistent: diets rich in fruits, and especially vegetables, can improve health and reduce illness and death from NCDs (Katz and Meller, 2014; Oyebode et al., 2014; USDA and HHS, 2015b, Part D, Chapter 2:45).

Nutrients are the chemical compounds necessary for the physical maintenance, work, and reproduction of all living plants and animals (Section 7.3), as well as for combating disease and other forms of stress. Essential nutrients must be obtained from our diet. For example, of the 21 amino acids needed by humans to make proteins, 9 are essential, while 12 are non-essential because they can be synthesized in the body. The fruits and vegetables grown in gardens are often good sources of many nutrients important for human health, including vitamins, minerals, and protein-building amino acids, as well as dietary fiber (see Appendix 1A, Table 1A.1). For example, dark green leafy vegetables like collard, amaranth, arugula, *chaya* (*Cnidoscolus aconitifolius*), and *molokheyyah* (*Corchorus olitorius*) are rich sources of iron, calcium, and vitamins C and A; seeds including dry bean, sunflower, and almond are good sources of iron, zinc, calcium, protein, and energy in the form of healthy carbohydrates and unsaturated fats. People in industrialized countries are often concerned about getting enough protein, especially if they reduce or eliminate meat or all animal foods, but in diets of diverse, whole plant food, that include whole grains, beans, nuts and legumes, adequate energy and protein requirements are easily met. Iron deficiency is present in about 10% of US women, and 25% of those who are pregnant, but garden produce can be a source of that nutrient as well (Miller, 2014). However, vegetable varieties selected for high yields in industrial agriculture may have reduced concentrations of some nutrients (Section 10.4), something to take into account if those nutrients are lacking in the diet.

In addition to nutrients, other compounds in garden produce are also beneficial. *Antioxidants* are compounds that prevent damage from the chemical process of *oxidation*, the loss of an electron from a molecule that alters its function, sometimes contributing to diseases including cancers, inflammation that can lead to gastrointestinal and cardiovascular diseases, and possibly mental illnesses such as dementia, depression, and anxiety (Bouayed et al., 2009). Oxidation in our cells can increase because of the presence of *free radicals*, atoms or molecules with an unpaired electron that therefore tend to oxidize other molecules, by removing or "stealing" their electrons. Sources of free radicals include normal body maintenance, energy use in physical exertion, and exposure to environmental sources, including cigarette smoke and ultraviolet light, and food contaminated with agrochemicals including organophosphate pesticides and bipyridyl herbicides (Abdollahi et al., 2004) that are consumed as residues on food. Foods themselves can also be sources of free radicals, for example when charred or browned.

Antioxidants are believed to help prevent cell damage because free radicals oxidize the antioxidants instead of our DNA or other important molecules, although other functions may also be important. A major class of antioxidants are *phytopigments*—the pigments produced by plants, which include the blue and purple anthocyanins, red and orange lycopenes, and the orange and yellow carotenoids, including beta carotene (pre-vitamin A). Other antioxidants are vitamins C (ascorbic acid) and E; minerals such as selenium and manganese; and phytoestrogens such as isoflavones and lignan. Most brightly colored green, purple, red and orange garden produce contains high levels of phytopigments and vitamins. Because many antioxidants are only effective against specific free radicals and/or in particular biological contexts, eating a diversity of whole fruits and vegetables provides more benefits than taking antioxidant supplements.

Some garden plants such as garlic, chiles, turmeric root, and broccoli can also provide compounds that appear to stimulate resiliency in our cells. This occurs through the process of *hormesis*, in which low doses of a compound are beneficial, even though the same compound may be toxic in higher doses. Plants have evolved to produce such compounds, including in response to heat, drought, or insect herbivory. For example, plants often produce toxins that are not deadly to humans but are sufficient to discourage insects from eating them. When we eat those plants, we consume low doses of these toxins, which stimulates a stress response in our brain cells, with the result that the cells are more resistant to processes that lead to cognitive decline, a benefit for cognitive health (Mattson, 2015). There is also evidence that the production of such hormesis-stimulating compounds is not only greater in some crop varieties, but also depends on how the plants are grown (Hooper et al., 2010), and may be greater in some "organic" produce (Box 1.3).

Garden foods are a rich source of *dietary fiber*, the indigestible part of plant foods (Lattimer and Haub, 2010) that's considered a non-nutrient because it is not directly metabolized by the body, but there is increasing evidence that the short chain fatty acids produced by fermentation of fiber in the gut have important health benefits (Koh et al., 2016; Morrison and Preston, 2016). Dietary fiber is often classified into two types: *soluble fiber* is dissolved by water in the gut (the colon or large intestine); *insoluble fiber* is not dissolved, but absorbs water in the gut, helps remove waste and makes some antioxidants available (Palafox-Carlos et al., 2011). Each type of fiber supports different forms of beneficial *fermentation* (the conversion of carbohydrates to acids or gases by bacteria) in the large intestine, that can increase the feeling of fullness and satisfaction after eating, and protect against diabetes and overweight (De Vadder et al., 2014). On average, fiber in foods is approximately one-third soluble and two-thirds insoluble (Wong et al., 2007), but with lots of variation, another reason for eating a diverse diet. Eating whole grains, vegetables, fruits, and legumes which are high in fiber is associated with maintaining healthy body weight (Du et al., 2010; Mozaffarian et al., 2011), increased health, and reduced disease and death (Park et al., 2011). And yet most people in industrialized countries don't eat enough fiber: in the US women on average eat only about 60% and men only 40% of the recommended dietary allowance (RDA) for fiber (USDA, 2010).

1.1.2. Fresh fruits and vegetables can replace unhealthy foods

While food is necessary for survival, its appearance, aroma, taste, texture, and cultural and personal meaning provide enjoyment and comfort. However, in industrialized countries we often eat too much—especially calories (energy), protein, salt, and fat—contributing to an epidemic of overweight and obesity, and debilitating NCDs. Whereas in the past, low body weight was an indicator of poverty, today in regions that have experienced a nutrition transition (Box 1.4), overweight and obesity are increasingly the indicators. Our diets are a major cause of poor health and early mortality, raising the question of why we eat what we do. The food environment (Box 1.2) plays a huge role, but there are other factors as well (Box 1.4).

A characteristic of the diets produced by the industrial agrifood system is the contrast between

Box 1.3. Is "organic" produce healthier?

All plants and animals are organic in the sense that they are composed primarily of carbon-containing (organic) compounds. But when talking about agriculture and food, "organic" has a wide range of other meanings, from broad definitions like that of IFOAM that includes ecological health and social justice, to the much more narrow definition of the USDA which is focused on what cannot be used in their production (synthetic fertilizers, herbicides, or pesticides) (Section 1.7.1).

A frequent topic of debate about farming and gardening is whether "organically" grown fruits and vegetables are healthier for consumers than their conventionally grown equivalents, in terms of the level of healthful compounds and toxic contaminants. At this time more research is needed to really understand the issue, and many factors are involved resulting in lots of variability. Still, a few findings are consistent. First, because no synthetic pesticides are used in organic agriculture, it is not surprising that residues of these are up to four times higher in conventional compared to organic produce (Baranski et al., 2014), and the proportion of organic produce with residues present (7%) is significantly lower than for non-organic produce (38%) (Smith-Spangler et al., 2012).

Second, when everything else is exactly the same there is little evidence for any clear difference in overall nutrient content (Dangour et al., 2009; Smith-Spangler et al., 2012; Baranski et al., 2014). However, organic production is not exactly the same as conventional production without those agrochemicals (see below).

Third, two recent meta-analyses found that organic produce contained significantly higher concentrations of *polyphenols* (Smith-Spangler et al., 2012;

Baranski et al., 2014;), because organically grown crops are thought to experience more environmental stress due to less powerful pest control and possibly lower applications of soil nutrients (i.e., no synthetic fertilizers). Polyphenols are compounds found in many fruits and vegetables, some of which act as antioxidants. Some polyphenols help prevent pests from eating the plants and so may support hormesis, and others may be more concentrated in certain growing environments, such as those with less available soil nitrogen (Brandt et al., 2011).

Fourth, because of different growing environments, the use of traditional varieties, and the development of new varieties specifically for organic production, as suggested above, it may be that those varieties have not experienced selection resulting in a lower concentration of nutrients and hormesis-stimulating compounds as is the case for some conventional varieties (Section 10.4) (Davis, 2009).

This means that gardening "organically" may provide you with fruits and vegetables that are healthier because of less contamination and possibly more antioxidants, nutrients, and other beneficial compounds, but these benefits are not automatic or universal to all organic produce. There are also many other individual and social benefits of gardening organically—you won't expose yourself and anyone working in or living near the garden to synthetic toxic compounds, and organic methods have been shown to promote biodiversity (Gabriel et al., 2013) and ecosystem health (Cavigelli et al., 2013) in agriculture, although sometimes with lower yield. And you won't be supporting the production of agrochemicals that often involves negative social, economic, and environmental consequences.

---

their energy content, and content of other nutrients (Appendix 1A, Table 1A.2). These diets contain lots of *empty calories* from energy-dense foods high in fat and sugar, and low in other nutrients or beneficial compounds like antioxidants and fiber (Box 1.4). Talking about energy can be confusing: a "Calorie" (with upper case "C"), is the "heat energy needed to raise 1 kilogram of water 1°C from 14.5° to 15.5° at 1 unit of atmospheric pressure," and is the unit on food labels (Nestle and Nesheim, 2012). One Calorie is equal to one kilocalorie (kcal), or 1,000 calories (with a lower case "c"). Even more confusing: the scientific standard unit of energy is the joule, with 1 kcal $\cong$ 4.2 kilojoules. Kilocalorie is the unit reported by the USDA and the one we use in

tables in this book; "calorie" is also used popularly and in this book as a generic reference to energy.

For many people, especially those with limited resources, empty-calorie, energy-dense foods with very little additional nutritional value, are abundant and inexpensive (Box 1.2), while healthier foods are harder to obtain and more expensive. Diets high in empty calories are major contributors to the combination of malnutrition, overweight, and NCDs now increasingly common in the global north and spreading rapidly elsewhere.

Fresh vegetables and fruit from the garden are nutrient rich compared to empty-calorie foods (Appendix 1A, Table 1A.2), and so for the many reasons outlined above they can contribute to healthier diets.

What Can Food Gardens Contribute? Gardens and Wellbeing

17

Box 1.4. Why do we eat what we eat?

The photos of the foods people eat in different places around the world in Menzel and D'Aluisio's book *Hungry planet* (Menzel and D'Aluisio, 2005) show how different human diets can be in different places, for different people. So, why do we eat what we eat? That is an incredibly interesting and difficult question, because it's so complex and so many variables are involved. A frequently cited study of nearly 3,000 respondents in the US (Glanz et al., 1998) found that when asked how they decide what to eat, people ranked the following five factors in order of most to least important: taste, cost, nutrition (Is it good for me?), convenience, and weight control. That study also found variation: cost and convenience were most important to younger and low-income individuals, while nutrition was more important to female, older, African American and Latinx individuals than others. To discover more about why we eat what we eat, we'll very briefly look at the evolution and history of the human diet.

For about the last 7 million years, the diet of our ancestors has been predominantly vegetarian (Ungar and Sponheimer, 2011): 97% of the diet of our relatives the chimpanzees is plants (Watts et al., 2012); on average 73% of energy in the diets of eight contemporary gatherer-hunter groups was from plant foods (Stanford, 1999), although there is tremendous variation, with the environment being the major influence. Human gut microbiota today reflect this long reliance on plant foods, leading some scientists to argue that, if anything, humans are biologically adapted to a primarily vegetarian diet (Dunn, 2012). Still, the advent of hunting by humans, providing sources of energy and protein-dense food, appears to be an important causal factor in the emergence of large brain size. Our gathering and hunting ancestors' environments contained many plant and animal species, and some researchers suggest that cuisine developed in part as a sociocultural system for identifying edible (vs. poisonous or unpalatable) species and methods of preparing them, and transmitting this knowledge to others (Armelagos, 2010).

The early Paleolithic environment (starting ~2.5 million years ago) selected for adaptation to periodic food scarcity and famine (Prentice, 2005), including a desire for energy- and protein-dense foods, which was adaptive when human ancestors were foragers, and those foods were scarce (Armelagos, 2010). Because of the scarcity of these foods, selection favored behaviors such as eating as much as possible when they were available, contributing to the

difficulty we have today in resisting these foods now that they are no longer scarce for most of us (Appelhans, 2009). This adaptation to scarcity is thought to have contributed to our sense of taste. While that adaptation appears to still affect our behavior and diets, evolution is ongoing and we are not the same physiologically as we were 2 million years ago. A clear example affecting our diets is the retention of genetic mutations—that occurred more than 7,000 years ago—which made some human populations capable of digesting dairy products after infancy (Ingram et al., 2009). There have been other mutations, such as ones providing greater ability to digest starch that adapted populations to another major lifestyle and dietary change: agriculture (Laland et al., 2010).

Our diets changed dramatically with increased consumption of grains, and then the advent of agriculture starting more than 12,000 years ago. This led to a great reduction in the diversity of species used for food (Harlan, 1992), accompanied by a larger proportion of starchy foods that increased dental cavities (Eshed et al., 2006), among other things. Agriculture also required tending crops, making us more sedentary, and along with the change in diet, probably led to an overall reduction in quality of life (Larsen, 2006). In this context, cuisine provided different ways to process and prepare a limited number of food species, like fermentation and drying that helped preserve harvests and make the diet more interesting.

The next major change in human diets came with the advent of industrial agriculture and food systems following World War II, and increasing consumption of fat, refined grains, added sugar, and animal products (meat, dairy, eggs) accompanied by an increase in diet-related NCDs. There is a well-documented correlation between increasing wealth, often measured as per capita *gross domestic product* (GDP, value of all goods and services produced in a country in a year), and increasing consumption of animal products and energy-dense foods (FAO, 2010), associated with an increase in NCDs (Tilman and Clark, 2014). Overall, this societal change is referred to as a *nutrition transition* between dietary patterns (Popkin, 1993, 2006), in this case accompanied by a change in the primary causes of sickness and death from under-consumption of nutritious food and communicable diseases, to over over-consumption of non-nutritious food and NCDs.

This nutrition transition might diversify diets initially, making more food products such as processed foods

Continued

Box 1.4. Continued.

made from a few crop species (e.g., wheat, maize, soy-bean, oil palm, rice) widely available, including where these crops weren't previously eaten. However, local crops are often marginalized in this process, which eventually decreases agricultural and dietary diversity (Khoury et al., 2014), and increases added refined starches, fats, and sugars (Ponce et al., 2006; Drescher et al., 2007).

Perhaps the most dramatic form of this nutrition transition occurs when people consuming healthier, traditional diets in their home countries migrate to industrialized countries. Diets often change within one generation, or even a few years, in a process of *dietary acculturation*. For example, second and third generation Mexican Americans in the US eat less fruits, vegetables, beans, and rice than does the first generation, and have a higher incidence of type 2 diabetes, especially if they are in low-income households (Afable-Munsuz et al., 2013), although higher disease incidence may also be due in part to better diagnosis in the US (Barcellos et al., 2012).

Garden foods can replace empty calories with the same *number* of calories (Fig. 1.2a), while providing lots more nutrients. Or garden foods can replace the same *volume* of empty calorie foods with vegetables and fruits, reducing calories and increasing nutrients, fiber, and other healthy compounds (Fig. 1.2b). Either way, eating more garden produce than the amount in the standard American diet (SAD) is healthier.

Gardens can help improve diets in so called *food deserts*, which are usually low-income neighborhoods dominated by "junk" food chains, with few markets selling high quality, fresh produce at affordable prices. However, urban farmer, educator, and food activist Karen Washington points out that the food desert label ignores the human resources and culture of these areas. She says these areas are more appropriately seen as symptoms of *food apartheid*— social organization of food and agriculture reflecting historical, structural inequities (Brones, 2018). Living in such an area increases the likelihood for being overweight, especially for lower income people (Reitzel et al., 2014). In these situations, food gardens controlled by local residents might provide their only access to affordable fresh produce.

But cutting back on highly processed energy-dense foods isn't easy, since we are evolutionarily adapted to seek high energy foods (Box 1.4), and because they are prepared with lots of salt, fat, and sugar to satisfy our taste, a primary reason we choose food. Hunger can signal a need for energy, but it can also be the result of a neurological and hormonal need for reward in the form of food with positive sensory and emotional associations, especially in people prone to being overweight (Zheng et al., 2009). That need results in a "hungry brain" that seeks food, regardless of whether the body needs it or not, one reason that losing weight can be so difficult.

The volume and form of foods affect a meal's *satiety*, how long it keeps you from wanting to eat again, and your *satiation*, the feeling of fullness and satisfaction that causes you to stop eating. Solids tend to have higher satiety than liquids, especially sweet liquids; eating an apple provides much greater satiation than does drinking the filtered juice of an identical apple with the same amount of calories (Rolls et al., 2004). One way to increase the satiety of meals, even when the total number of calories is lower, is to include more foods with higher water and fiber content (Rolls et al., 2004). Fiber increases satiety by increasing food volume, and perhaps also by slowing the release of sugar (glucose) into the bloodstream during digestion (Livesey et al., 2008).

Most raw vegetables and fruits have a water content of about 80% or more by weight, and are high in fiber and nutrients, with ratios of nutrients per calorie, and volume per calorie much higher than most other foods, especially processed foods (Appendix 1A, Table 1A.2). This is one reason why, especially in combination with physical activity, garden foods can help reduce obesity and overweight (Ledoux et al., 2011).

Despite evidence of the multiple benefits of eating lots of vegetables and fruits, few people eat the minimum amount recommended. For example, the recommendation for a 2,000 kcal per day diet in the US is 2.5 cups of vegetables and 2 cups of fruit. But one study found only 9% and 13% of US adults eat recommended amounts of vegetables and fruits, respectively (Moore and Thompson, 2015). However, community food gardeners eat more vegetables and fruits than their neighbors who garden at home, or do not garden at all (Alaimo et al., 2008; Litt et al., 2011). In the City of Seattle, Washington, the P-Patch Community Garden Program comprises about 31 acres gardened by over 2,850 households, with 71% of these low

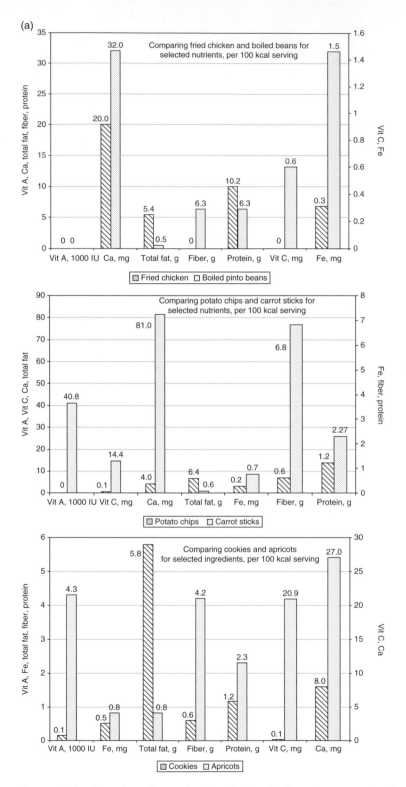

Figure 1.2a. Comparing a standard American diet meal with garden foods of equal energy value. Data source: USDA ARS (2018)

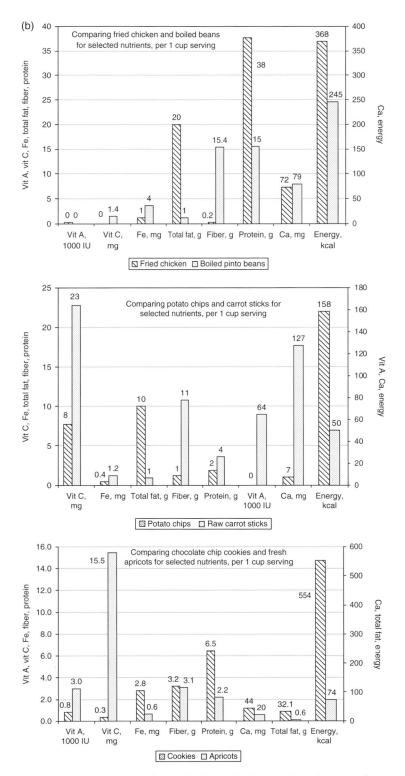

Figure 1.2b. Comparing a standard American diet meal with garden foods of equal volume. Data source: USDA ARS (2018)

income, and with 23% of gardeners people of color (P-Patch, 2014). In addition, 40% of P-Patch gardening households are contributing an average of nearly 1,100 kg (2,400 lb) of produce each month to local food banks and hot meal programs. The evidence for the effect of home gardens on nutrition is less clear (Girard et al., 2012), though some studies do support a positive association (Talukder et al., 2010; Cameron et al., 2012).

1.1.3. Flavor from the garden

The *flavor* of a food is the result of its taste combined with other sensory information such as aroma, feel in the mouth, and even the way the food looks. Flavor plays a big role in how appealing we find foods, and we learn to associate certain flavors with experiences and even identities (Mattes, 2012). Because of our evolutionary history (Box 1.4), the taste of empty-calorie foods high in fat, salt and sugar, typical of most processed fast foods, is appealing, and we crave these foods for their emotional reward, especially when experiencing anxiety or depression (Platte et al., 2013). They stimulate our appetite for more of those foods, as well as seeming to raise the threshold for satisfaction from those tastes as we become habituated to increasing levels of them in our diet. For example, the more salt in your diet, the more salt you need to eat to get the same salty-taste reward (Kim and Lee, 2009). Similarly, diets higher in fats reduce the satiety of those fats, encouraging you to eat more and more fats to achieve the same level of satisfaction (Stewart and Keast, 2012), and the same effect may be true for sweet foods. Now there is also evidence that obesity itself causes inflammation that decreases the number of taste buds, inhibits the growth of new taste buds and is a "likely cause of taste dysfunction in obese populations" (Kaufman et al., 2018). Tastes can affect diets in other ways; for example, drinking sweetened beverages instead of water discourages children from eating vegetables, because those taste bitter in comparison to sugary drinks (Cornwell and McAlister, 2013).

Gardens are sources of flavors that make foods appetizing. Herbs such as arugula, basil, cilantro, fenugreek, garlic chives, *papaloquelite*, or marjoram, and other flavorants from the garden such as chiles, citrus zest and juice, and rose petals, can make food more appetizing and personally and culturally meaningful. This can be especially helpful when trying to reduce the use of salt, sugar, and fat in the diet, when people try new vegetables and fruits, and for supporting important cultural traditions. Gardens can also add to the visual appeal of foods, for example by adding bright orange or red sliced *tuna* (fruit of *Opuntia* cacti), a sprinkling of chopped greens, or edible petals from calendula, nasturtium, or fejoa blossoms.

Tastes are also formed by experience. A food environment with aggressive food industry advertising contributes to a cycle of experience and behavior that makes it very difficult to have a healthy diet. Gardens offering alternative experiences and associated with educational programs can help address some of these obstacles. Schools, community centers, and other groups can organize programs demonstrating simple preparation, and appetizing presentation of foods made with garden produce, especially foods community members are already making, or remember. Such programs may be the first time that people taste these vegetables and fruits, and can change attitudes. Vegetables and fruits can go from being unfamiliar or unappealing to ingredients people want to include in their meals. However, while a review of research on the effect of garden-based interventions among children and youth in the global north from 2005–2015 suggests they do result in increased fruit and vegetable consumption, more rigorous research is needed to test this relationship (Savoie-Roskos et al., 2017).

Evaluation of a popular school garden program for students from fourth and fifth grades in three Berkeley, California schools showed significant increases in consumption of fruits and vegetables in and out of school for students in schools with comprehensive programs, including cooking, compared with those in schools with less-developed programs. However, this difference disappeared as students transitioned to middle schools (sixth grade) and high schools that did not have this program (Rauzon et al., 2010). This suggests that broader institutional and social change is needed to sustain the positive results of these programs.

Other strategies can also help. Recently, researchers presented high-school students in Texas with information about "junk" food companies, including how those companies' strategies to grow their profits include recruiting teenage consumers. This knowledge activated common adolescent values of autonomy, social justice, and concern for their peers, resulting in reduced junk food consumption by those youth compared with a control group that only received information about food nutrition (Bryan et al., 2016).

1.2. Physical Activity

Work in the garden can be long and strenuous, like a morning of digging garden beds into dense soil, or turning a big compost pile. Or garden work can be brief and leisurely, like harvesting a few kumquats, picking some herbs for a dish you are cooking, or removing cutworms from the soil around young plants. Either way, gardening can bring healthy physical movement into our daily routines and improve physical fitness and health.

Lack of physical activity can lead to declines in physical and mental health, and in the industrialized world we spend much of our time sitting (Owen et al., 2009). Time spent sitting is positively correlated with body weight, and with NCDs like type 2 diabetes; the more you sit, the more likely you are to be overweight and sick (Katzmarzyk, 2010). Surveys of how working Americans spend their time found that in 24 hours the vast majority of the time we are sleeping or sedentary (Tudor-Locke et al., 2011) (Fig. 1.3). The combination of inactivity and overweight impairs our health, capacity for satisfying work, and happiness. But the good news is that even light physical activity

throughout the day can help prevent these negative outcomes (Katzmarzyk, 2010). We do not have to be devoted athletes to be healthy, minimize NCDs, and avoid falls and related accidents. Still, even vigorous activity can't make up for all the time we spend sitting (Dunlop et al., 2014), we just need to sit less because doing nearly *anything* physical is better for us than sitting!

Some garden activities can improve aerobic *physical fitness*, quantified as VO_2 *max*, the maximum volume of oxygen a person can acquire during the most strenuous activity. VO_2 max is a measure of the capacity of the lungs and cardiovascular system to capture oxygen and deliver it to the working muscle tissue where it is used to oxidize glucose to produce energy, a process called *respiration* (Section 5.4). VO_2 max is increased through extended periods of moderate to vigorous physical activity that gets you breathing harder and raises the number of heart beats per minute, increasing the strength and capacity of the cardiovascular and respiratory systems. While gardening is typically classified as moderate exercise (Matthews et al., 2008), the physical fitness benefits of gardening will depend on what work is actually being done, for how long, and by whom. For the elderly (Park et al., 2008), ill, overweight, and very young, even the lightest garden tasks can increase heart rate and oxygen capture and delivery,

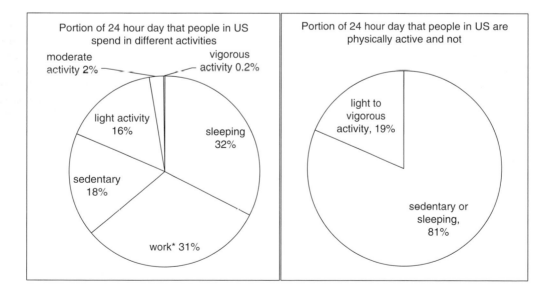

Figure 1.3. Physical activity in the US

*For 79% of population "work" is sedentary
Tudor-Locke et al. (2011)

What Can Food Gardens Contribute? Gardens and Wellbeing

23

and improve physical fitness, especially when there are few other opportunities for physical activity.

Garden tasks like digging, carrying tools, or buckets of compost or water are weight-bearing or impact exercise, and so contribute to greater bone strength, for example among older people who may be most at risk. Physical activity is also important for maintaining or improving the motor skills, balance, and agility that are needed for healthy daily activity throughout life, and it can make us feel better about ourselves, and those around us.

1.3. Psychological and Social Benefits

Human beings are social animals. We evolved in groups of related and unrelated cooperating individuals that formed large social networks (Hill et al., 2011), and our brains and emotional lives have been formed by hundreds of thousands of years of living together. Greater population size and urbanization, increasing scarcity of resources, and the growing recognition that in a globalized world the biophysical and sociocultural aspects of our lives are interconnected, all mean that improving our capacity for positive social interactions is more important than ever.

For most of us, feeling that we are contributing and are appreciated members of a social group is important for our self-regard and happiness, and has positive effects on health (Fredrickson et al., 2013). And yet, life today in the industrialized world is often characterized by social isolation, making us lonely, reliant on social media, and vulnerable (Fig. 1.4). Much of our personal lives is centered within the walls of our homes, or inside automobiles; often our free time is spent in activities requiring no face-to-face interactions with others. For example, adults in the US spend an average of 2.8 hours per day watching TV, with hours being greater among those with low income and low educational status (USBLS, 2017). In an older study, children spent an average of 4.5 hours per day watching TV, plus another almost 3 hours per day with other forms of personal media (internet, computer games) (Rideout et al., 2010). This sort of social isolation is known to have negative psychological consequences (Turkle, 2011) which are especially acute when we are physically inactive. Being socially isolated, or feeling that way, is now recognized as predictive of early mortality—especially among those under 65, and scientists anticipate

Figure 1.4. Just me and my TV (or social media)

loneliness will become a major public health problem in the coming decades (Holt-Lunstad, 2018).

Obviously, food gardens are not a solution to all of the personal and social problems posed by our lifestyles, but the positive contributions of physical activity, experiences with nature, other people, problem solving, and mastery of tasks have been documented in a number of studies.

1.3.1. Direct psychological benefits of physical activity

The positive correlation between physical activity and mental health is very strong across all ages (Penedo and Dahn, 2005). Even as little as 20 minutes of activity each week can make a real difference in how we feel, and the more activity, the better we feel (Hamer et al., 2009). Physical activity has this effect because it helps our bodies function better with fewer limitations, and less discomfort and disease. It also makes us feel better because it stimulates the release of chemicals in the brain like dopamine that create positive feelings and attitudes (Ströhe, 2009). Indeed, a rapidly growing number of researchers (e.g., Rhodes et al., 2017), medical practitioners, and health promotion organizations worldwide agree that physical activity is the single most powerful and wide-reaching "treatment" available for many of the physical and psychological problems common today, and is recommended along with the improved diets we discussed above.

1.3.2. Positive effects of experiences in the garden

Nearly all human evolutionary history has been spent in natural environments, and this may be one reason that some interaction with nature seems important for our happiness (Wilson, 1984; White et al., 2013). Interacting with the natural world—including the soils, plants, and animals in the garden—has positive effects on mood, self-regard, and attitude (Pretty et al., 2005; Barton and Pretty, 2010), and stimulates the brain. Even periods as short as five minutes in a green environment can be beneficial. Gardeners themselves report that their gardening provides them with physical, emotional, and social benefits (Waliczekz et al., 1996; Egerer et al., 2018), and a more positive mood and health compared to non-gardeners (Waliczek et al., 2005), reasons that

food gardens are sometimes referred to as "therapeutic landscapes" by public health workers (Hale et al., 2011).

Because of the many factors that vary over time and space, gardening can be a constant source of mental stimulation. For example, observing changes during a plant's life cycle, the effects of changing weather, and basic gardening tasks such as figuring out the best way to stake a tomato plant using materials on hand, without having to buy anything. As with other diversions from our daily routine, focusing on the garden can provide relief from worries and obsessive thought patterns that often have negative consequences for mood and social interactions. Observations and learning while working in food gardens can also contribute to greater awareness of biological and ecological processes that in turn have positive effects on gardeners' understanding of their own health and the health of the environment (Hale et al., 2011). Over time, working in the garden can provide a sense of mastery with a resulting positive effect for gardeners who practice and hone skills such as seed planting and pruning, and harvesting. As a place where we must constantly problem solve, gardens provide the opportunity to practice responding to small challenges, strengthening our capacity to meet greater adversity when it comes along.

Food gardens can be a way for new community members, including immigrants and refugees, to feel connected and comforted through activities that are familiar, that make them feel useful (Wen Li et al., 2010), and help them feel more at home when they can grow food plants that are familiar to them. Cultural identity, including the plants, foods, and activities that help define who you are, is important for everyone, not just new residents. For example, when controlled by people themselves and not outsiders, gardens growing traditional crops and varieties can be a part of individual or community strategies to cope with the historical trauma affecting many groups (Mohatt et al., 2014; Ramírez, 2015). These positive effects of food gardens can be especially important when age, poverty, illness, fear, and other conditions isolate people and limit their opportunities for positive, stimulating activities and interactions.

Food gardens can also provide social benefits, though some are easier to quantify than others. Positive interactions among community members can be rare, and difficult to initiate, since we frequently live among strangers. Especially in the poorest neighborhoods, gardens can beautify and humanize shared spaces, making positive interactions with natural and social environments easier (Voicu and Been, 2008). A study of community gardens in Denver found that "community gardeners talked about the garden as a way to awaken the senses and support a more holistic way to contemplate health and wellness," and that gardeners' perception of how pleasant their neighborhood is was positively correlated with fruit and vegetable consumption, and with community gardening compared to home gardening, or no gardening (Litt et al., 2011, 1470).

Visible home or community gardens may be one of the few places to see neighbors outdoors. Since we all eat and share some interest in food, gardens can be an easy topic for interactions among people who might not otherwise have anything to say to each other, including people of different cultures or backgrounds (Shinew et al., 2004). The social benefits of food gardens are enhanced when they are community gardens. Participating in community gardens increases face-to-face interactions, and can create mutual trust and social cohesion, including between generational, "racial" and cultural groups, and resource levels, as was reported by gardeners in a small study of community gardens in Minneapolis (Grewell, 2015). Community gardens in New York City are believed to have contributed to greater social cohesion and support in gardeners' daily lives, and some of these gardens were the basis of neighborhood organizing and mutual support during the disaster brought by hurricane Sandy in 2012 (Chan et al., 2017). These relationships may support wider engagement and reduce conflict, although we can't assume this is always true (Section 1.6).

Community gardening experience is a way people can learn and practice important group-based skills such as decision making, creating, observing and enforcing rules and social norms, and engaging in community issues. It can also be the basis for experiential science education (Krasny and Tidball, 2017). These gardens may also offer the opportunity for members to experience individual social roles such as leaders, followers, and learners (Teig et al., 2009). Altogether, community gardening can be a chance to practice many of the prosocial skills needed to create healthy, positive, safe communities.

1.4. Environmental Benefits

Food gardens can provide benefits by decreasing our environmental impact in two ways. First, supporting healthy ecosystems by reducing negative impacts on the environment from human activities in general. Second, by producing and delivering food with less negative environmental impact than the same food from the industrial agricultural system. However, like all garden benefits, these environmental ones are not automatic—they depend on how gardens are managed.

1.4.1. Gardens supporting healthy environments

Among the changes we are experiencing in the twenty-first century are the dramatic negative environmental and climate consequences of human activity, including feeding ourselves (Chapter 2). One approach to understanding the importance of the biophysical environment to humanity, and encouraging practices that stop and reverse harm, is to quantify the benefits it provides to society. *Ecosystem services* (ES) are the different ways in which the environment contributes to the existence and support of human societies and can be given a monetary value. ES are a popular way to bring environmental and resource issues into mainstream economic and political discussions—by determining how much water filtering by the soil, or a morning birdsong, is worth in dollars—but this is a subjective process and often limited by lack of data (Norgaard, 2010). That is, there are many facets to ecosystem function that we simply do not understand and therefore cannot assess the "value" of. Even if our knowledge was complete, ES values reflect the beliefs and priorities of society, especially governments and the powerful. An important philosophical question is whether ES values should only be determined by value to humans—and to which humans—the anthropocentric view, or whether other species and parts of nature have value independent of their utility for humans—the ecocentric view. In addition, philosopher Michael Sandel argues that monetizing things like carbon dioxide (CO_2) emissions or ES, has an effect beyond benefit:cost calculations by fundamentally changing our attitudes toward those resources or processes, and each other (Sandel, 2012, 72). By assigning them a monetary value ES become commodities with a monetary "worth," that can be owned or transferred or replaced, all ideas that contradict the reality of ecological and social processes. We agree, and briefly describe ES not as a way to calculate economic value, but because ES are so widely used for appreciating the significance of the environment, including gardens.

ES are often classified into the following four groups: support (photosynthesis, soil formation);

regulation (water filtration, pollination, carbon storage affecting climate regulation, biodiversity conservation); provision (water, domesticated and wild plants for food, fiber and medicine); and culture (aesthetic, educational, recreational values) (Millennium Ecosystem Assessment, 2005). Food gardens, like other green areas, can provide a combination of all these services—support and regulation, in addition to the provision and culture services we usually think of. For example, fruit trees and other large perennials can capture and store carbon, absorb some soil pollutants, block the movement of airborne pollutants, and provide cooling and shade (Pataki et al., 2011). When replacing paved urban areas (*hardscapes*) like parking lots, gardens can increase infiltration of rainwater, especially important in areas of low rainfall and high temperatures (Pataki et al., 2011). Replacing hardscapes with gardens can also provide cooling as the result of increased evapotranspiration (Section 8.2.3), shading, and increased *albedo* or reflectance of sunlight. All of these reduce the *urban heat island* effect, the phenomenon of urban microclimates being warmer than surrounding areas (Section 4.1.3). Gardens can also attract and protect pollinators and other wildlife. For example, in both agricultural and ornamental landscapes, wildlife diversity is increased by the plant genetic diversity and planting complexity often found in polycultures (Section 6.2.3), with tall and short plants, different growth habits and lifecycles (Goddard et al., 2010).

Of course, the way gardens are managed can result in providing either positive ecosystem services or negative "disservices" (Cameron et al., 2012). Distinguishing between these requires ecological thinking and careful consideration of benefits:costs. Trees planted in the garden may take 3–10 years to become carbon neutral—that is, begin to have a net benefit on carbon emissions, sequestering as much carbon in the plant as was released due to the energy used in growing and transporting the tree to the garden, and soil disturbance when planting (Cameron et al., 2012). Similarly, composting organic waste in the garden instead of exporting it may result in either lower or higher greenhouse gas emissions, depending on how the composting is done and what the alternatives are (Cleveland et al., 2017) (Section 7.6). Gardens in industrial countries can be very resource intensive when they include high levels of external inputs, such as purchased transplants, pesticides, soil amendments like imported guano or peat or compost, and machinery like rototillers. Poor water management in a garden that uses lots of piped water means that despite the ES from air cooling via increased shade, albedo, evapotranspiration and water infiltration, and the hardscape reduction the garden may be providing, the impact of pumping and delivering that amount of water could result in a net increase in greenhouse gas emissions.

1.4.2. Gardens reducing the environmental impact of food production

Yields in terms of harvested food per unit area are often greater from smaller areas, like gardens, compared to commercial-scale fields (Netting, 1993; Barrett et al., 2010, 88), primarily because small areas are more carefully managed. For the same reason, gardens may also be more efficient in terms of harvest per unit of inputs like water or compost, but because of the time spent, they are usually less efficient in terms of harvest per unit of labor. How time spent in the garden is valued, and how the environmental and social costs of the fossil fuels that increase labor efficiency in industrial agriculture are calculated, makes a big difference in estimates of efficiency.

Gardens can also have lower negative environmental impacts than larger-scale agriculture because they use less energy and resources per unit of food for production, storage, cooling, and packaging, the proportion of wasted food is generally lower, and transportation is eliminated. For example, in one model of home gardens in southern California, the greenhouse gas emissions of purchased vegetables (1 kg CO_2 equivalent per kg of vegetables per year) was more than twice that of vegetables harvested from the garden (Cleveland et al., 2017).

1.5. Economic Benefits

The word *economy* derives from the Greek *oikonomia* meaning "household management." This reminds us that even though we tend to think that the economic benefits of gardens are financial ones like the market value of fruits and vegetables grown, other benefits that gardens contribute to household and community wellbeing could also be included, such as recreation, increased knowledge, pleasant spaces, maintenance of preferred crop varieties, improved nutrition and health, and ES. But, as with ES, the question is, how are

What Can Food Gardens Contribute? Gardens and Wellbeing

27

non-monetary benefits valued? We suggest that sometimes, instead of absolute numbers, valuation can be relative. However it is assessed, valuation should be done locally by those affected, because it is very context specific and subjective. For example, the value of a garden that is a pleasant public space providing enjoyable physical activity may be much greater in a poor neighborhood or housing project than a middle-class suburb.

In this section we focus on the more easily monetized economic benefits: indirect benefits from savings when garden produce replaces food that would otherwise be purchased, and direct benefits from selling or trading garden produce. For these, monetary costs need to be subtracted from benefits in order to estimate the net economic benefit of food gardens.

While household and community gardeners may not want to take the time to estimate monetary economic benefits, this can be an important exercise when food gardens are being compared with alternative uses of resources like land and water, or alternative ways of achieving benefits, like community centers or nutritional programs.

1.5.1. The net economic benefit of food gardens

Calculating the net economic benefit of the garden requires subtracting total monetary cost of inputs like time, water and compost from total monetary value of outputs, like the vegetables, fruits, herbs, and other things harvested. An important input is the gardener's time, but calculating its cost is very subjective. The *opportunity cost* approach is favored by most economists and involves comparison of the benefits of one choice with the benefits of another available choice. For example, according to this approach, the chard grown by a 30-year old earning $200 per hour as a software engineer would have a much higher cost than chard grown by an older, unemployed man because in theory her time is "worth" much more than his.

However, using monetary returns from alternative ways of spending your time assumes that maximizing income is always the priority, and does not include other ways of valuing time or people. For example, an urban extension agent in Tucson discouraged us from working with a community garden at a housing project, telling us that gardens are only for wealthy hobbyists, because growing food at home is more expensive than buying it at the store. This may have been because he assumed, without providing any evidence, that resource-intensive methods would be used, or that time spent working in the garden would be valued at gardeners' rates of pay in their regular jobs. It may also be because he assumed that the value of gardens could only be estimated from the money saved in grocery bills. This encounter inspired us to document the net economic benefit of two gardens, and we think we proved him wrong (Cleveland et al., 1985) (Section 1.5.2). In addition, for most people, time in the garden has other values, and may even be seen as recreation, something people often pay for.

Calculating the economic value per unit of individual inputs or outputs can also be useful. For example, the retail value of different vegetables grown per m^2 of garden area or per m^3 of water applied, or the cost of water applied per m^2 of gardened area. An advantage of per unit value calculations compared to total value calculations is that it makes it easier to compare different gardens, seasons, gardening methods, and other variables, including other ways of production. Per unit calculations are also very useful when trying to make your garden more efficient for specific variables like water, space, or your time (Section 8.5).

1.5.2. Savings through direct consumption

Saving money on food bills is a major motivation for food gardening in the US (NGA, 2014). The amount of money that people spend on food is relatively *inelastic*, that is, once a basic quantity of food is obtained, the amount does not change much when the price of food or household income changes, because food is essential. We say relatively inelastic because consumption, including of food, in industrialized countries has been expanding beyond basic needs as consumers seek to increase not only the quantity they consume, but also quality or prestige, such as gourmet, exotic, or novelty foods (Witt, 2017). In comparison, spending on recreation is very *elastic*, changing a lot depending on its price or household income. Because the amount of food spending is relatively inelastic, low-resource households spend a greater proportion of their resources buying food. For example, in the US, 15.7% of the lowest income households' expenses are for food, compared to 11.2% for households in the highest income bracket (USBLS, 2016).

Interestingly, even though saving money is a major motivation for food gardening in the US, there have been very few systematic studies of the financial costs and benefits of food gardening. One reason is that the overall trend in home gardening in the US had been downward for many years (Putnam, 2000), although there have been spikes of increased gardening, for example during World War I, the depression of the 1930s, and World War II (Lawson, 2005). In 1919, per capita consumption of home grown vegetables was 59 kg (131 lb) per year, in 1998 it was 5 kg (11 lb), and then data stopped being collected (Putnam, 2000, Fig. 16). However, survey data show that gardening in the US is now increasing; there was an estimated total of 42 million households with food gardens in 2013, a 17% increase from 2008 (NGA, 2014). One estimate is that for every dollar spent on garden inputs approximately $8 worth of fruits and vegetables is produced (NGA, 2009). However, these data are from an industry-sponsored survey and assume the use of many purchased inputs, but also do not include the value of land, water or labor, and assume gardening to be recreational.

The economic savings that gardens can provide depend on how this is calculated, including the value given to different inputs and outputs. For some of these, such as purchased inputs, the market value is typically accepted as representing the financial costs or savings to the gardener. For example, the direct costs of water ($ per m^3 or per ft^3), or purchased seeds ($ per m^2 or per ft^2 planted), are relatively easy to calculate. Putting a monetary value on inputs like homemade compost, saved seeds, and labor is much more complicated. Putting a monetary value on the food harvested can also be tricky, because there may not be good equivalents in local stores.

Most studies of the value of garden production have been extrapolations based on individual plant yield. Using species-specific per plant yield estimates calculated from unpublished data from organic garden advocates (Rodale Institute, Ecology Action), a study of 144 Philadelphia community gardens in the late 1980s estimated their garden produce had an average annual value of $160 per garden (Blair et al., 1991). More recently (2008), a small sample of participating gardeners, also in Philadelphia, recorded produce weight per crop for individual crops which was then used to estimate that 226 gardens produced a mean of $35.94 of produce m^{-2} ($3.34 ft^{-2}) (Vitiello and Nairn, 2009). Similarly in New York City, data from a sample of gardeners provided yields and/or number of plants of a crop

in an area, and these data were used to estimate that 67 gardens produced an average of $32.28 of produce m^{-2} ($3 ft^{-2}) in 2010 (Gittleman et al., 2012). A study in San Jose, California weighed the produce of 10 experienced community gardeners, and found an average yield of 1.78 kg m^{-2} (0.75 lb ft^{-2}), valued at $12.16 m^{-2} ($1.13 ft^{-2}) (Algert et al., 2014). None of these studies documented the costs of inputs, including water.

With resources becoming scarcer, and many growing environments becoming more stressful, the cost of inputs such as piped water or imported compost will become important in calculating the economic viability of gardens, so thinking in terms of ecological networks and life cycles is essential. In our study of two urban desert gardens in Tucson, Arizona (Cleveland et al., 1985) we measured all inputs and outputs over a total of five garden years, using market rate equivalents, and excluding high prices in estimating value of harvest. For example, although we often grew basil that sells for a high price per pound, we used the price of more common leafy greens like lettuce to avoid over-valuing our produce. We found average net returns of $3.67 and $5.44 m^{-2} ($0.33 and $0.49 ft^{-2}) of growing area, including the cost of water, compost and seeds. The average yield was 1.75 kg m^{-2} yr^{-1}, and returns to labor were $0.72 and $1.11 $hour^{-1}$, but the gardens were not intended to maximize production or labor income—and the time spent was considered to be leisure, not drudgery. In Section 3.2.2 we discuss doing a food garden input:output study.

1.5.3. Increased income or trade value

Food gardens can provide supplemental income or goods from selling or trading garden produce. Opportunities for this are increasing as farmers' markets become more popular, and cities are changing regulations to allow gardeners to legally sell their produce, for example in San Francisco (Section 4.2.2). Barter networks and exchanges are also becoming popular, such as the public "crop swaps" the San Diego Seed Library has started organizing.

As with household savings from gardens, the net value of the income or trade depends on the costs of inputs. Other benefits can be hard to quantify but are quite important. For example, trade may be the only way to obtain special foods or goods grown or made by others, and participation in a market can be a source of positive feelings about self and a valued social activity for gardeners.

What Can Food Gardens Contribute? Gardens and Wellbeing

29

Increasing income by marketing garden produce may be a motivation for some gardeners. However, low-income gardeners may find marketing difficult if the price they need to charge for their produce is too high for people in their community to afford (Alkon and Mares, 2012).

1.6. Food Gardens and Food Justice

Justice, including food justice, is one of the most challenging benefits to achieve, yet in many ways the most important, and there is a growing number of inspiring garden projects focused on justice. Without justice, all other benefits and resources are at risk for anyone not a part of the power structure. Thinking about justice, food, and gardens is part of a larger discussion of food security, food sovereignty, and social equity, and demands that we move beyond equal treatment, and address the historical, structural, economic, and other enduring sources of inequity.

Discussions of food and hunger are slowly changing from a focus on food security to food sovereignty and food justice. *Food security* is focused on eliminating hunger by ensuring people are fed, however it has become clear that this has not eliminated many of the reasons people are hungry and does not address injustice. *Food sovereignty* means that people and communities not only have a secure source of food, but that they have control over their food system, and that it provides a healthy, just, dignified life. The idea of food sovereignty was first defined by La Via Campesina in 1996 as control of the food system based on social justice (LVC, 2008, 61, 147–148). Started as a movement among farmers and peasants in the global south, the relevance and importance of food sovereignty to industrialized country food systems has become clear as well. Reasons for this include the division of food supplies into expensive healthy diets for the wealthy, and the cheap, unhealthy diets for the poor, the enormous negative health consequences for low-income people, and their lack of power in the food system. The food justice movement emphasizes the need for equity and fair treatment of everyone in the food system (Gottlieb and Joshi, 2010; Food Chain Workers Alliance, 2012).

The UN recognized the human right to food in 2004 (FAO, 2005), and shifting the emphasis from a right to food to a right to sovereignty over food links the food system strongly to other human rights, and not simply the right to consume minimally adequate diets.

1.6.1. Gardens reinforcing inequity

We have described lots of potential benefits of food gardens, but we also know that gardens are not magical. Despite the many potential benefits outlined above, there are problems for which gardens alone are not a sufficient or appropriate response. And sometimes gardens don't help at all, and can even do harm. In those cases, acknowledging this and supporting communities in setting their own priorities, as well as trying to understand why gardens are insufficient or inappropriate, helps address the problem in different, better ways. Indeed, there are times when food gardens are part of the problem, and this is why listening to what people say, what they want, their worries and dreams, is so essential. But the situation is complicated; for example, gardens in the US necessarily function as part of the larger inequitable and unjust society, dominated by neoliberal economic policies. Therefore, many gardens, garden projects, and programs may advance justice in some ways, and injustice in other ways (McClintock, 2014), another example of tradeoffs.

Obviously, denying individuals, groups, or communities access to gardens or resources that are available to others is unjust, but in some cases so is providing access to gardens. For example, gardens have been used to support the exploitation of laborers by requiring them to grow some of the food they need because they are not paid enough to obtain it any other way. These are gardens used to subsidize inequity, and they have been a part of feudal and colonial strategies for centuries. It was the case with garden plots given to African slaves on sugar plantations in the "New World," which they were allowed to work only in what time they had left after long hours laboring in their masters' fields and gardens (Carney and Rosomoff, 2009, 108–109, 131–135). A similar situation existed for Native Americans after the invasion by Europeans, for example in California's Roman Catholic missions (Castillo, 2015). In the US, programs encouraging "victory" gardens when the national priority was war funding have been seen by some this way (Pudup, 2008). Similarly, providing a housing project with a garden space is not a substitute for better living conditions or space for children to play, and does not make up for the lack of affordable grocery stores with good quality fresh produce in the neighborhood.

The positive associations with gardens can also be co-opted for other ends. In Vancouver, British

Columbia, Canada, a private development company established a "community garden" on a vacant lot it acquired, later admitting the purpose was to keep homeless people off the land and give the company a good image; the garden was also used to place a large billboard advertising their gentrifying development in a low-income neighborhood (Quastel, 2009).

Gardens and garden projects that are seen as rationalizing and perpetuating inequity will understandably result in resentment, and participation may simply be the result of desperation. Community gardens may also create or reinforce community divisions when one group takes control, monopolizing resources or decision making, increasing animosity, and exacerbating instead of diminishing inequity (Glover, 2004).

Inequity arising from gardens is not always intentional. Community food gardens have been shown to improve property values by removing blight, providing recreation, and encouraging a sense of community (Voicu and Been, 2008). But unless precautions are taken, these improvements may not have positive results for people in the neighborhood, because they can lead to increased tax rates and gentrification, pushing those residents out, as observed for urban green spaces generally in examples in China and the US (Wolch et al., 2014). When a community garden was established by a well-meaning outside organization in a low-income, African American neighborhood in Seattle, residents perceived it as replicating unjust historical power relations, removing that space from community access and control (Ramírez, 2015).

1.6.2. Gardens and hidden agendas

Some popular garden projects today have been interpreted as serving an implicit sociocultural or economic agenda, in a sense indoctrinating participants into practices and values that do not necessarily serve their best interests (Pudup, 2008). We have seen this in garden projects around the world where a failure to appreciate, let alone even acknowledge, the existence of local gardens, led to formally trained technicians and agencies imposing their ideas of "proper" gardens, crops, and inputs. Even though it may not be intentional, this approach reflects a lack of respect and an ignorance of local knowledge and environments, with the end result being inappropriate gardens that dry up simultaneously with project funding.

When the priority is exploitation, making a profit, avoiding social responsibility, or bolstering some other hidden agenda, the full benefits of food gardens will not be realized because decisions will ultimately be based on those priorities, and not the wellbeing of the people and communities involved. And most importantly, these hidden and not so hidden agendas take control and decision making away from participants and their communities. Being thoughtful and explicit about the assumptions and goals of garden projects involving community members and outsiders helps in working towards genuine collaboration and participation, and in avoiding manipulation and exploitation.

1.7. Resources

All the websites listed below were verified on May 3, 2018.

1.7.1. Nutrition and physical activity

The USDA's online, searchable Food Composition Database provides nutrient content of many foods, including for different preparations and quantities. The most recent release came out in April 2018. https://ndb.nal.usda.gov/ndb/

The USDA's Dietary Guidelines for Americans provide recommendations for diets, including vegetarian diets (USDA and HHS, 2015a). However, some of its recommendations may be more influenced by the food industry than scientific knowledge, for example its recommendation to increase dairy consumption.

The US National Health and Nutrition Examination Survey (NHANES) can be explored online. https://www.cdc.gov/nchs/nhanes/index.htm

The US Institute of Medicine develops the dietary reference intakes (DRIs) for different nutrients and different age and gender groups. These data, and more information about the DRIs are available here: https://www.nal.usda.gov/fnic/dietary-reference-intakes

Harvard University's School of Public Health has a number of webpages with accessible information about diets, health, and physical activity based on published scientific research. https://www.hsph.harvard.edu/nutritionsource/

The American College of Lifestyle Medicine is a global professional association for medical practitioners emphasizing the capacity of lifestyle habits, including healthy diet, physical activity, and not smoking to address many major health problems. They offer training, and their scientific evidence page has links to abstracts of relevant current research.
http://www.lifestylemedicine.org/

The US Government's Office of Disease Prevention and Health Promotion publishes Physical Activity Guidelines that include chapters for different parts of the population, such as Chapter 4, "Active adults." Physical fitness is described and the contribution of different types of gardening activity is included in their recommendations of ways to achieve healthy physical activity.
https://health.gov/paguidelines/guidelines/

New guidelines published in 2018 are less detailed.
https://health.gov/paguidelines/second-edition/

Oldways is a nonprofit organization that receives funding from some trade organizations. Nevertheless, it has some interesting graphics and is a good source of culturally relevant food pyramids (African heritage, Latinx, Mediterranean, Asian, vegetarian/vegan).
http://oldwayspt.org/

The weight of the nation is an excellent documentary produced by HBO and the Institute of Medicine (IOM) describing the evolution, causes, and costs of the current obesity epidemic in the US, based on a report by the IOM (IOM, 2012).

The US Department of Agriculture's National Organic Program (USDA NOP) regulates "organic" production in that country. USDA NOP information, including links to pages with regulations, allowed and prohibited materials and practices, can be found at this website:
https://www.usda.gov/topics/organic

For information on organic standards and production globally: IFOAM Organics International in collaboration with other organizations has defined organic agriculture very broadly, and publishes the Common Objectives and Requirements of Organic Standards (COROS) (https://www.ifoam.bio/en/coros). Since 2000 the Research Institute of Organic Agriculture and IFOAM have published

the yearbook *The world of organic agriculture* with data on organic agriculture from countries worldwide.
https://www.organic-world.net/yearbook.html

1.7.2. Food justice

The UN's Special Rapporteur on the Human Right to Food has been at the forefront of bringing the discussion of food and humane food production as a human right to international attention.
https://www.ohchr.org/EN/Issues/Food/Pages/FoodIndex.aspx
The current Rapporteur's blog is here
https://hilalelver.org/

The website of the Special Rapporteur 2008–2014 has links to many documents he produced regarding food, agriculture, biodiversity, food sovereignty, and more.
http://www.srfood.org/

The USDA's Food Access Research Atlas has data by census tract and subpopulation for the US that can be downloaded and used to make maps.
https://www.ers.usda.gov/data-products/food-access-research-atlas.aspx

There are a growing number of organizations and projects devoted to food growing and justice that offer training, mentoring, social networking, and other forms of support. For example, Black Urban Growers and the Native American Food Sovereignty Alliance.
https://www.blackurbangrowers.org/
https://nativefoodalliance.org/

1.8. References

Abdollahi, M., Ranjbar, A., Shadnia, S., Nikfar, S. & Rezaie, A. (2004) Pesticides and oxidative stress: A review. *Medical Science Monitor: International Medical Journal of Experimental and Clinical Research*, 10, RA141-7.

Afable-Munsuz, A., Gregorich, S. E., Markides, K. S. & Pérez-Stable, E. J. (2013) Diabetes risk in older Mexican Americans: Effects of language acculturation, generation and socioeconomic status. *Journal of Cross-Cultural Gerontology*, 28, 35–373.

Alaimo, K., Packnett, E., Miles, R. A. & Kruger, D. J. (2008) Fruit and vegetable intake among urban community gardeners. *Journal of Nutrition Education and Behavior*, 40, 94–101.

Algert, S. J., Baameur, A. & Renvall, M. J. (2014) Vegetable output and cost savings of community gardens in San Jose, California. *Journal of the Academy of Nutrition and Dietetics*, 114, 1072–1076, DOI: 10.1016/j.jand.2014.02.030.

Alkon, A. & Mares, T. (2012) Food sovereignty in US food movements: Radical visions and neoliberal constraints. *Agriculture and Human Values*, 29, 347–359, DOI: 10.1007/s10460-012-9356-z.

Appelhans, B. M. (2009) Neurobehavioral inhibition of reward-driven feeding: Implications for dieting and obesity. *Obesity*, 17, 640–647, DOI: 10.1038/oby. 2008.638.

Archer, E., Hand, G. A. & Blair, S. N. (2013) Validity of U.S. Nutritional Surveillance: National Health and Nutrition Examination Survey caloric energy intake data. *Plos One*, 8, e76632, DOI: 10.1371/journal. pone.0076632.

Armelagos, G. J. (2010) The omnivore's dilemma: The evolution of the brain and the determinants of food choice. *Journal of Anthropological Research*, 66, 161–186.

Aune, D., Norat, T., Romundstad, P. & Vatten, L. J. (2013) Whole grain and refined grain consumption and the risk of type 2 diabetes: A systematic review and dose–response meta-analysis of cohort studies. *European Journal of Epidemiology*, 28, 845–858.

Aune, D., Giovannucci, E., Boffetta, P., Fadnes, L. T., Keum, N., Norat, T., Greenwood, D. C., Riboli, E., Vatten, L. J. & Tonstad, S. (2017) Fruit and vegetable intake and the risk of cardiovascular disease, total cancer and all-cause mortality—a systematic review and dose-response meta-analysis of prospective studies. *International Journal of Epidemiology*, 46, 1029–1056, DOI: 10.1093/ije/dyw319.

Baranski, M., Srednicka-Tober, D., Volakakis, N., Seal, C., Sanderson, R., Stewart, G. B., Benbrook, C., Biavati, B., Markellou, E. & Giotis, C. (2014) Higher antioxidant and lower cadmium concentrations and lower incidence of pesticide residues in organically grown crops: A systematic literature review and meta-analyses. *British Journal of Nutrition*, 112, 794–811.

Barcellos, S. H., Goldman, D. P. & Smith, J. P. (2012) Undiagnosed disease, especially diabetes, casts doubt on some of reported health "advantage" of recent Mexican immigrants. *Health Affairs*, 31, 2727–2737.

Barrett, C. B., Bellemare, M. F. & Hou, J. Y. (2010) Reconsidering conventional explanations of the inverse productivity-size relationship. *World Development*, 38, 88–97, DOI: 10.1016/j.worlddev.2009.06.002.

Barton, J. & Pretty, J. (2010) What is the best dose of nature and green exercise for improving mental health? A multi-study analysis. *Environmental Science & Technology*, 44, 3947–3955, DOI: 10.1021/ es903183r.

Blair, D., Giesecke, C. C. & Sherman, S. (1991) A dietary, social and economic-evaluation of the Philadelphia urban gardening project. *Journal of Nutrition Education*, 23, 161–167.

Bouayed, J., Rammal, H. & Soulimani, R. (2009) Oxidative stress and anxiety: Relationship and cellular pathways. *Oxidative Medicine and Cellular Longevity*, 2, 63–67.

Brandt, K., Leifert, C., Sanderson, R. & Seal, C. (2011) Agroecosystem management and nutritional quality of plant foods: The case of organic fruits and vegetables. *Critical Reviews in Plant Sciences*, 30, 177–197.

Bremer, A. A., Mietus-Snyder, M. & Lustig, R. H. (2012) Toward a unifying hypothesis of metabolic syndrome. *Pediatrics*, 129, 557–570, DOI: 10.1542/peds. 2011–2912.

Brones, A. (2018) Karen Washington: It's not a food desert, it's food apartheid. *Guernica*, May 7, available at: https://www.guernicamag.com/karen-washington-its-not-a-food-desert-its-food-apartheid/ (accessed Oct. 4, 2018).

Bryan, C. J., Yeager, D. S., Hinojosa, C. P., Chabot, A., Bergen, H., Kawamura, M. & Steubing, F. (2016) Harnessing adolescent values to motivate healthier eating. *Proceedings of the National Academy of Sciences*, DOI: 10.1073/pnas.1604586113.

Cameron, R. W. F., Blanuša, T., Taylor, J. E., Salisbury, A., Halstead, A. J., Henricot, B. & Thompson, K. (2012) The domestic garden—its contribution to urban green infrastructure. *Urban Forestry & Urban Greening*, 11, 129–137, DOI: http://dx.doi.org/10.1016/j.ufug.2012. 01.002.

Capewell, S. & Graham, H. (2010) Will cardiovascular disease prevention widen health inequalities? *Plos Medicine*, 7.

Carney, J. & Rosomoff, R. N. (2009) *In the shadow of slavery: Africa's botanical legacy in the Atlantic world*, University of California Press, Berkeley, CA.

Castillo, E. (2015) *Crown of thorns: The enslavement of California's Indians by the Spanish missions*. Craven Street Books, Fresno, CA.

Cavigelli, M. A., Mirsky, S. B., Teasdale, J. R., Spargo, J. T. & Doran, J. (2013) Organic grain cropping systems to enhance ecosystem services. *Renewable Agriculture and Food Systems*, 28, 145–159, DOI: 10.1017/s1742170512000439.

Chan, J., DuBois, B. B., Nemec, K. T., Francis, C. A. & Hoagland, K. D. (2017) Community gardens as urban social–ecological refuges in the global north. In WinklerPrins, A. M. G. A. (ed.) *Global urban agriculture*, 229–241. CAB International, Wallingford, UK.

Cleveland, D. A., Orum, T. V. & Ferguson, N. F. (1985) Economic value of home vegetable gardens in an urban desert environment. *Hortscience*, 20, 694–696.

Cleveland, D. A., Phares, N., Nightingale, K. D., Weatherby, R. L., Radis, W., Ballard, J., Campagna, M., Kurtz, D., Livingston, K., Riechers, G., et al. (2017)

The potential for urban household vegetable gardens to reduce greenhouse gas emissions. *Landscape and Urban Planning*, 157, 365–374, DOI: http://dx.doi.org/10.1016/j.landurbplan.2016.07.008.

Cornwell, T. B. & McAlister, A. R. (2013) Contingent choice: Exploring the relationship between sweetened beverages and vegetable consumption. *Appetite*, 62, 203–208, DOI: http://dx.doi.org/10.1016/j.appet.2012.05.001.

Craig, W. J. (2010) Nutrition concerns and health effects of vegetarian diets. *Nutrition in Clinical Practice*, 25, 613–620, DOI: 10.1177/0884533610385707.

Dangour, A. D., Dodhia, S. K., Hayter, A., Allen, E., Lock, K. & Uauy, R. (2009) Nutritional quality of organic foods: A systematic review. *The American Journal of Clinical Nutrition*, 90, 680–685, DOI: 10.3945/ajcn.2009.28041.

Davis, D. R. (2009) Declining fruit and vegetable nutrient composition: what is the evidence? *HortScience*, 44, 15–19.

De Schutter, O. (2014) Unhealthy diets greater threat to health than tobacco: UN expert calls for global regulation, available at: http://www.srfood.org/en/unhealthy-diets-greater-threat-to-health-than-tobacco-un-expert-calls-for-global-regulation (accessed Oct. 22, 2018).

De Vadder, F., Kovatcheva-Datchary, P., Goncalves, D., Vinera, J., Zitoun, C., Duchampt, A., Bäckhed, F. & Mithieux, G. (2014) Microbiota-generated metabolites promote metabolic benefits via gut-brain neural circuits. *Cell*, 156, 84–96.

Drescher, L. S., Thiele, S. & Mensink, G. B. M. (2007) A new index to measure healthy food diversity better reflects a healthy diet than traditional measures. *The Journal of Nutrition*, 137, 647–651.

Drewnowski, A. (2009) Obesity, diets, and social inequalities. *Nutrition Reviews*, 67, S36–S39, DOI: 10.1111/j.1753-4887.2009.00157.x.

Drewnowski, A. (2010) The cost of US foods as related to their nutritive value. *American Journal of Clinical Nutrition*, 92, 1181–1188, DOI: 10.3945/ajcn.2010.29300.

Du, H., van der A, D. L., Boshuizen, H. C., Forouhi, N. G., Wareham, N. J., Halkjær, J., Tjønneland, A., Overvad, K., Jakobsen, M. U., Boeing, H., et al. (2010) Dietary fiber and subsequent changes in body weight and waist circumference in European men and women. *The American Journal of Clinical Nutrition*, 91, 329–336, DOI: 10.3945/ajcn.2009.28191.

Dunlop, D., Song, J., Arnston, E., Semanik, P., Lee, J., Chang, R. & Hootman, J. (2014) Sedentary time in US older adults associated with disability in activities of daily living independent of physical activity. *Journal of Physical Activity & Health*, 12, 93–101.

Dunn, R. (2012) Human ancestors were nearly all vegetarians. *Scientific American*, available at: https://blogs.scientificamerican.com/guest-blog/human-ancestors-were-nearly-all-vegetarians/ (accessed June 30, 2013).

Egerer, M. H., Philpott, S. M., Bichier, P., Jha, S., Liere, H. & Lin, B. B. (2018) Gardener well-being along social and biophysical landscape gradients. *Sustainability*, 10, 96, DOI:10.3390/su10010096.

Eshed, V., Gopher, A. & Hershkovitz, I. (2006) Tooth wear and dental pathology at the advent of agriculture: New evidence from the Levant. *American Journal of Physical Anthropology*, 130, 145–159.

Esselstyn, C. B., Jr, Gendy, G., Doyle, J., Golubic, M. & Roizen, M. F. (2014) A way to reverse CAD? *Journal of Family Practice*, 63, 356–364b.

EWG (Environmental Working Group) (2017) EWG farm subsidy database. EWG, Washington, DC. Available at: https://farm.ewg.org/index.php (accessed Aug. 5, 2017).

FAO (Food and Agriculture Organization) (2005) *Voluntary guidelines to support the progressive realization of the right to adequate food in the context of national food security*. FAO, Rome, Italy, available at: http://www.fao.org/3/a-y7937e.pdf (accessed Oct. 5, 2018).

FAO (UN Food and Agriculture Organization) (2010) Global trends and future challenges for the work of the organization. Web annex. In: *Addressing food crises—towards the elaboration of an agenda for action in food security in countries in protracted crisis, High-Level Expert Forum (HLEF), Introduction—setting the context of 36th CFS recommendations on further analysis and actions on food security in protracted crisis. SOFI 2010 report*, available at: www.fao.org/docrep/meeting/025/md784e.pdf (accessed July 20, 2018).

Food Chain Workers Alliance (2012) *The hands that feed us: Challenges and opportunites for workers along the food chain*. Food Chain Workers Alliance, Los Angeles, CA.

Fredrickson, B. L., Grewen, K. M., Coffey, K. A., Algoe, S. B., Firestine, A. M., Arevalo, J. M. G., Ma, J. & Cole, S. W. (2013) A functional genomic perspective on human well-being. *Proceedings of the National Academy of Sciences of the United States of America*, 110, 13684–13689, DOI: 10.1073/pnas.1305419110.

Gabriel, D., Sait, S. M., Kunin, W. E. & Benton, T. G. (2013) Food production vs. biodiversity: Comparing organic and conventional agriculture. *Journal of Applied Ecology*, 50, 355–364, DOI: 10.1111/1365-2664.12035.

Girard, A. W., Self, J. L., McAuliffe, C. & Olude, O. (2012) The effects of household food production strategies on the health and nutrition outcomes of women and young children: A systematic review. *Paediatric and Perinatal Epidemiology*, 26, 205–222, DOI: 10.1111/j.1365-3016.2012.01282.x.

Gittleman, M., Jordan, K. & Brelsford, E. (2012) Using citizen science to quantify community garden crop yields. *Cities and the environment (CATE)*, 5, article 4.

Glanz, K., Basil, M., Maibach, E., Goldberg, J. & Snyder, D. A. N. (1998) Why Americans eat what they do: Taste, nutrition, cost, convenience, and weight control concerns as influences on food consumption. *Journal of the American Dietetic Association*, 98, 1118–1126.

Glover, T. D. (2004) Social capital in the lived experiences of community gardeners. *Leisure Sciences*, 26, 143–162.

Goddard, M. A., Dougill, A. J. & Benton, T. G. (2010) Scaling up from gardens: Biodiversity conservation in urban environments. *Trends in Ecology & Evolution*, 25, 90–98.

Gottlieb, R. & Joshi, A. (2010) *Food justice (food, health, and the environment)*. MIT Press, Cambridge, MA.

Grewell, R. (2015) *Urban Farm and Garden Alliance Nelson Report UPDATE*. Kris Nelson Community-Based Research Program, Center for Urban and Regional Affairs, University of Minnesota, Minneapolis, MN, available at: https://conservancy.umn.edu/bitstream/handle/11299/178552/KNCBR%201405.pdf?sequence=1&isAllowed=y (accessed Oct. 5, 2018).

Guthrie, J. & Lin, B.-H. (2014) Healthy vegetables undermined by the company they keep. *Amber Waves*, available at: https://www.ers.usda.gov/amber-waves/2014/may/healthy-vegetables-undermined-by-the-company-they-keep/ (accessed Oct. 22, 2018).

Hale, J., Knapp, C., Bardwell, L., Buchenau, M., Marshall, J., Sancar, F. & Litt, J. S. (2011) Connecting food environments and health through the relational nature of aesthetics: Gaining insight through the community gardening experience. *Social Science & Medicine*, 72, 1853–1863, DOI: http://dx.doi.org/10.1016/j.socscimed.2011.03.044.

Hamer, M., Stamatakis, E. & Steptoe, A. (2009) Dose-response relationship between physical activity and mental health: the Scottish Health Survey. *British Journal of Sports Medicine*, 43, 1111–1114, DOI: 10.1136/bjsm.2008.046243.

Harlan, J. R. (1992) *Crops and man*, 2nd edn. American Society of Agronomy, Inc. and Crop Science Society of America, Inc., Madison, WI.

Harris, J. L. & Fox, T. (2014) Food and beverage marketing in schools: Putting student health at the head of the class. *JAMA Pediatrics*, 168, 206–208.

Harris, J. L., Shehan, C., Gross, R., Kumanyika, S., Lassiter, V., Ramirez, A. G. & Gallion, K. (2015) *Food advertising targeted to Hispanic and Black youth: Contributing to health disparities*. The Rudd Center For Food Policy & Obesity, University of Connecticut, Hartford, CT, available at: http://www.uconnruddcenter.org/files/Pdfs/272-7 Rudd_Targeted Marketing Report_Release_081115%5B1%5D.pdf (accessed Oct. 5, 2018).

Hastings, G. (2012) Why corporate power is a public health priority. *British Medical Journal*, 345.

Hayes, S. C., Villatte, M., Levin, M. & Hildebrandt, M. (2011) Open, aware, and active: Contextual approaches as an emerging trend in the behavioral and cognitive therapies. *Annual Review of Clinical Psychology*, 7, 141–168.

Heinrich Böll Foundation (2014) *The meat atlas*. Heinrich Böll Foundation and Friends of the Earth Europe, Berlin, Germany and Brussels, Belgium.

Hill, K. R., Walker, R. S., Bozicevic, M., Eder, J., Headland, T., Hewlett, B., Hurtado, A. M., Marlowe, F. W., Wiessner, P. & Wood, B. (2011) Co-residence patterns in hunter-gatherer societies show unique human social structure. *Science*, 331, 1286-1289, DOI: 10.1126/science.1199071.

Holt-Lunstad, J. (2018) Why social relationships are important for physical health: a systems approach to understanding and modifying risk and protection. *Annual Review of Psychology*, 69, 437–458, DOI: 10.1146/annurev-psych-122216-011902.

Hooper, P. L., Hooper, P. L., Tytell, M. & Vígh, L. (2010) Xenohormesis: Health benefits from an eon of plant stress response evolution. *Cell Stress and Chaperones*, 15, 761–770.

Hu, D., Huang, J., Wang, Y., Zhang, D. & Qu, Y. (2014) Fruits and vegetables consumption and risk of stroke: A meta-analysis of prospective cohort studies. *Stroke*, 45, 1613–1619, DOI: 10.1161/strokeaha.114.004836.

IAASTD (International Assessment of Agricultural Knowledge, Science and Technology for Development) (2009) *Synthesis report with executive summary: A synthesis of the global and sub-global IAASTD reports*. Island Press, Washington, DC.

IARC (International Agency for Research on Cancer) (2003) *Fruits and vegetables,* Vol. 8. *IARC Handbooks of Cancer Prevention*, IARC, Lyon, France.

Ingram, C. J. E., Mulcare, C. A., Itan, Y., Thomas, M. G. & Swallow, D. M. (2009) Lactose digestion and the evolutionary genetics of lactase persistence. *Human Genetics*, 124, 579–591, DOI: 10.1007/s00439-008-0593-6.

IOM (Institute of Medicine of the National Academies) (2012) *Accelerating progress in obesity prevention: Solving the weight of the nation*. The National Academies Press, Washington, DC.

Jelliffe, D. B. (1972) Commerciogenic malnutrition? *Nutrition Reviews*, 30, 199–205, DOI: 10.1111/j.1753-4887.1972.tb04042.x.

Johnson, W. (2008) *Gardening at the dragon's gate: At work in the wild and cultivated world*. Bantam, New York.

Katz, D. L. & Meller, S. (2014) Can we say what diet is best for health? *Annual Review of Public Health*, 35, 83–103, DOI: 10.1146/annurev-publhealth-032013-182351.

Katzmarzyk, P. T. (2010) Physical activity, sedentary behavior, and health: Paradigm paralysis or paradigm

What Can Food Gardens Contribute? Gardens and Wellbeing

35

shift? *Diabetes*, 59, 2717–2725, DOI: 10.2337/db10-0822.

Kaufman, A., Choo, E., Koh, A. & Dando, R. (2018) Inflammation arising from obesity reduces taste bud abundance and inhibits renewal. *PLOS Biology*, 16, e2001959, DOI: 10.1371/journal.pbio.2001959.

Khoury, C. K., Bjorkman, A. D., Dempewolf, H., Ramirez-Villegas, J., Guarino, L., Jarvis, A., Rieseberg, L. H. & Struik, P. C. (2014) Increasing homogeneity in global food supplies and the implications for food security. *Proceedings of the National Academy of Sciences*, DOI: 10.1073/pnas.1313490111.

Killingsworth, M. A. & Gilbert, D. T. (2010) A wandering mind is an unhappy mind. *Science*, 330, 932–932.

Kim, G. H. & Lee, H. M. (2009) Frequent consumption of certain fast foods may be associated with an enhanced preference for salt taste. *Journal of Human Nutrition and Dietetics*, 22, 475–480. DOI: 10.1111/j.1365-277X.2009.00984.x.

Koh, A., De Vadder, F., Kovatcheva-Datchary, P. & Bäckhed, F. (2016) From dietary fiber to host physiology: Short-chain fatty acids as key bacterial metabolites. *Cell*, 165, 1332–1345, DOI: https://doi.org/10.1016/j.cell.2016.05.041.

Krasny, M. E. & Tidball, K. G. (2017) Community gardens as contexts for science, stewardship, and civic action learning. *Urban Horticulture: Ecology, Landscape, and Agriculture*, 267.

Laland, K. N., Odling-Smee, J. & Myles, S. (2010) How culture shaped the human genome: Bringing genetics and the human sciences together. *Nature Reviews Genetics*, 11, 137–148.

Larsen, C. S. (2006) The agricultural revolution as environmental catastrophe: Implications for health and lifestyle in the Holocene. *Quaternary International*, 150, 12–20, DOI: 10.1016/j.quaint.2006.01.004.

Lattimer, J. M. & Haub, M. D. (2010) Effects of dietary fiber and its components on metabolic health. *Nutrients*, 2, 1266–1289.

Lawson, L. J. (2005) *City bountiful: A century of commuity gardening in America*. University of California Press, Berkeley, CA.

Ledoux, T. A., Hingle, M. D. & Baranowski, T. (2011) Relationship of fruit and vegetable intake with adiposity: A systematic review. *Obesity Reviews*, 12, e143-e150, DOI: 10.1111/j.1467-789X.2010.00786.x.

Levenstein, H. A. (1996) The politics of nutrition in North America. *Neuroscience & Biobehavioral Reviews*, 20, 75–78, DOI: https://doi.org/10.1016/0149-7634(95)00036-E.

Li, D. (2014) Effect of the vegetarian diet on non-communicable diseases. *Journal of the Science of Food and Agriculture*, 94, 169–173, DOI: 10.1002/jsfa.6362.

Li, Y., Pan, A., Wang, D. D., Liu, X., Dhana, K., Franco, O. H., Kaptoge, S., Di Angelantonio, E., Stampfer, M., Willett, W. C., et al. (2018) Impact of healthy lifestyle factors on life expectancies in the US population. *Circulation*, 137, DOI: 10.1161/circulationaha.117.032047.

Litt, J. S., Soobader, M.-J., Turbin, M. S., Hale, J. W., Buchenau, M. & Marshall, J. A. (2011) The influence of social involvement, neighborhood aesthetics, and community garden participation on fruit and vegetable consumption. *American Journal of Public Health*, 101, 1466–1473.

Livesey, G., Taylor, R., Hulshof, T. & Howlett, J. (2008) Glycemic response and health—a systematic review and meta-analysis: Relations between dietary glycemic properties and health outcomes. *The American Journal of Clinical Nutrition*, 87, 258S–268S.

LVC (La Vía Campesina) (2008) *La via Campesina policy documents*. 5th conference, Mozambique, October 16–23, available at: https://viacampesina.org/en/la-via-campesina-policy-documents/ (accessed July 28, 2018).

Malik, V. S., Popkin, B. M., Bray, G. A., Després, J.-P., Willett, W. C. & Hu, F. B. (2010) Sugar-sweetened beverages and risk of metabolic syndrome and type 2 diabetes: A meta-analysis. *Diabetes Care*, 33, 2477–2483, DOI: 10.2337/dc10-1079.

Mason, A. E., Epel, E. S., Aschbacher, K., Lustig, R. H., Acree, M., Kristeller, J., Cohn, M., Dallman, M., Moran, P. J., Bacchetti, P., et al. (2016) Reduced reward-driven eating accounts for the impact of a mindfulness-based diet and exercise intervention on weight loss: Data from the SHINE randomized controlled trial. *Appetite*, 100, 86–93, DOI: 10.1016/j.appet.2016.02.009.

Mattes, R. D. (2012) Spices and energy balance. *Physiology & Behavior*, 107, 584–590.

Matthews, C. E., Chen, K. Y., Freedson, P. S., Buchowski, M. S., Beech, B. M., Pate, R. R. & Troiano, R. P. (2008) Amount of time spent in sedentary behaviors in the United States, 2003–2004. *American Journal of Epidemiology*, 167, 875–881, DOI: 10.1093/aje/kwm390.

Mattson, M. P. (2015) What doesn't kill you. *Scientific American*, 313, 40–45.

McClintock, N. (2014) Radical, reformist, and garden-variety neoliberal: Coming to terms with urban agriculture's contradictions. *Local Environment*, 19, 147–171, DOI: 10.1080/13549839.2012.752797.

Menzel, P. & D'Aluisio, F. (2005) *Hungry planet: What the world eats*. Ten Speed Press, Berkeley, CA.

Millennium Ecosystem Assessment (2005) *Ecosystems and human well-being: Synthesis*. World Resources Institute, Washington, DC.

Miller, E. M. (2014) Iron status and reproduction in US women: National Health and Nutrition Examination Survey, 1999–2006. *PLoS ONE*, 9, e112216, DOI: 10.1371/journal.pone.0112216.

Mohatt, N. V., Thompson, A. B., Thai, N. D. & Tebes, J. K. (2014) Historical trauma as public narrative: A conceptual review of how history impacts present-day health. *Social Science & Medicine*, 106, 128–136, DOI: http://dx.doi.org/10.1016/j.socscimed.2014.01.043.

Moore, L. V. & Thompson, F. E. (2015) Adults meeting fruit and vegetable intake recommendations—United States. *Morbidity and Mortality Weekly Report (MMWR)*, 64, 709–713.

Morrison, D. J. & Preston, T. (2016) Formation of short chain fatty acids by the gut microbiota and their impact on human metabolism. *Gut Microbes*, 7, 189–200, DOI: 10.1080/19490976.2015.1134082.

Mozaffarian, D., Hao, T., Rimm, E. B., Willett, W. C. & Hu, F. B. (2011) Changes in diet and lifestyle and long-term weight gain in women and men. *New England Journal of Medicine*, 364, 2392–2404, DOI: doi:10.1056/NEJMoa1014296.

Nestle, M. (2002) *Food politics: How the food industry influences nutrition and health*. University of California Press, Berkeley, CA.

Nestle, M. & Nesheim, M. (2012) *Why calories count: From science to politics*, Vol. 33. University of California Press, Berkeley, CA.

Netting, R. M. (1993) *Smallholders, householders: Farm families and the ecology of intensive, sustainable agriculture*. Stanford University Press, Stanford, CA.

NGA (National Gardening Association) (2009) *The impact of home and community gardening in America*, National Gardening Association, South Burlington, VT. Available at: https://garden.org/learn/articles/view/3126/ (accessed Oct. 18, 2018).

NGA (National Gardening Association) (2014) Garden to Table: A 5-year look at food gardening in America. National Gardening Association, available at: http://www.hagstromreport.com/assets/2014/2014_0402_NGA-Garden-to-Table.pdf (accessed March 3, 2016).

Norgaard, R. B. (2010) Ecosystem services: From eye-opening metaphor to complexity blinder. *Ecological Economics*, 69, 1219–1227, DOI: 10.1016/j.ecolecon.2009.11.009.

Owen, N., Bauman, A. & Brown, W. (2009) Too much sitting: A novel and important predictor of chronic disease risk? *British Journal of Sports Medicine*, 43, 81–83, DOI: 10.1136/bjsm.2008.055269.

Oyebode, O., Gordon-Dseagu, V., Walker, A. & Mindell, J. S. (2014) Fruit and vegetable consumption and all-cause, cancer and CVD mortality: Analysis of Health Survey for England data. *Journal of Epidemiology and Community Health*, DOI: 10.1136/jech-2013-203500.

P-Patch (2014) P-Patch narrative information sheet. Seattle Department of Neighborhoods, Seattle, WA, available at: http://www.seattle.gov/Documents/Departments/Neighborhoods/PPatch/Narrative-Information-Sheet.pdf (accessed July 29, 2018).

Palafox-Carlos, H., Ayala-Zavala, J. F. & González-Aguilar, G. A. (2011) The role of dietary fiber in the bioaccessibility and bioavailability of fruit and vegetable antioxidants. *Journal of Food Science*, 76, R6–R15, DOI: 10.1111/j.1750-3841.2010.01957.x.

Park, S.-A., Shoemaker, C. A. & Haub, M. D. (2008) A preliminary investigation on exercise intensities of gardening tasks in older adults. *Perceptual and Motor Skills*, 107, 974–980, DOI: 10.2466/pms.107.3.974-980.

Park, Y., Subar, A. F., Hollenbeck, A. & Schatzkin, A. (2011) Dietary fiber intake and mortality in the NIH-AARP diet and health study. *Archives of Internal Medicine*, 171, 1061–1068, DOI: 10.1001/archinternmed.2011.18.

Pataki, D. E., Carreiro, M. M., Cherrier, J., Grulke, N. E., Jennings, V., Pincetl, S., Pouyat, R. V., Whitlow, T. H. & Zipperer, W. C. (2011) Coupling biogeochemical cycles in urban environments: Ecosystem services, green solutions, and misconceptions. *Frontiers in Ecology and the Environment*, 9, 27–36, DOI: 10.1890/090220.

Penedo, F. J. & Dahn, J. R. (2005) Exercise and well-being: a review of mental and physical health benefits associated with physical activity. *Current Opinion in Psychiatry*, 18, 189–193.

Platte, P., Herbert, C., Pauli, P. & Breslin, P. A. S. (2013) Oral perceptions of fat and taste stimuli are modulated by affect and mood induction. *PLoS ONE*, 8, e65006, DOI: 10.1371/journal.pone.0065006.

Ponce, X., Ramirez, E. & Delisle, H. (2006) A more diversified diet among Mexican men may also be more atherogenic. *The Journal of Nutrition*, 136, 2921–2927.

Popkin, B. M. (1993) Nutritional patterns and transitions. *Population and Development Review*, 19, 138–157.

Popkin, B. M. (2006) Technology, transport, globalization and the nutrition transition food policy. *Food Policy*, 31, 554–569, DOI: 10.1016/j.foodpol.2006.02.008.

Prentice, A. M. (2005) Early influences on human energy regulation: Thrifty genotypes and thrifty phenotypes. *Physiology & Behavior*, 86, 640–645, DOI: http://dx.doi.org/10.1016/j.physbeh.2005.08.055.

Pretty, J., Griffin, M., Peacock, J., Hine, R., Sellens, M. & South, N. (2005) A countryside for health and wellbeing: The physical and mental health benefits of green exercise, executive summary. Countryside Recreation Network, Sheffield Hallam University, Sheffield, UK,

Pudup, M. B. (2008) It takes a garden: Cultivating citizen-subjects in organized garden projects. *Geoforum*, 39, 1228–1240.

Putnam, J. (2000) Major trends in U.S. food supply, 1909–99. *Food Review*, 23, 8–15.

Quastel, N. (2009) Political ecologies of gentrification. *Urban Geography*, 30, 694–725.

Ramírez, M. M. (2015) The elusive inclusive: Black food geographies and racialized food spaces. *Antipode*, 47, 748–769.

Rauzon, S., Wang, M., Studer, N. & Crawford, P. (2010) *Changing students' knowledge, attitudes and behavior*

What Can Food Gardens Contribute? Gardens and Wellbeing

37

in relation to food: An evaluation of the school lunch initiative. Final Report. A report by the Dr. Robert C. and Veronica Atkins Center for Weight and Health, University of California at Berkeley, CA. Commissioned by the Chez Panisse Foundation, available at: https://edibleschoolyard.org/resource/evaluation-school-lunch-initiative (accessed Oct. 22, 2018).

Reitzel, L. R., Regan, S. D., Nguyen, N., Cromley, E. K., Strong, L. L., Wetter, D. W. & McNeill, L. H. (2014) Density and proximity of fast food restaurants and body mass index among African Americans. *American Journal of Public Health*, 104, 110–116, DOI: 10.2105/AJPH.2012.301140.

Rhodes, R. E., Janssen, I., Bredin, S. S. D., Warburton, D. E. R. & Bauman, A. (2017) Physical activity: Health impact, prevalence, correlates and interventions. *Psychology & Health*, 32, 942–975, DOI: 10.1080/08870446.2017.1325486.

Rideout, V. J., Foehr, U. G. & Roberts, D. F. (2010) *Generation M2. Media in the lives of 8- to 18-year-olds*, Henry J. Kaiser Family Foundation, available at: https://www.kff.org/other/event/generation-m2-media-in-the-lives-of/ (accessed July 29, 2018).

Rolls, B. J., Roe, L. S. & Meengs, J. S. (2004) Salad and satiety: Energy density and portion size of a first-course salad affect energy intake at lunch. *Journal of the American Dietetic Association*, 104, 1570–1576, DOI: 10.1016/j.jada.2004.07.001.

Sandel, M. J. (2012) *What money can't buy*. Farrar, Straus and Giroux, New York, NY.

Savoie-Roskos, M. R., Wengreen, H. & Durward, C. (2017) Increasing fruit and vegetable intake among children and youth through gardening-based interventions: A systematic review. *Journal of the Academy of Nutrition and Dietetics*, 117, 240–250, DOI: https://doi.org/10.1016/j.jand.2016.10.014.

Scarborough, P., Allender, S., Clarke, D., Wickramasinghe, K. & Rayner, M. (2012) Modelling the health impact of environmentally sustainable dietary scenarios in the UK. *European Journal of Clinical Nutrition*, 66, 710–715, DOI: 10.1038/ejcn.2012.34.

Shinew, K. J., Glover, T. D. & Parry, D. C. (2004) Leisure spaces as potential sites for interracial interaction: Community gardens in urban areas. *Journal of Leisure Research*, 36, 336–355, DOI: 10.1080/00222216.2004.11950027.

Siegel, K. R., McKeever Bullard, K., Imperatore, G., Kahn, H. S., Stein, A. D., Ali, M. K. & Narayan, K. M. (2016) Association of higher consumption of foods derived from subsidized commodities with adverse cardiometabolic risk among US adults. *JAMA Internal Medicine*, 176, 1124–1132, DOI: 10.1001/jamainternmed.2016.2410.

Smith-Spangler, C., Brandeau, M. L., Hunter, G. E., Bavinger, J. C., Pearson, M., Eschbach, P. J., Sundaram, V., Liu, H., Schirmer, P. & Stave, C. (2012) Are organic foods safer or healthier than conventional alternatives? A systematic review. *Annals of Internal Medicine*, 157, 348–366.

Stanford, C. B. (1999) *The hunting apes: Meat eating and the origins of human behavior*. Princeton University Press, Princeton, NJ.

Stewart, J. E. & Keast, R. S. J. (2012) Recent fat intake modulates fat taste sensitivity in lean and overweight subjects. *Intenational Journal of Obesity*, 36, 834–842.

Ströhe, A. (2009) Physical activity, exercise, depression and anxiety disorders. *Journal of Neural Transmission*, 116, 777–784.

Talukder, A., Haselow, N. J., Osei, A. K., Villate, E., Reario, D., Kroeun, H., SokHoing, L., Uddin, A., Dhunge, S. & Quinn, V. (2010) Homestead food production model contributes to improved household food security and nutrition status of young children and women in poor populations: Lessons learned from scaling-up programs in Asia (Bangladesh, Cambodia, Nepal and Philippines). *Field Actions Science Reports*, Special Issue 1, 1–9.

Teig, E., Amulya, J., Bardwell, L., Buchenau, M., Marshall, J. A. & Litt, J. S. (2009) Collective efficacy in Denver, Colorado: Strengthening neighborhoods and health through community gardens. *Health & Place*, 15, 1115–1122.

Tilman, D. & Clark, M. (2014) Global diets link environmental sustainability and human health. *Nature*, 515, 518–522, DOI: 10.1038/nature13959.

Tudor-Locke, C., Leonardi, C., Johnson, W. D. & Katzmarzyk, P. T. (2011) Time spent in physical activity and sedentary behaviors on the working day: The American time use survey. *Journal of Occupational & Environmental Medicine*, 53, 1382–1387.

Turkle, S. (2011) *Alone together: Why we expect more from technology and less from each other*. Basic Books, New York, NY.

Ungar, P. S. & Sponheimer, M. (2011) The diets of early hominins. *Science*, 334, 190–193, DOI: 10.1126/science.1207701.

USBLS (U.S. Bureau of Labor Statistics) (2016) Consumer expenditures midyear update—July 2014 through June 2015 average, available at: http://www.bls.gov/news.release/cesmy.nr0.htm (accessed Sep. 21, 2016).

USBLS (U.S. Bureau of Labor Statistics) (2017) American time use summary. US Department of Labor, Washington, D.C, available at: https://www.bls.gov/news.release/atus.t11a.htm (accessed July 29, 2018).

USDA (United States Department of Agriculture) (2010) *Dietary guidelines for Americans 2010*, available at: http://www.cnpp.usda.gov/Publications/DietaryGuidelines/2010/PolicyDoc/PolicyDoc.pdf (accessed March 23, 2018).

USDA ARS (2018) USDA Food Composition Databases [Online]. US Department of Agriculture, Agricultural Research Service, Nutrient Data Laboratory, Available at: https://ndb.nal.usda.gov/ndb/ (accessed March 14, 2019).

USDA & HHS (Department of Agriculture and U.S. Department of Health and Human Services) (2015a) *201–2020 Dietary guidelines for Americans*, 8th edn. USDA, HHS, Washington, DC, available at: http://health.gov/dietaryguidelines/2015/guidelines/ (acessed Oct. 5, 2018).

USDA & HHS (US Departments of Agriculture and Health and Human Services) (2015b) *Scientific report of the 2015 Dietary Guidelines Advisory Committee*. USDA, HHS, Washington, DC, available at: http://www.health.gov/dietaryguidelines/2015-scientific-report/ (accssed Oct. 5, 2018).

Veerman, J. L., Van Beeck, E. F., Barendregt, J. J. & Mackenbach, J. P. (2009) By how much would limiting TV food advertising reduce childhood obesity? *The European Journal of Public Health*, ckp039.

Vitiello, D. & Nairn, M. (2009) *Community gardening in Philadelphia: 2008 harvest report*. Penn Planning and Urban Studies, University of Pennsylvania, PA. Available at: https://sites.google.com/site/harvestreportsite/philadelphia-report (accessed July 29, 2018).

Voicu, I. & Been, V. (2008) The effect of community gardens on neighboring property values. *Real Estate Economics*, 36, 241–283.

Waliczekz, T., Mattson, R. & Zajicek, J. (1996) Benefits of community gardening on quality-of-life issues. *Journal of Environmental Horticulture*, 14, 194–198.

Waliczek, T. M., Zajicek, J. M. & Lineberger, R. D. (2005) The influence of gardening activities on consumer perceptions of life satisfaction. *HortScience*, 40, 1360–1365.

Watts, D. P., Potts, K. B., Lwanga, J. S. & Mitani, J. C. (2012) Diet of chimpanzees (*Pan troglodytes schweinfurthii*) at Ngogo, Kibale National Park, Uganda, 1. Diet composition and diversity. *American Journal of Primatology*, 74, 114–129.

Wen Li, W., Hodgetts, D. & Ho, E. (2010) Gardens, transitions and identity reconstruction among older Chinese immigrants to New Zealand. *Journal of Health Psychology*, 15, 786–796, DOI: 10.1177/1359105310368179.

White, M. P., Alcock, I., Wheeler, B. W. & Depledge, M. H. (2013) Would you be happier living in a greener urban area? A fixed-effects analysis of panel data. *Psychological Science*, 24, 920–928, DOI: 10.1177/0956797612464659.

Wilson, E. O. (1984) *Biophilia*. Harvard University Press, Cambridge, MA.

Witt, U. (2017) The evolution of consumption and its welfare effects. *Journal of Evolutionary Economics*, 27, 273–293.

Wolch, J. R., Byrne, J. & Newell, J. P. (2014) Urban green space, public health, and environmental justice: The challenge of making cities "just green enough." *Landscape and Urban Planning*, 125, 234–244, DOI: http://dx.doi.org/10.1016/j.landurbplan.2014.01.017.

Wong, R., Palloni, A. & Soldo, B. J. (2007) Wealth in middle and old age in Mexico: The role of international migration. *International Migration Review*, 41, 127–151, DOI: 10.1111/j.1747-7379.2007.00059.x.

Zheng, H., Lenard, N., Shin, A. & Berthoud, H.-R. (2009) Appetite control and energy balance regulation in the modern world: Reward-driven brain overrides repletion signals. *International Journal of Obesity*, 33, S8–S13.

What Can Food Gardens Contribute? Gardens and Wellbeing

39

Appendix 1A Nutrients and Energy in Garden Foods

Table 1A.1. Nutrient requirements and nutrient content of garden produce

	Vitamins			Minerals (mg)					Macronutrients and fiber (g)				Energy (kcal)
	A (IU)[a]	C (mg)	E (mg)	Ca	Fe	K	Mg	Zn	Protein	Fiber	Fat, total	Sugars	
RDA (recommended dietary allowance, units day⁻¹)													
RDA for active 40-year-old female	14,000	75	15	1000[b]	18	90[b]	320	8	46	25[b]	ND	130[c]	[d]
RDA for active-40-year old male	18,000	90	15	1000[b]	8	120[b]	420	11	56	38[b]	ND	130[c]	[d]
Garden produce (nutrient content of one cup prepared volume, with gram equivalent given)													
Almonds, raw whole (143 g)	3	0.0	36.65	385	5.31	1048	386	4.46	30.24	17.9	71.40	6.22	828
Black beans, boiled (172 g)	0	0.0	NA	385	3.61	611	120	1.93	15.24	15.0	0.93	0.55	227
Carrots, raw, grated (110 g)	18,377	6.5	0.73	36	0.33	352	13	0.26	1.02	3.1	0.26	5.21	45
Chard, boiled and drained (175 g)	10,717	31.5	3.31	102	3.95	961	150	0.58	3.29	3.7	0.14	1.93	35
Cilantro, fresh, raw leaves (16 g)	1,080	4.3	0.40	11	0.28	83	4	0.08	0.34	0.4	0.08	0.14	4
Collards, boiled and drained (190 g)	14,440	34.6	1.67	268	2.15	222	40	0.44	5.15	7.6	1.37	0.76	63
Green beans, boiled (125 g)	800	12.1	0.56	55	0.81	182	22	0.31	2.36	4.0	0.35	1.94	44
Jujube fruit, fresh, raw (100 g)	40	69.0	NA	21	0.48	250	10	0.05	1.20	NA	0.20	NA	79
Loquat, fresh, raw, chopped (149 g)	2,277	1.5	NA	24	0.42	396	19	0.07	0.64	2.5	0.30	NA	70
Nopales, cooked (141 g)	660	7.9	0.00	244	0.74	291	70	0.31	2.01	3.0	0.07	1.67	22
Olives, cured, green (100 g)	400	0.0	3.81	52	0.49	42	11	0.04	1.03	3.3	15.22	0.54	145
Persimmon, fresh, raw (100 g)	1,620	7.5	0.73	8	0.15	161	9	0.11	0.58	3.6	0.19	12.53	70
Picante red chile pepper, raw (150 g)	1,440	215.6	1.04	21	1.54	483	34	0.39	2.80	2.2	0.66	7.95	60
Pomegranate, fresh, arils only (174g)	0	17.7	1.04	17	0.52	411	21	0.61	2.91	7.0	2.04	23.79	144
Tomatoes, raw, sliced (180 g)	1,520	24.7	0.97	18	0.49	427	20	0.31	1.58	2.2	0.36	4.73	32
Winter acorn squash, baked (205 g)	877	22.1	NA	90	1.91	896	88	0.35	2.30	9.0	0.29	NA	115
Yard long beans, boiled (104 g)	480	16.8	NA	46	1.02	302	44	0.37	2.63	NA	0.10	NA	49
Examples of animal products for comparison (nutrient content of one cup prepared volume, with gram equivalent given)													
Cheddar cheese, shredded (113 g)	1,403	0	0.80	802	0.16	86	31	4.11	25.84	0	37.64	0.54	455
Chicken breast, roasted (140 g)	130	0	0.38	20	1.50	343	38	1.43	41.72	0	10.89	0.00	276
Cooked ground beef (100 g)	22	0	0.43	25	2.67	353	22	6.19	25.07	0	14.53	0.00	240
Low fat, 1% milk, vitamin A & D fortified (244 g)	478	0	0.02	305	0.07	366	27	1.02	8.22	0	2.37	12.69	102

RDAs based on Otten et al. (2006); nutrient content based on USDA ARS (2018).

IU = international units; RDA = recommended dietary allowance. "Average daily dietary nutrient intake level sufficient to meet the nutrient requirement of nearly all (>97%) of individuals in a particular life stage and gender group" (Otten et al., 2006); ND = not determined; NA = not available.

[a]20 IUs of beta carotene ≅ 1 RAE (retinol activity equivalent) (https://ods.od.nih.gov/factsheets/VitaminA-HealthProfessional/). Nutritionists use RAE, but we use IUs in order to visualize and compare nutrients between figures and tables, and because IUs are the units typically given on food labels.

[b]Given only as adequate intake (AI), sufficient minimum intake for health.

[c]Carbohydrates (starches and sugars).

[d]Depends on height, physical activity, and other factors.

Table 1A.2. Comparison of some energy-dense foods with some nutrient-dense foods from the garden

Food (g equivalent)	Water (g)	Protein (g)	Total fat (g)	Carbo-hydrate (g)	Sugars (g)	Total fiber (g)	Energy (kcal)	Vitamins			Minerals (mg)				
								A (IU)	C (mg)	E (mg)	Ca	Fe	Mg	K	Na
Content kcal⁻¹															
Energy-dense foods															
French fries (0.35 g)	0.14	0.01	0.05	0.15	0.00	0.0	1	0	0.0	0.01	0	0.00	0	2	1
Potato chips (0.17 g)	0.00	0.01	0.06	0.09	0.00	0.0	1	0	0.0	0.02	0	0.00	0	2	1
Sugar doughnut (0.23 g)	0.04	0.01	0.05	0.12	NA	0.0	1	0	0.0	NA	0	0.00	0	0	1
Cola soda drink (2.44 g)	2.14	0.00	0.01	0.25	0.24	0.0	1	0	0.0	0.00	0	0.00	0	0	0
Nutrient-dense foods from the garden															
Raw carrot sticks (2.40 g)	2.12	0.02	0.01	0.23	0.11	0.1	1	402	0.1	0.02	1	0.10	0	8	2
Boiled, fresh green beans (2.90 g)	2.59	0.05	0.01	0.23	0.11	0.1	1	18	0.3	0.01	1	0.02	0	4	0
Cooked collards (3.00 g)	2.71	0.08	0.02	0.17	0.01	0.1	1	228	0.5	0.03	4	0.03	0	4	0
Boiled amaranth leaves (4.80 g)	4.39	0.10	0.01	0.20	NA	NA	1	133	2.0	NA	10	0.11	3	31	1
Content cup⁻¹ volume															
Energy-dense foods															
French fries (117 g)	45.10	4.01	17.23	48.48	0.35	4.4	365	0	5.5	1.95	21	0.95	41	677	246
Potato chips (227 g)	4.22	14.51	77.13	122.19	0.75	7.0	1208	0	49.0	23.72	48	2.91	143	2715	1196
Sugar doughnut (45 g)	8.82	2.43	10.30	22.86	NA	0.7	192	4	0.0	NA	27	0.48	8	46	181
Cola soda drink (246 g)	219.82	0.00	0.61	25.49	24.45	0.0	103	0	0.0	0.00	2	0.05	0	12	7
Nutrient-dense foods from the garden															
Raw carrot sticks (122 g)	107.70	1.13	0.29	11.69	5.78	3.4	50	20,381	7.2	0.81	40	0.37	15	390	84
Boiled, fresh green beans (125 g)	111.53	2.36	0.35	9.85	4.54	4.0	44	791	12.1	0.57	55	0.81	22	182	1
Cooked collards (190 g)	171.34	5.15	1.37	10.73	0.76	7.6	63	14,440	34.6	1.67	268	2.15	40	222	28
Boiled amaranth leaves (132 g)	121.00	2.79	0.24	5.43	NA	NA	28	3656	54.3	NA	276	2.98	73	846	28

(USDA ARS, 2018)
IU = international units; NA = not available.

References

Otten, J. J., Hellwig, J. P. & Meyers, L. D. (2006) *Dietary reference intakes: The essential guide to nutrient requirements*. National Academies Press, Atlanta, GA.

USDA ARS (2018) USDA food composition databases. US Department of Agriculture, Agricultural Research Service, Nutrient Data Laboratory, Beltsville, MD, available at: https://ndb.nal.usda.gov/ndb/ (accessed Oct. 18, 2018).

2 Changes Coming to your Garden

D. Soleri, S.E. Smith, D.A. Cleveland

Chapter 2 in a nutshell.

- Four major trends are affecting gardeners and food gardens:
  - declining quantity and quality of many garden resources;
  - changing climate;
  - rising inequity in society;
  - an aging population, that is more urban, diverse, and unhealthy than 50 years ago.
- To respond to these changes in ways that will support our environmental and social goals, we need to distinguish between familiar variation, such as year-to-year variation in the date of first and last frost, and directional trends we haven't experienced before, such as increasing average temperatures over years.
- The best responses to trends help us adapt to them, and help to slow or reverse negative trends.

Every year, many people start food gardening for the first time, and it's all new to them. But gardening year to year is never exactly the same, even if you have gardened for a long time, because weather, soils, plants, insects, your schedules, your neighbors, your goals, and many other variables change from year to year. This year is always different, in some way, from last year, so as gardeners we are always learning and adapting. However, while year-to-year familiar variation is the norm in gardening, there are long-term directional changes, or trends, occurring now, that are the result of human actions. These trends are bringing conditions we have not experienced before, which means that even seasoned gardeners will have to find new responses.

In this chapter we step back to see the larger picture of the connection between our gardens and what is happening in our communities and environments. We look at how change in the form of familiar variation is different from change in the form of trends, and then outline the four major trends affecting food gardens and gardeners. Chapter 3 is about basic strategies for responding to these changes in ways that can avoid or minimize any negative impacts on you, your garden, community, and environment.

2.1. Change: Is it Variation or a Trend?

We have all experienced change in many different forms, it's inseparable from life! In this chapter we focus on change as a trend over time, and not the periodic, *familiar variation* that is always part of gardening. Variation in a variable like the amount of annual rainfall can be large across years, but if the pattern of variation is familiar over years, for example, if the mean millimeters of annual rainfall, and the standard deviation of the mean, remain the same over time, there is no directional change. But when there is cumulative directional change over time, a *trend* is present (Fig. 2.1). The effect of trends on gardening can be mostly negative, like the increase in temperature described below, or positive, like the increasing

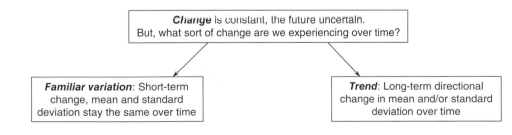

Figure 2.1. Variation and trend

support for community gardens by state legislatures (NGA, 2014). We focus on negative trends and how gardeners can respond successfully to them—responding to positive trends is much easier.

Confusion between variation (familiar, recurrent change) and a trend (novel, directional change) has been a problem for public understanding, including of anthropogenic climate change (ACC). For example, short-term variation can be mistaken as evidence of no long-term trend, and this has been used by climate change deniers to feed public doubt and stall actions needed to address causes of climate trends like greenhouse gas emissions (Oreskes et al., 1994; Oreskes and Conway, 2010). This is what happened in 2014 when a spell of unusually cold weather in the northcentral US and parts of Britain was claimed by climate change deniers as

evidence that global warming is a hoax; in fact, it appears that extreme cold was likely related to climate change (Overland et al., 2016).

In Fig. 2.2 we show the difference between variation and trend for mean maximum annual temperature in Sacramento, California from 1880–2017 based on observations, and 2018–2099, based on climate projections (see Box 2.3 for definitions). Year-to-year variation is present in both periods, but there is a statistically significant positive trend (upward slope of the line on a graph, representing the ratio of changes on the y axis to those on the x axis for the line) for both (Box 2.1). For example, looking only at a few data points such as 1934, 24°C (75.2°F) and 1963, 22°C (71.6°F) or 2010, 23°C (73.4°F) (see circled points on Fig 2.2), you would not discern a warming trend—the more recent years were both

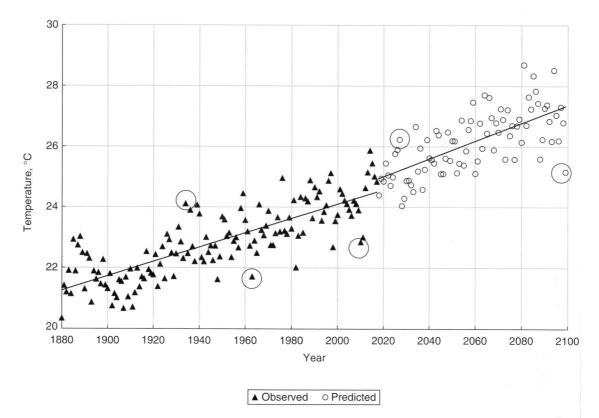

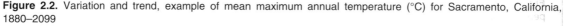

Figure 2.2. Variation and trend, example of mean maximum annual temperature (°C) for Sacramento, California, 1880–2099

Linear trends are presented for two periods: observed 1880–2017, significantly different from zero, P <0.001, r = 0.802, slope = 0.0232; and projected 2018–2099, significantly different from zero, P <0.001, r = 0.681, slope = 0.0295. Linear trend for the entire period (1880–2099) observed plus projected: significantly different from zero, P <0.001, r = 0.925, slope = 0.0282. Circled points discussed in text. Projections based on moderate emissions scenario (RCP 4.5) and HadGEM2-ES model. Cal-adapt (2018); SCENIC (2018)

Box 2.1. Visualizing trends

Figures 2.2 and 2.3 illustrate how trendlines can help visualize and identify trends in variables through time. Trendlines, which can be straight lines, are drawn to minimize the average distance from each of the data points and the line, and are often referred to as "best-fit" lines. If the trend is not significant over time (x axis), that is, there is no trend, the trendline will be close to horizontal with a slope close to 0. If the value of the y axis variable increases significantly through time, the trendline will move upward, and the slope will increase, and be positive. Likewise, if the value of the variable decreases, the trend line will move downward, and the slope will decrease, and be negative. For example, a slope of 0.0237 for the line depicting mean maximum annual temperature from 1880–2017 in Sacramento (Fig. 2.2) indicates that temperature (y axis on the graph) increased by about 0.0237°C (0.043°F) per year (x axis on the graph) over this period. Over the entire 137 years this change is nearly 3.25°C (5.85°F).

The trendline for August maximum temperature at El Centro, California (Fig. 2.3) during the 68-year period from 1950–2017 is nearly horizontal over time, with a mean value of 41.4°C (106.5°F) for this period. The significance of trendlines can be assessed by the *correlation coefficient* (*r*), a statistic calculated using correlation analysis, which measures the strength of the relationship between change in the x axis variable and change in the y axis variable. Correlation coefficients will take on the sign of the slope. When the data points are relatively close to the line, the correlation coefficient will be closer to its maximum values of 1.0 or –0.1. A correlation coefficient of 0 indicates that variation in the x variable is completely unrelated to that for the other variable (y axis). Calculating correlation coefficients involves simultaneous assessment of both the trend (slope) and the variation, defined as the standard deviation around the mean. With relatively high levels of variation, trends must be more pronounced (larger positive or negative slope) in order to be statistically significant, and therefore different than zero. The statistical significance of a correlation coefficient is also affected by the number of data points involved. Given equal amounts of variation, a slope is more likely to be considered significantly different from zero if it is based on more, compared to fewer, data points.

The correlation coefficient for 1950–2017 at El Centro (mean: 41.4°C, 106.5°F) is –0.118, and this coefficient is not statistically different than zero (*P* = 0.338), and we would conclude that there is no relationship between the two variables (August maximum

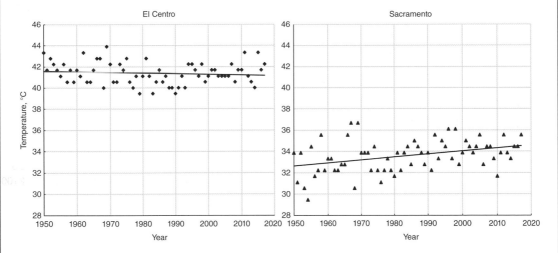

Figure 2.3. Observed mean maximum August temperature (°C) at El Centro, CA and Sacramento, CA for the period 1950–2017

Slopes of linear trendlines: El Centro, –0.0062 (not significantly different from zero, *P* = 0.338, *r* = –0.118); Sacramento, 0.0282 (significantly different from zero, *P* = 0.001, *r* = 0.370). Projections based on moderate emissions scenario (RCP 4.5) and HadGEM2-ES model.
Cal-adapt (2018); SCENIC (2018)

Continued

Box 2.1. Continued.

temperature and time) over this time period at this location. In contrast, the trend line for August maximum temperature over the same period at Sacramento (mean: 33.6°C, 92.4°F) is sloping upward with a correlation coefficient of 0.370, meaning the slope is positive. When analyzed, together, the strength of the correlation coefficient, the amount of variation with regard to the mean, and the number of data points ($n = 68$), indicate that the Sacramento data slope is significantly different from 0 ($P = 0.001$).

cooler than 1934. However, including more years and thus data points makes the larger trend visible, despite some year-to-year variation that can seem to contradict the trend. Variation is also present in the projected data points—see for example 2027, 26°C (78.8°F), and 2099, 25°C (77°F).

Experienced gardeners and farmers are aware of the range of familiar variation that affects them, for example in the date of first frost, annual rainfall, or the size of aphid populations, and often have effective ways to deal with it. But trends are directional and present new challenges that may require new strategies (Section 3.1). Trends are especially challenging when they are big, and they can be "big" in two ways: the amount or magnitude of change, and the rate of change (Box 2.2).

What makes the twenty-first century unusual is not only that environmental and social trends are occurring at magnitudes and rates humans haven't experienced before, but that we now have the ability to document and analyze these changes to see where they are heading, and to understand what is driving them. We also have an understanding of thresholds that we should avoid, including for greenhouse gas emissions, biodiversity loss, water pollution, use of water resources (Foley et al., 2011), and income and social inequity. This is powerful information if we act on it. Our dual challenge is, first, to understand negative changes so that we can respond to them in ways that support rewarding lives, and healthy communities and environments, and, second, to *mitigate* them as much as possible, that is, stop and even reverse the behaviors that create and drive these changes.

If you have lots of resources and are not concerned with mitigation, it can be easy to respond to some trends in the short term. For example, you could respond to increasing summer temperatures or decreased rainfall by watering your garden more and paying higher water bills. But using more piped water, and the energy required to bring it to you, exacerbates problems of resource depletion, climate change, and inequity,

undermining positive environmental and social goals. Instead, there are other, more prosocial ways to respond that can also mitigate change, for example switching to crop varieties and species that use less water, using water in the garden more efficiently, and supporting policies that reduce ACC and social inequity. Helping food gardeners develop practices that have positive synergies between their responses to change and their goals and values is what *FGCW* is about.

In the rest of this chapter we highlight four significant trends that are having major effects on our food gardening: increasing resource scarcity; changing climate; increasing inequity; and the changing character of populations in the global north, using the US as an example. In Chapter 3 we'll discuss ways of responding to these trends.

2.2. Critical Resources are Becoming Scarcer

Resources are becoming scarcer both because of the increasing demand from growth in population and per capita consumption, and from the loss of resources due to pollution and over-exploitation. Our food system has often used water, soil, organic matter, and nutrients for growing food without concern for their limits, or for the waste and pollution it generates. Ecological thinking means recognizing that resources are limited, and that our waste doesn't simply disappear—realities we have ignored in the way we feed, clothe, transport, and house ourselves.

A major source of these problems is the habits developed through most of human history when human population size and consumption levels were much lower than in recent decades. These habits are now unsustainable, especially in the industrialized countries, yet are reinforced by political and economic pressures that ignore long-term consequences in favor of short-term growth and profit. It is good for the short-term profits of a few to get people to consume more! For example, there has been cheap fossil fuel for pumping surface and groundwater to irrigate water-intensive crops, lawns, and gardens in hot, dry areas, like the deserts of the US Southwest. This has encouraged the rapid depletion of water resources, leading to severe shortages in a region now living with drought.

In addition, the effects of resource scarcity are not evenly felt. Scarcity means that resource availability follows existing, historical patterns of structural inequity, with finite resources often used for the benefit of a few individuals at the expense of society, and especially those with less economic and political power. This situation is being made worse by ACC, as shown by the massive pumping of groundwater for large-scale crop production in California's Central Valley during the 2012–2016 drought, and exacerbated by a switch from row to tree crops, together resulting in large reductions in groundwater (Xiao et al., 2017). This pumping has also resulted in land subsidence as aquifers were depleted, causing billions of dollars of damage to bridges, roads and aqueducts, some of which are public infrastructure. It has also caused wells on which many low-income farmworker communities rely for drinking water to go dry, while drilling deeper, and treating the water contaminated by agricultural chemicals, is prohibitively expensive or impossible (Lohan, 2017).

Globalization of the food system means that resource inequities can easily occur across national borders. For example, when large, private, irrigated farms grow vegetables and fruits in Mexico and ship them to the US, water tables drop in Mexico, making access to water more difficult for people who live there, and rely on the same source of water for home use, including food gardens. The exported produce embodies exported *virtual water* in the form of all the water used to grow the crops, which can't be used to feed people in Mexico. For example, in 2009, northwest Mexico alone exported about 1 km^3 of virtual groundwater per year just in produce sent to the US (Scott, 2013). In Table 2.3 at the end of this chapter we summarize environmental and resource changes, many of which are trends, and their effects on food gardens.

2.3. Anthropogenic Climate Change is Happening

We live in what many are calling the *Anthropocene epoch* (from *anthropo-*, human, *-cene*, new), a proposed

---

Box 2.2. Ways trends can be big

First, the *magnitude* or absolute amount of change over a period of time could be much greater than in previous experience, sometimes surpassing a critical threshold, for example, higher mean temperatures as a result of ACC will lead to increased plant water requirements, and could become too high for some trees to set fruit (Section 5.8.2). ACC will also mean less precipitation in many regions, so there will also be less water available for plants to use (Cook et al., 2014). Because of these trends, the chance that severe drought will occur will rise, for example in the US Southwest from <12% in the twentieth century, to ≥80% by 2050–2099 for multidecadal drought under high (RCP 8.5) and >70% under moderate (RCP 4.5) IPCC future emissions scenarios (see Box 2.3), a level never experienced before (Cook et al., 2015). That is, droughts more extreme than anything in our historical record—even more severe than the worst "megadrought" (an extreme and long-lasting drought), documented in tree-ring data from 1,100–1,300 years ago. Such a drought will become extremely likely, using the IPCC's classification. This will lead to a large increase in the probability that plant needs will not be met, resulting in water stress and reduced yield (Sections 5.5, 5.6).

Second, trends can also be big in their *rate* of change over time, as indicated by the slope of the trendline. For example, the average global surface temperatures have been rising for some time due to anthropogenic *greenhouse gas emissions*—emissions into the atmosphere of gases such as methane and carbon dioxide, which absorb long-wave radiation from the Earth and re-radiate some back to it, contributing to global warming. The rate of temperature rise has also been accelerating—the slope of the trendline has been increasing. The average rate of increase in global annual temperatures (including ocean and terrestrial surfaces) has gone from 0.04°C (0.03°F) per decade over the 70 years from 1880–1950, a trendline with a slope of 0.0037, to 0.13°C (0.23°F) per decade over the 64 years from 1951–2015, a trendline with a slope of 0.0129 (NOAA NCEI, 2016). As the amount of change per unit of time increases over time, it produces the "hockey stick" shaped graphs that characterize many indicators of the Anthropocene epoch (Section 2.3).

---

new geological epoch (period of time) (SQS, 2018), and a term already in wide use informally (Ruddiman et al., 2015). The name is based on evidence that many of the Earth's biophysical processes are being affected by human activity, including atmospheric composition and chemistry leading to ACC (Vince, 2011; Monastersky, 2015). Greenhouse gases have been released into the Earth's atmosphere at increasing rates from the mining and burning of fossil fuels, agricultural production, and land use change, resulting in higher atmospheric concentrations of those gases, and rising average surface temperatures (IPCC, 2014b). If these trends are not reversed, they threaten to alter the climate to an extent that will be catastrophic for humans and many other species (Hansen et al., 2013, 2017).

As mentioned above, agriculture is one of the human activities most responsible for these changes, starting with deforestation and cultivation beginning up to 7,000 years ago, and greatly intensifying with the Industrial Revolution starting in the late 1700s (Kaplan et al., 2011; Ruddiman, 2013). Currently, the entire food system, including agriculture, probably accounts for up to one third or more of the greenhouse gas emissions that drive ACC

(Garnett, 2011), through carbon dioxide (CO_2) emissions from fossil fuel use, and from decomposition and combustion of organic matter in vegetation and soils through extensive land clearing and cultivation; methane (CH_4) from domestic ruminant animals (e.g., cows, sheep, goats), food decomposition in landfills and irrigated rice fields; and nitrous oxide (N_2O) from nitrogen fertilizers, especially synthetic ones (Ruddiman et al., 2015).

There are clear trends occurring in many of the components of climate, like temperature and precipitation, and those trends are large in magnitude. In many, the rate of change is accelerating. At one time these changes may have seemed abstract and easy to ignore, but they are increasingly affecting the lives of most people around the world every day. The assessment and interpretation of climate changes globally is the work of the Intergovernmental Panel on Climate Change (IPCC) that has developed a set of terms to describe the scientific understanding of these trends (Box 2.3).

2.3.1. Climate change and plants

Predicting the net impact of ACC on plants in detail for specific locations, including garden crops, is

---

Box 2.3. Describing ACC predictions

The IPCC was established in 1988 for reporting, evaluating, interpreting, and communicating scientific findings about ACC, and it engages thousands of scientists worldwide to do this. It has released five assessment reports thus far, all of which can be viewed and downloaded from the IPCC website (Section 2.7.2). Arriving at useful estimates of climate change, and projections for the future, is challenging because climate is complex, with lots of interacting variables and many unknowns. Because of this there are two basic parts to IPCC descriptions of ACC. First, how *likely* is a change, based on current scientific research? Likelihood is a quantitative description, expressed as probabilities from 0–100%. Second, how *confident* are scientists about predictions of change? Confidence is a qualitative descriptor based on the quality, amount, and consistency of the data, plus the level of expert agreement in interpreting those data (IPCC, 2014a). For example, the IPCC found with medium confidence that in the "northern hemisphere 1983–2012 was likely (66–100% probability) the warmest 30-year period of the last 1400 years" (Stocker et al., 2013).

Understanding how ACC will affect the future is especially difficult because it will depend upon interacting biophysical and socioeconomic conditions at particular locations: current and future greenhouse gas emissions, normal (non-anthropogenic) climate variability, and both global and local interactions of these and other factors (Melillo et al., 2014, 61–62). Because of this, different *scenarios* are developed that are scientifically plausible models of the future under particular conditions (IPCC, 2014d); for example, if greenhouse gas emissions remain at current levels or become substantially lower. Scenarios can then be used to construct *projections*, alternate predictions of the future under each scenario, and probabilities for each of those projections can be calculated.

Since the 1990s the IPCC has developed a number of scenarios focused on different issues, and the most well known have been the emissions scenarios based on different levels of greenhouse gas emissions from human activities, and the effect of socioeconomic variables. The IPCC *Assessment Report 5* (AR5, the most recent), uses four scenarios (now called representative concentration pathways, RCPs) that include both anthropogenic emissions as well as non-anthropogenic sources of greenhouse gases such as volcanoes (IPCC, 2013, 29).

---

challenging. CO_2 molecules in the atmosphere are taken up by plants and used in the process of photosynthesis (Section 5.4), which incorporates the carbon (C) into carbohydrates that store energy from the sun in their chemical bonds. Plants then release this energy for use through respiration, during which the C compounds are oxidized, with some C used in building and maintaining plant tissue, and some C released back into the atmosphere as CO_2. C can remain in plant tissues for a long time, especially in large woody plants like trees, and in some types of soil organic matter (Section 7.5.1), and can also remain out of the atmosphere for centuries or millennia in *peat* (partially decomposed plants, Box 6.4), and in fossilized plants that form natural gas, coal, and oil. Large-scale mining and burning of these fuels has released substantial amounts of that fossil C back into the atmosphere at unprecedented rates, and is a major driver of ACC (for more on the C cycle see Chapter 7).

Much of the impact of ACC on food gardens in areas that are already warm and dry will be due to higher temperatures, less precipitation, and more extreme weather events, but increased CO_2 concentration in the atmosphere will also affect plants directly (Box 2.4). With so many interacting variables, the net effect of ACC on plants is difficult to predict in detail, but in general it will make growing food gardens more challenging. That's why *FGCW* includes lots of discussion about how to mitigate and adapt to the effects of ACC.

2.3.2. Trends in temperatures and precipitation

The global climate system is complex, and changes are taking many different forms, but warming is the most obvious. Since the beginning of verifiable global weather record-keeping in the late nineteenth century, there's been an increase of 1.1°C (2.0°F) in average combined land and water surface temperature (NASA, 2017). Global temperatures are both higher (magnitude), and generally rising faster (rate) than in the past. Much of this warming has occurred recently, "with 16 of the 17 warmest years on record occurring since 2001," with five of the hottest occurring since 2010 (NASA, 2018), and 2016 the warmest year so far, "with globally averaged temperatures 0.99°C (1.78°F) warmer than the mid-20th century mean" (NASA, 2017). Overall, higher average temperatures will have a

larger effect on gardens than higher CO_2 concentration, and this is already evident (see Table 2.4).

A 2009 study estimated that increasing temperatures mean that globally, on average, plants adapted to a particular local climate will need to move about 0.4 km (0.25 mi) toward the poles every year to remain in the climate to which they are best suited (Loarie et al., 2009), although this estimate would probably be higher now. While this is based on non-domesticated plants growing in the wild, the crop plants in our gardens are also being affected by rising temperatures, changing precipitation and growing seasons, and climatic extremes, and many gardeners will need to adjust or change strategies (Karl et al., 2009), including finding new species or varieties. This is one reason you see changes in the USDA's Plant Hardiness Zone Map for the US, used by many US gardeners (Section 2.7.2).

With global warming, *isotherms* (the continuous geographic contour lines that mark locations having the same temperature on the same date) on both sides of the equator are moving toward the Earth's poles, as reflected in the new map. Plant hardiness zones are defined by isotherms based on the average annual coldest temperatures at a location over a designated period (Daly et al., 2012). In the current map, based on data from 1976–2005, the designations are in half zones, each a step of 5°F (i.e., zone 7a = 0–5°F, 7b = 5–10°, 8a = 10–15°, 8b =15–20°, etc.). Compared to the previous map (1976–1990), the new plant hardiness map shows 56% of the area of the continental US has shifted one half zone or more toward warmer temperatures, including 55% of land in the west, and over 59% of the midwestern agricultural belt (Daly et al., 2012), although some of the change is due to improved data quality.

The overall warming and extreme temperatures of ACC will have significant impacts on farmers, gardeners, and our global food system, moving species of crops and other organisms into new areas, but also altering the timing of flowering, fruiting, harvest, and length of crop life cycles (Parker and Abatzoglou, 2016). For example, on average, the frost-free period in the southwestern US has increased by two to three weeks since 1991 (Melillo et al., 2014).

Higher temperatures worldwide contribute to increased water consumption due to evapotranspiration because as air warms, its water-holding capacity increases, and so more water evaporates from plants and soil. For example, much of the region of the US historically considered arid or

Box 2.4. Increased atmospheric CO_2

Some plant scientists have suggested that higher average atmospheric CO_2 concentration could mean greater physiological water use efficiency (WUE), especially for C_3 plants (Box 5.1), because their stomata would not need to be open as long to acquire CO_2 for photosynthesis (Section 5.4), and so not lose as much water to evaporation through the stomata. It also seemed possible that increased CO_2 would reduce photorespiration, which replaces photosynthesis when a plant lacks CO_2. Photorespiration is less energy efficient than photosynthesis because it uses stored plant energy to capture solar energy, which is why it was hypothesized that ACC might be good for food production. For example, yields of C_3 crops like soybean, rice, wheat, and peanut could increase up to 30% with atmospheric CO_2 levels of 660 ppm (parts per million), at least in some environments (summarized in Backlund et al., 2008, 34–37). Of course, atmospheric CO_2 levels that high would cause climate change so catastrophic for humanity that good cereal and legume yields would not be a consolation.

More research is needed, but for a number of possible reasons (Uddling et al., 2018) it now appears that the benefit to crop production from higher atmospheric CO_2 concentration may be small or even nonexistent because: a) higher temperatures may cancel WUE advantages (Box 5.1) by increasing evapotranspiration; b) plants seem to acclimate to higher CO_2 concentrations (Hatfield and Prueger, 2011), resulting in much smaller yield increases over time; c) plant response is in part due to soil nitrogen availability and how that's affected by increased atmospheric CO_2 which is not yet understood (Reich et al., 2018); and finally, d) photorespiration helps plants assimilate nitrogen, so reduced photorespiration could decrease crop nitrogen content (Bloom et al., 2014).

Less photorespiration could lead to reduced absorption of nitrogen from the soil, limiting protein production because nitrogen is an essential component of the amino acids that make up proteins. For example, with current methods and crop varieties, average protein content of the staple crops wheat, rice, barley, and potato could decrease by about 8% if CO_2 levels increase to over 548 ppm as currently projected if current trends continue (Bloom et al., 2014). This level of elevated atmospheric CO_2 would also reduce zinc and iron content and therefore the nutritional quality of major C_3 food crops rice, wheat, and soybean. Worldwide between 2015 and 2050 those reductions would contribute about 125.8 million disability-adjusted life years (DALYs), that is, productive years lost to premature disability or death (Weyant et al., 2018), and disproportionately impact the global south. ACC mitigation policies to prevent the problem would be almost twice as effective at addressing nutrient deficiencies compared to smaller scale public health measures to treat those same deficiencies, an example of the importance of finding appropriate scales for organizing responses to ACC and other trends.

---

semi-arid is experiencing higher temperatures, and also receiving less precipitation, which contribute to an increased probability of drought. Some of this represents the familiar, long-term cycles of variation in precipitation that are expected in the region, but it's also due to the trend of increasing temperatures and decreasing precipitation due to ACC (Bevelander et al., 2014; Mao et al., 2015).

With ACC, parts of northern and southwestern Africa, southwestern Australia, southern Europe, and much of the western US will receive less precipitation (IPCC, 2014c), meaning gardeners will have to supply more of the water their gardens need through irrigation. But changing precipitation also impacts sources of irrigation water. In the Sierra Nevada mountains of California, the major source of water for that state, warmer temperatures mean less precipitation falling as snow, and an earlier spring snow melt. Consequently, instead of precipitation being held as snowpack that melts slowly through the spring,

soaking into the soil and being available as stream flow that fills reservoirs, the melt is happening earlier and more rapidly with more of that water lost to evaporation and flow to the ocean (Garfin et al., 2014). In addition, variable and intense short-duration precipitation events are increasing. Some reservoirs don't have the capacity to store the large amounts of water from early snowmelt and intense rainfall. This can also damage dams and threaten surrounding areas as in winter 2017 with the Oroville dam in California. It also means less surface water is stored and available for irrigation later in the growing season when it's hotter. These large-scale changes affect gardeners by decreasing the availability and quality of piped water, and increasing its cost.

Higher temperatures, less precipitation, and a greater demand for water can also result in more salt build-up (Section 8.8.5), from both greater salt concentration in irrigation water, and higher irrigation and evaporation rates in the garden.

2.4. Inequity

Worldwide, since the middle of the twentieth century, there is less sickness and death from communicable diseases, less hunger, and increased access to shelter and clean water (UNU-WIDER, 2014). However, especially in recent decades, access to these improvements has become increasingly uneven within and between nations, driven by rising economic inequity, with large numbers of people in or near poverty. For example, there is evidence that the growing number of violent conflicts and resulting refugee crises are a direct consequence of environmental degradation, ACC, and systematic inequity (e.g., Kelley et al., 2015). And inequity affects food gardens too; Table 2.5 at the end of this chapter summarizes some potential effects of the increased economic inequity in the US on food gardens, and Box 2.5 describes how it is increasing.

Many of the needs that can be met by food gardens are especially urgent as a result of social and economic inequity, for example, lack of food, especially good quality food. In 2016, 31.6% of US households with incomes below 185% of the poverty threshold were food insecure, compared with only 5.6% of households at or above 185% of the poverty threshold. Poverty is also positively associated with more overweight and obesity and the non-communicable diseases (NCDs) that accompany them (Levine, 2011). The poor quality of food most low-income households can afford is a key part of this problem, due in part to government subsidies of junk food ingredients (Siegel et al., 2016), and fresh vegetables and fruits being hard to find and expensive in poorer neighborhoods (Section 1.1.2). Evidence is also accumulating for a relationship between inequity and poor health, measured by incidence of disease or violence (Pickett and Wilkinson, 2015). Similarly, a history of inequity, discrimination, and violence experienced by a community results in deep and persistent psychological and social harm, including historical trauma (Brave Heart, 2003; Mohatt et al., 2014).

Gardens can support equity by increasing the availability of higher quality food, as well as providing psychological and social benefits. But without an intentional, prosocial orientation (Introduction) those benefits may not occur, and gardens can be used to perpetuate social, educational, economic, and other inequities (Section 1.6).

Inequity is also associated with harmful environmental conditions (Cushing et al., 2015). Evidence for this is clearest for problems of local environmental quality including contamination of water, air, and soil. While studies show populations of color have the highest risk, everyone's risk can increase the more segregated neighborhoods are (Morello-Frosch and Jesdale, 2006), perhaps because no comprehensive environmental quality programs are created or enforced. Sometimes we see those consequences of inequity because environmental quality is only ensured for a powerful minority, or they "opt out," for example through having the power to keep polluters from locating near their neighborhoods, or avoiding poor quality tap water by buying bottled water. There is increasing evidence that inequity in multiple forms, such as environmental injustice, poverty, social violence, and predisposition to disease occur together and amplify each other (Solomon et al., 2016).

Finally, inequity also seems to fuel itself, making people distrustful, seeing each other as members of different groups, rivals for limited resources, and so less willing to work together to find prosocial, environmentally beneficial responses (Section 3.3). Thus, for many reasons, disregarding inequity, and leaving the wealthy to control and consume a vastly disproportionate amount of resources and benefits harms everyone (Fig. 2.5).

Ecological thinking helps when trying to understand the environmental consequences of inequity for food gardens. In ecological systems, activities require resources obtained from a *source* environment, and produce waste that is returned to *sink* environments. If extraction and waste exceed the ecosystem's ability to maintain itself, then long-term environmental degradation results. This often happens when powerful groups take a disproportionate amount of resources from sources, and discharge a disproportionate amount of waste into sinks, compared to those who are less powerful, and who also depend on those environments. In other words, there is environmental injustice. An example we already mentioned is the Central Valley of California, where drought and economic pressures on farmers mean that large-scale agriculture uses much of the available water from a large regional watershed (source), and then pollutes local water tables (sinks) with runoff containing nitrates, pesticides, and other contaminates. The result is that farmworker communities providing the labor for that agriculture have to spend some of their modest incomes buying bottled water because their tap water is unsafe to drink (Francis and Firestone, 2011) (Section 2.2).

Good source environments for food gardens have high quality resources, like soil, water, compost, and air, or have reliable access to them. But these environments and resources are often controlled by the privileged, powerful members of society. This

Box 2.5. Increasing global inequity, the example of the US

An inequitable distribution of income and wealth is present in all regions of the world but varies from Europe, where the wealthiest 10% of the population receive 37% of national income, to the Middle East where they receive 61% (WID.world, 2018). In Canada and the US the wealthiest 10%'s share is 47% of national income. The average income of the richest 10% of Americans (~31 million people) has grown 144% since 1960, whereas incomes for the remaining 90% of the population have grown only 22% during the same period (WID.world, 2018), and by 2014 the wealthiest 10% of the US population had incomes 19 times greater than the incomes of the poorest 50%, on average (Fig. 2.4).

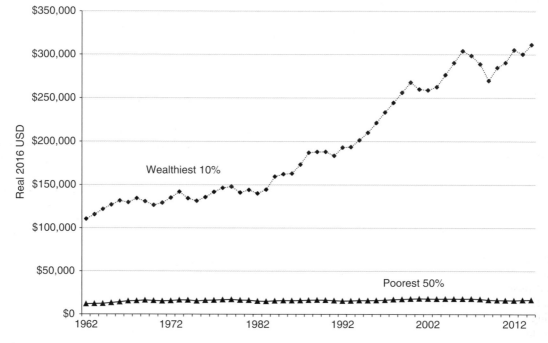

Figure 2.4. Trends in mean incomes of wealthiest 10% compared to poorest 50% of individuals in the US population, 1962–2014 (pre-tax national mean incomes)

WID.world (2018)

Many different people are experiencing stagnation or decline in their incomes, but these economic differences are not simply due to current events—this inequity reflects and exacerbates disparities resulting from our ongoing heritage of structural racism. The inequity between people of color compared to whites in the US can be seen in economic differences such as the lower median household incomes, higher poverty rates, greater impact of economic recession, as well as higher disease burden, shorter life expectancy, higher rates of incarceration, poorer quality of schools and other factors, many of which persist despite improving aggregate national level trends (Table 2.1).

Table 2.1. Percentage of "racial" groups in the US population below the poverty level, 1976–2015

Year	All	Asian	African American	"Hispanic"	White, not "Hispanic"	Native American[a]
1976	11.8	NA	31.1	24.7	8.1	
2000	11.3	9.9	22.5	21.5	7.4	27.0
2015	13.5	11.4	24.0	21.4	9.1	

[a]Macartney et al. (2013), data only available for 2007–2011
For all except Native American (USCB, 2017c, Table 6). NA, not available

Figure 2.5. The control and consumption of large amounts of the Earth's resources by the very wealthy threatens everyone's wellbeing

means that many people most in need have few resources, and are relegated to contaminated sink environments, where it's difficult to grow food, and where there are health problems. Some predominantly Latinx neighborhoods in Los Angeles have had their soils contaminated for decades with lead and other toxic substances by the Exide battery recycling plant, and the situation has been largely ignored by state and local regulators (Nazaryan, 2016). Since the plant was finally closed in 2015, clean-up operations continue to be extremely slow, and fines imposed on Exide provide only a fraction of the costs (Pulido et al., 2016), with residents still afraid to eat food from their gardens, or let their kids play on bare soil. They have accused the government of racist neglect (Barboza and Poston, 2018). Organizing and working together was a critical factor in finally getting the battery recycling plant closed, and will be key for effective clean-up, as well as for resolving similar situations of environmental injustice (Section 3.3).

2.5. Who we are: Demographic Trends, the US Example

Who we are influences our interest in food gardens, the kinds of gardens and plants we want, and our ability to create and maintain our gardens. In the US, like the rest of the world, the population in the twenty first century is changing in ways that have created both challenges to, and opportunities for, food gardens (Table 2.6).

Compared to 50 years ago, the US, and most industrialized countries, have a larger, more urban, more diverse, older, and less healthy population. This can make finding a place to garden a challenge, and tailoring the garden to your particular needs more important. Growing urban populations and a demand for garden space has stimulated new laws to protect garden areas, and encourage landowners to make space available for gardening (Section 4.2.2). Resource and legal centers for helping urban gardeners navigate city regulations, and which advocate for garden-friendly changes are opening in many cities (e.g., the Sustainable Economies Law Center in Oakland, California—SELC, 2018).

These changes can strengthen the importance of gardens for meeting food, health, psychological, and social needs. For example, as refugees new to a country find comfort in growing and eating familiar foods, their gardens and farms may even become a source of income if they can market some of their produce. This is the case for a project in Arizona where international refugees work together to farm and sell their produce through community supported agriculture programs (Gonzales et al., 2013).

As we described earlier in this chapter, and in Chapter 1, more and more people worldwide are becoming sick with NCDs, and food gardens can be part of a positive response to this epidemic. Still, like all social, environmental, and climate challenges we are facing, enduring inequity and discrimination make the burden of NCDs disproportionately heavy on those with the least power and fewest resources. For reasons outlined in Sections 1.1 and 1.6, those historically facing discrimination and injustice make up a disproportionate part of the overweight and obese population (Table 2.2). Gardens can play a key

Table 2.2. Proportion of overweight and obese adult females in different US groups, 2011–2014

Group	Overweight (BMI ≥25), %	Obese (BMI ≥30), %
All US	66.2	38.1
White, non-"Hispanic"	63.5	35.3
"Hispanic, Mexican origin"	80.3	49.6
African American	82.0	56.5
Income 100–200% of federal poverty level	73.4	42.6

BMI = body mass index, a weight to height ratio used as an indicator of weight status. Adults 20 years and older, age adjusted; overweight includes obese.
NCHS (2014)

role in changing diets, creating environments that encourage more enjoyable physical activity, and building community action to advocate for the policy changes needed to curb food industry advertising and sales of the foods that fuel the obesity and NCD epidemics. But no matter what, solving this and our other problems requires changing pervasive inequity.

Finally, there has been a steady increase in the number of US households growing food gardens; an estimated 42 million US households had food gardens in 2013—35% of all US households—up 17% from 2008 (NGA, 2014). This increase has occurred for several reasons including greater urbanization, economic hardship, and the rise of the "food movement." Saving money was a reason for growing food for 54% of the people surveyed for the National Gardening Association study, evidence of how important it is to garden in ways that minimize purchased inputs. Other notable changes between 2008 and 2013 were that the increase in the number of gardening households was greatest for the 5 million households with five or more people, and that food gardening rose 38% among households with annual incomes of less than $35,000 USD. Food gardening may also address the psychological stress of unemployment and loss of income, which is made worse when social services are lacking. Spending time outdoors working with plants and soil, especially in the company of other gardeners, can be very positive if you are un- or underemployed (Okvat and Zautra, 2011), although gardens can also be used to rationalize or maintain inequity (Section 1.6).

2.6. Summary Tables of Major Changes

Tables 2.3–2.6 provide an outline of major environmental, climate, equity, and demographic changes currently underway, the section of *FGCW* where each is discussed in more depth if relevant, how the change affects food gardens, and references. These tables demonstrate why many current changes will have locally specific consequences, and substantial variation across locations.

Table 2.3. Environmental resource changes and their effects on food gardens

Environmental resource	Description	Actual or potential effect on food gardens	References
Resources in general (Sections 2.2, 2.4)	Lack of stewardship of sources of inputs important for food gardens, including depletion and pollution of water, soil, air, biodiversity	Degradation or loss of essential garden inputs, making gardening more difficult and costly, increasing scarcity and cost of resources	McLean (2016)
Sinks in general (Sections 2.4, 4.2.1)	Air, water, and soil contamination of gardens and garden resources	Degradation or loss of garden areas, making gardening hazardous, more difficult, increasing scarcity and cost of resources Gardening possible, but produce is poor quality or contaminated Extra precautions and expenses required for gardening and using produce	Clark et al. (2006, 2015)
Crop genetic resources (Section 10.4)	Ongoing consolidation of large, multinational seed corporations	Decreased crop genetic diversity Homogenization of germplasm, inputs, and seed institutions for some vegetables produced commercially Reduced options for growing environments, strategies, end uses, adaptation to change	Howard (2009)
	Dominance of intellectual property regimes (patents, grower contracts) by corporate seed industry	Public plant breeders, farmers, gardeners lack access and right to share crop genetic diversity, reducing capacity to breed publicly available, regional, or specialty varieties	Kloppenburg (2014); Howard (2015)
	More named, commercial varieties available from smaller, regional seed companies; level of diversity represented by those varieties unknown, but overall more varieties and species than from multinational seed corporations	Increased options for sources of germplasm, planting, and adaptation to change; potentially more dietary, end use and marketing options	Heald and Chapman (2009); OSSI (2018)

Continued

Chapter 2

Table 2.3. Continued.

Environmental resource	Description	Actual or potential effect on food gardens	References
	Increased interest in and value of "heirloom," specialty, and locally adapted seed	Increase in new institutions for alternative access to and movement of crop diversity	Howard (2015)
Land access (Sections 1.6, 4.2.2)	Increasing competition for urban land	Obtaining and securing tenure to land for gardens more difficult due to prohibitive costs, including rent, taxes, insurance; pressure to convert food gardens to ornamental green space, or buildings	Voicu and Been (2008)
Water (Section 2.2, Chapter 8)	Reduced supply due to: increased rural and urban demand; decreased precipitation due to ACC; decreased recharge of groundwater; contamination	Increased likelihood of drought and water stress; species and varieties no longer adapted in some areas; increased cost; decreased productivity	Melillo et al. (2014)
	Reduced quality due to reduced supply and increased evaporation which concentrate salts and contaminants	Decreased productivity	Connor et al. (2012)
Soil (Chapter 7, Section 8.8.5)	Loss due to wind and water erosion from tillage, increased precipitation intensity and drying	Management required to stop loss; soil replacement may be needed	Melillo et al. (2014)
	Reduced quality due to salinization, contamination, loss of organic matter	Productivity decreased if no remediation; soil replacement may be needed	

Table 2.4. Anthropogenic climate changes and their effects on food gardens

Climate change	Description	Actual or potential effect on food gardens	References
Increased atmospheric CO_2 (Section 2.3.1)	C_3 plants: more CO_2 → less time stomata need to be open (less stomatal conductance) → less H_2O loss from stoma (transpiration) → greater physiological WUE	Growth and productivity ↑, but may be limited by ACC impact on soil nutrients, see below about CO_2 + higher temperatures	Long and Ort (2010); Hatfield and Prueger (2011); Reich et al. (2018)
	C_3 plants: less CO_2 deficit → less photorespiration	Growth and productivity ↑, but see possible tradeoff below with CO_2 + higher temperatures; N may be limiting	Long and Ort (2010); Hatfield and Prueger (2011); Reich et al. (2018)
	C_3 plants: more CO_2 → less transpiration, → less cooling of the plant, potentially reaching temperatures restricting growth	Growth and productivity ↓, shading is more important	Long and Ort (2010)
	Equal or greater vegetative growth + reduced nutrient assimilation, including N → reduced nutrient concentrations	Nutritional quality ↓ in some crops	Bloom et al. (2014); Myers et al. (2014)
	C_4 plants: little net effect	Growth and productivity unchanged	Long and Ort (2010); Hatfield and Prueger (2011)

Continued

Table 2.4. Continued.

Climate change	Description	Actual or potential effect on food gardens	References
Higher average temperatures (Sections 2.3.2, 5.6)	Faster rate of development in plants, shorter lifespan, less total photosynthesis per individual plant	Productivity = or ↓, especially for fruits, nuts, seeds, roots, bulbs and tubers	Ainsworth and Ort (2010); Long and Ort (2010); Hatfield and Prueger (2011)
	More water stress due to more evapotranspiration	Productivity ↓ if no additional water applied	Long and Ort (2010); Hatfield and Prueger (2011)
	Pests, pathogens and weeds can spread to new areas, have increased number of generations per year	Productivity ↓	Melillo et al. (2014)
	Greater risk of heat stroke, especially for elderly, ill, very young	Productivity ↓ because reduced ability to work in garden, and work less productive	Melillo et al. (2014); Hansen and Sato (2016)
	Increased soil temperatures may → faster organic matter decomposition	Productivity ↑ or ↓	Davidson and Janssens (2006); von Lützow and Kögel-Knabner (2009)
	Drier soil → slower organic matter decomposition	Productivity ↑ or ↓	
Increased atmospheric CO_2 + higher air temperatures (Section 2.3)	Eliminates advantage of greater physiological WUE with higher CO_2 levels because higher air temperatures → more evapotranspiration from soil and plants, water becomes limiting; higher CO_2 concentration with limited water → more stomata closure and high plant tissue temperatures	Productivity = or ↓	Long and Ort (2010); Hatfield and Prueger (2011)
Higher nighttime temperatures (Section 5.8.2)	Winter: minimum chilling hours for fruit set of some perennial fruit crops not met	Productivity ↓	Luedeling et al. (2011)
	Summer: inhibits reproductive growth, seed and fruit set of some species and varieties, e.g., tomato	Productivity ↓	Melillo et al. (2014)
	Expansion of some cold-sensitive species and varieties into higher latitudes and elevations	Productivity ↑	Melillo et al. (2014)
Shorter, milder cold season	Extended growing season and cycles for some species and varieties (opposite is true for higher summer temperatures, see higher average temperatures above)	Productivity ↑	Melillo et al. (2014)
	Temperate pest and pathogen populations more prolific due to greater number of generations possible, not controlled by low temperatures	Productivity ↓	Melillo et al. (2014)
Earlier onset of warm spring temperatures	Extended growing season; in temperate areas changing fit between temperature and daylength could limit temperate-origin photoperiod-sensitive crops; new opportunities for cold and photoperiod-sensitive crops originally from low latitudes	Productivity ↑ or ↓	Long and Ort (2010)

Continued

Table 2.4. Continued.

Climate change	Description	Actual or potential effect on food gardens	References
More extreme warm to cold temperature fluctuations in late winter, spring	More cold temperature damage to prolific growth encouraged by early warm temperatures	Productivity ↓	Gu et al. (2008)
Less snowpack, faster melt	Less infiltration and recharge, less dilution of salts and contaminants; less available surface and groundwater	Productivity ↓, piped water more costly, lower quality	Melillo et al. (2014)
More intense and variable precipitation	Flooding, crop damage, less water infiltration and less *in situ* storage	Productivity ↓, water harvesting and storage more difficult, less water in the soil	Melillo et al. (2014)

C_3, plants with photosynthesis capturing carbon in three carbon molecule; C_4, plants with photosynthesis capturing carbon in four carbon molecule; CO_2, carbon dioxide; ACC, anthropogenic climate change; H_2O, water; N, nitrogen; WUE, water use efficiency, g C fixed g^{-1} H_2O lost

Table 2.5. Changes in economic indicators of inequity in the US, and their effects on food gardens, the example of the US

Indicator	Description	Actual or potential effect on food gardens	References
Poverty	Proportion of people living at or below 125% of poverty level in 2015 is much lower than 1960, but the number of people is higher 1960 30.4%, 54.6M people 2000 15.6%, 43.6M 2015 17.9%, 56.9M		USCB (2017c, Table 3)
Food insecurity	Proportion of food insecure HHs has increased 2000 10.5%, 11.1M US HHs 2007 11.0%, 13.0M 2015 12.7%, 15.8M	Increased need for gardens, but fewer resources to start them Potential garden benefits are	Coleman-Jensen et al. (2016)
Unemployment	Individuals 16–65 years old unemployed in US 2000 4.0%, 5.7M people 2013 7.4%, 11.5M 2015 5.3%, 8.3M	increasingly important Gardens using minimal	USBLS (2018)
Income inequity	Inequity has risen, financial and other resources increasingly captured by a wealthy minority, while majority become poorer. Share of total income going to richest 10% of adults in US: 1960 35.6%, 11.5M people 2000 43.9%, 20.9M 2015 47.0%, 24.5M	purchased inputs important Increased risk of gardens being promoted for the wrong reasons, including avoiding greater equity in social policies (Section 1.6)	WID.world (2018)
Proportion of income for basic needs among poorest	Proportion of total expenditures spent on food increases with decreasing income, 2015: 15.7% in 20% of HHs with lowest income 12.6% in 20% of HHs with middle income 11.2% in 20% of HHs with highest income		

M = million, HHs = households

Table 2.6. Changes in who we are, and their effects on food gardens, the example of the US

Change	Description	Actual or potential effect on food gardens	References
Aging	Proportion of US population ≥65 years old is rising 1960 9.2%, 17M people 2000 12.4%, 35M 2015 14.9%, 48M 2060 23.6%, 98M, projected	People may have more time and greater need for garden benefits than when younger	USCB (2017d, Table 9)
Recent migrants and refugees	Proportion of US population not born in US, including undocumented 1960 5.4%, 10M people 2000 11.1%, 31M 2015 13.5%, 43M 2060 18.8%, 78M	Access to resources for gardening, e.g., land, especially limited Diverse experiences and cultures, including seeds and foods; cultural identity, social integration important	Gibson and Jung (2006); USCB (2017d, Table 2)
Diversity	No longer one dominant demographic group. "Non-Hispanic white" as proportion of US population[a] 1980 79.6%, 180M 2000 71.4%, 196M 2015 63.5%, 204M 2060 48.2%, 201M	Access to resources for gardening especially limited, e.g., land Diverse experiences and cultures, including seeds and foods; cultural identity, social integration important	Gibson and Jung (2006); USCB (2017d)
Urbanization	Proportion of US population in "urban clusters" (population >2500) 1960 63.1%, 113.1M people 2000 79.0%, 222.4M 2010 80.7%, 249.3M	Urban gardening and demand for urban gardening land increasing	USCB (2018)
Different kinds of HHs	Number of shared HHs in US increased 11.4% from 2007 to 2015: 2007 17.0% US HHs, 61.7M adults 2015 19.2% HHs, 75.3M adults, 31.1% of all US adults	Multiple families, individuals, can be different ages, genders, schedules, economic needs, all affecting experience, time and interest in gardens	Mykyta and Macartney (2012); USCB (2017b)
	Children living in single parent HHs as proportion of all children 1960 9.1%, 5.8M children 2000 26.7%, 19.2M 2015 26.8%, 19.8M	Large proportion low income; single mother HHs are poorest, have greatest need for financial, nutritional and other potential garden benefits	USCB (2017a, Table CH-1)

M = million, HH = households
[a]1980 was first year data disaggregated

2.7. Resources

All the websites listed below were verified on May 8, 2018.

2.7.1. Trend and variation

This very short video clearly demonstrates the difference between trend and variation:
https://www.youtube.com/watch?v=e0vj-0imOLw.

2.7.2. Climate

The Food Climate Research Network (FRCN) at the University of Oxford, UK, curates an up-to-date collection of resources on the food–climate connection, including its own published reports, scientific documents, and blog posts.
https://www.fcrn.org.uk/

The new USDA Plant Hardiness Zone Map is available online and can be searched by zip code and other queries.
http://planthardiness.ars.usda.gov/PHZMWeb/

The Intergovernmental Panel on Climate Change (IPCC) is a Nobel-prize winning international scientific body founded by the UN and the World

Meteorological Organization. It is open to all governments that are members of the UN, with the goal of producing balanced, science-based climate change information directly useful for policy. Large international groups of scientific experts (>3500 for the last report) develop periodic reports that are approved by governments and then released to the public; the fifth report (*AR5*) was released starting in 2013, *AR6* will be finalized in 2021.
http://www.ipcc.ch/

Climate Watch "is an online platform designed to empower policymakers, researchers, media and other stakeholders with the open climate data, visualizations and resources they need to gather insights on national and global progress on climate change" (from the website). Much of the data are nationally determined contributions (NDCs) to greenhouse gas emissions. Data are from government or international agencies (e.g., NASA, FAO, UN Framework Convention on Climate Change, and others). Climate Watch is managed by the World Resources Institute, with support from diverse national and international governmental and university institutions and some large, private technology firms such as Google.
https://www.climatewatchdata.org/

Cal-Adapt at UC Berkeley, with support from the state government and Google, is a climate change database for the state of California, including temperatures, precipitation, sea level, snowpack, and wildfires, and can be queried at state or local levels for tailored data sets, and has simulations of change over time with different climate change models and scenarios. Figures 2.2, 2.3 and examples in Section 3.1 were generated using tools and data available at this site.
http://cal-adapt.org/tools/

The 4th US National Climate Assessment is a government report on current and projected climate change and its consequences in the US, published November 2018.
https://www.globalchange.gov/nca4

"A community climate and weather journal," ISeeChange is a partnership among scientists (from NASA, NOAA and other institutions), community members, and US public radio stations to document changes noticed on the ground by the public, and investigate them in the context of larger scale information such as satellite data sets, to help us understand how ACC is being experienced locally. There are free public apps on the website. This is one of a growing number of such "citizen" science initiatives worldwide through which the public uses smartphones to document and report local occurrences of variables (Section 3.2.1).
https://www.iseechange.org/

2.7.3. Water

The Water Footprint Network is a public, private, and non-governmental organization (NGO) supported site with water data, including for virtual water flows, and calculators for personal, national, product-specific, and corporate water footprints.
http://waterfootprint.org/en/
See also Pacific Institute, Section 8.10.1.

2.7.4. Equity

A powerful resource for data and visualizing the rapid rise in inequity is the World Inequity Database, assembled and maintained by a team of over 100 researchers from more than 70 countries and paid for with public and non-profit support. The database uses tax and other data, often dating back to the early twentieth century. Inequality data are available for 58 countries, with others in preparation. You can generate graphics and other summaries that make it easier to grasp the trend toward greater inequality worldwide (Fig. 2.4 is an example).
http://wid.world/

The UN's Special Rapporteur on the Human Right to Food has been at the forefront of bringing the discussion of food and humane food systems as a human right to international attention, including discussion of food sovereignty, the role of gender, and the impact of land grabs and forced urbanization on the global south. See Section 1.7.2 for that and other resources regarding equity and food.

The International Panel of Experts on Sustainable Food Systems (IPES Food) supports evidence based policy debate about alternative food systems, including gardens through interdisciplinary research and policy work.
http://www.ipes-food.org/

2.8. References

Ainsworth, E. A. & Ort, D. R. (2010) How do we improve crop production in a warming world? *Plant Physiology*, 154, 526–530. DOI: 10.1104/pp.110.161349.

Backlund, P., Janetos, A., Schimel, D. & Walsh, M. (2008) *The effects of climate change on agriculture, land resources, water resources, and biodiversity in the United States*. Climate Change Science Program, US Environmental Protection Agency, available at: https://www.fs.usda.gov/treesearch/pubs/32781 (accessed Oct. 5, 2018).

Barboza, T. & Poston, B. (2018) The Exide plant in Vernon closed 3 years ago. The vast majority of lead-contaminated properties remain uncleaned. *Los Angeles Times*, April 26, available at: (http://www.latimes.com/local/lanow/la-me-exide-cleanup-20180426-story.html (accessed Oct. 5, 2018).

Bevelander, K. E., Kaipainen, K., Swain, R., Dohle, S., Bongard, J. C., Hines, P. D. H. & Wansink, B. (2014) Crowdsourcing novel childhood predictors of adult obesity. *PLoS ONE*, 9, e87756, DOI: 10.1371/journal.pone.0087756.

Bloom, A. J., Burger, M., Kimball, B. A. & Pinter, J. P. J. (2014) Nitrate assimilation is inhibited by elevated CO_2 in field-grown wheat. *Nature Climate Change*, 4, 477–480, DOI: 10.1038/nclimate2183.

Brave Heart, M. Y. H. (2003) The historical trauma response among natives and its relationship with substance abuse: A Lakota illustration. *Journal of Psychoactive Drugs*, 35, 7–13, DOI: 10.1080/02791072.2003.10399988.

Cal-adapt (2018) Cal-adapt. Berkeley, CA, available at: http://cal-adapt.org/ (accessed May 13, 2018).

Clark, H. F., Brabander, D. J. & Erdil, R. M. (2006) Sources, sinks, and exposure pathways of lead in urban garden soil. *Journal of Environmental Quality*, 35, 2066–2074, DOI: 10.2134/jeq2005.0464.

Clarke, L. W., Jenerette, G. D. & Bain, D. J. (2015) Urban legacies and soil management affect the concentration and speciation of trace metals in Los Angeles community garden soils. *Environmental Pollution*, 197, 1–12, DOI: http://dx.doi.org/10.1016/j.envpol.2014.11.015.

Coleman-Jensen, A., Rabbitt, M. P., Gregory, C. A. & Singh, A. (2016) *Household food security in the United States in 2015, ERR-215*. US Department of Agriculture, Economic Research Service, available at: https://www.ers.usda.gov/publications/pub-details/?pubid=79760 (accessed Oct. 5, 2018).

Connor, J. D., Schwabe, K., King, D. & Knapp, K. (2012) Irrigated agriculture and climate change: The influence of water supply variability and salinity on adaptation. *Ecological Economics*, 77, 149–157, DOI: http://dx.doi.org/10.1016/j.ecolecon.2012.02.021.

Cook, B. I., Smerdon, J. E., Seager, R. & Coats, S. (2014) Global warming and 21st century drying. *Climate Dynamics*, 43, 2607–2627.

Cook, B. I., Ault, T. R. & Smerdon, J. E. (2015) Unprecedented 21st century drought risk in the American Southwest and Central Plains. *Science Advances*, 1, e1400082.

Cushing, L., Morello-Frosch, R., Wander, M. & Pastor, M. (2015) The haves, the have-nots, and the health of everyone: The relationship between social inequality and environmental quality. *Annual Review of Public Health*, 36, 193–209, DOI: doi:10.1146/annurev-publhealth-031914–122646.

Daly, C., Widrlechner, M. P., Halbleib, M. D., Smith, J. I. & Gibson, W. P. (2012) Development of a new USDA plant hardiness zone map for the United States. *Journal of Applied Meteorology and Climatology*, 51, 242–264.

Davidson, E. A. & Janssens, I. A. (2006) Temperature sensitivity of soil carbon decomposition and feedbacks to climate change. *Nature*, 440, 165–173.

Foley, J. A., Ramankutty, N., Brauman, K. A., Cassidy, E. S., Gerber, J. S., Johnston, M., Mueller, N. D., O'Connell, C., Ray, D. K., West, P.C., et al. (2011) Solutions for a cultivated planet. *Nature*, 478, 337–342, DOI: 10.1038/nature10452.

Francis, R. & Firestone, L. (2011) Implementing the human right to water in California's Central Valley: Building a democratic voice through community engagement in water policy decision making. *Willamette Law Review*, 47, 495–537.

Garfin, G., Franco, G., Blanco, H., Comrie, A., Gonzalez, P., Piechota, T., Smyth, R. & Waskom, R. (2014) Chapter 20: Southwest. In Melillo, J. M., Richmond , T. C. & Yohe, G. W. (eds) *Climate change impacts in the United States: The third national climate assessment. U.S. global change research program*, 462–486. US Government Printing Office, Washington, DC, DOI: 10.7930/J08G8HMN.

Garnett, T. (2011) Where are the best opportunities for reducing greenhouse gas emissions in the food system (including the food chain)? *Food Policy*, 36, S23–S32, DOI: 10.1016/j.foodpol.2010.10.010.

Gibson, C. & Jung, K. (2006) *Historical census statistics on the foreign-born population of the United States: 1850–2000,*. US Census Bureau, Population Division, Washington, DC, available at: https://www.census.gov/population/www/documentation/twps0081/twps0081.html (accessed Oct. 5, 2018).

Gonzales, V., Forrest, N. & Balos, N. (2013) Refugee farmers and the social enterprise model in the American southwest. *Journal of Community Positive Practices*, 13, 32–54.

Gu, L., Hanson, P. J., Post, W. M., Kaiser, D. P., Yang, B., Nemani, R., Pallardy, S. G. & Meyers, T. (2008) The 2007 eastern US spring freeze: Increased cold damage in a warming world? *AIBS Bulletin*, 58, 253–262.

Hansen, J. & Sato, M. (2016) Regional climate change and national responsibilities. *Environmental Research Letters*, 11, 034009.

Hansen, J., Kharecha, P., Sato, M., Masson-Delmotte, V., Ackerman, F., Beerling, D. J., Hearty, P. J., Hoegh-Guldberg, O., Hsu, S.-L., Parmesan, C., et al. (2013) Assessing "Dangerous climate change": required reduction of carbon emissions to protect young people, future generations and nature. *PLoS ONE*, 8, e81648, DOI: 10.1371/journal.pone.0081648.

Hansen, J., Sato, M., Kharecha, P., von Schuckmann, K., Beerling, D. J., Cao, J. J., Marcott, S., Masson-Delmotte, V., Prather, M. J., Rohling, E. J., et al. (2017) Young people's burden: Requirement of negative CO_2 emissions. *Earth System Dynamics*, 8, 577–616, DOI: 10.5194/esd-8-577-2017.

Hatfield, J. L. & Prueger, J. H. (2011) Agroecology: Implications for plant response to climate change. In Yadav, S., Redden, R. J., Jerry L. Hatfield, Lotze-Campen, H. & Hall, A. E. (eds) *Crop adaptation to climate change*, 27–43. Wiley-Blackwell, DOI: 10.1002/9780470960929.

Heald, P. J. & Chapman, S. (2009) Crop diversity report card for the twentieth century: Diversity bust or diversity boom? *SSRN*, DOI:10.2139/ssrn.1462917, available at: https://papers.ssrn.com/sol3/papers.cfm?abstract_id=1462917 (accessed July 31, 2018).

Howard, P. H. (2009) Visualizing food system concentration and consolidation. *Southern Rural Sociology*, 24, 87–110.

Howard, P. H. (2015) Intellectual property and consolidation in the seed industry. *Crop Science*, 55, 2489–2495.

IPCC (Intergovernmental Panel on Climate Change) (2013) *The physical science basis. Summary for policymakers*, IPCC, available at: http://www.climatechange2013.org/images/report/WG1AR5_SPM_FINAL.pdf (accessed Oct. 5, 2018).

IPCC (2014a) Summary for policymakers. In Field, C. B., Barros, V. R., Dokken, D. J., Mach, K. J., Mastrandrea, M. D., Bilir, T. E., Chatterjee, M., Ebi, K. L., Estrada, Y. O., Genova, R. C., et al. (eds) *Climate change 2014: Impacts, adaptation, and vulnerability. Part A: Global and sectoral aspects. Contribution of working group II to the Fifth Assessment Report of the Intergovernmental Panel on Climate Change*, 1–32. Cambridge University Press, Cambridge, UK and New York, NY.

IPCC (2014b) *Climate change 2014: Mitigation of climate change. Contribution of working group III to the Fifth Assessment. Report of the Intergovernmental Panel on Climate Change*. Cambridge University Press, Cambridge, UK and New York, NY.

IPCC (2014c) *Climate change 2014: Impacts, adaptation, and vulnerability. Part B: Regional aspects. Contribution of working group II to the Fifth Assessment Report of the Intergovernmental Panel on Climate Change*. Cambridge University Press, Cambridge, UK and New York, NY.

IPCC (2014d) Annex II: Glossary. In Barros, V. R., Field, C. B., Dokken, D. J., Mastrandrea, M. D., Mach, K. J., Bilir, T. E., Chatterjee, M., Ebi, K. L., Estrada, Y. O., Genova, R. C., et al. (eds) *Climate change 2014: Impacts, adaptation, and vulnerability. Part B: Regional aspects. Contribution of working group II to the Fifth Assessment Report of the Intergovernmental Panel on Climate Change*,

1757–1776. Cambridge University Press, Cambridge, UK and New York, NY.

Kaplan, J. O., Krumhardt, K. M., Ellis, E. C., Ruddiman, W. F., Lemmen, C. & Goldewijk, K. K. (2011) Holocene carbon emissions as a result of anthropogenic land cover change. *Holocene*, 21, 775–791, DOI: 10.1177/0959683610386983.

Karl, T. R., Melillo, J. M., Peterson, T. C. & Hassol, S. J. (eds) (2009) *Global climate change impacts in the United States*. Cambridge University Press, UK.

Kelley, C. P., Mohtadi, S., Cane, M. A., Seager, R. & Kushnir, Y. (2015) Climate change in the Fertile Crescent and implications of the recent Syrian drought. *Proceedings of the National Academy of Sciences*, 112, 3241–3246.

Kloppenburg, J. (2014) Re-purposing the master's tools: The open source seed initiative and the struggle for seed sovereignty. *Journal of Peasant Studies*, 1–22.

Levine, J. A. (2011) Poverty and obesity in the U.S. *Diabetes*, 60, 2667–2668, DOI: 10.2337/db11-1118.

Loarie, S. R., Duffy, P. B., Hamilton, H., Asner, G. P., Field, C. B. & Ackerly, D. D. (2009) The velocity of climate change. *Nature*, 462, 1052–1055.

Lohan, T. (2017) Systemic failure: Why 1 million Californians lack safe drinking water. News Deeply/Water Deeply, available at: https://www.newsdeeply.com/water/articles/2017/07/05/systemic-failure-why-1-million-californians-lack-safe-drinking-water (accessed May 12, 2018).

Long, S. P. & Ort, D. R. (2010) More than taking the heat: Crops and global change. *Current Opinion in Plant Biology*, 13, 240–247.

Luedeling, E., Girvetz, E. H., Semenov, M. A. & Brown, P. H. (2011) Climate change affects winter chill for temperate fruit and nut trees. *PLoS ONE*, 6, e20155.

Macartney, S., Bishaw, A. & Fontenot, K. (2013) Poverty rates for selected detailed race and Hispanic groups by state and place: 2007–2011. *American Community Survey Briefs*, 2.

Mao, Y., Nijssen, B. & Lettenmaier, D. P. (2015) Is climate change implicated in the 2013–2014 California drought? A hydrologic perspective. *Geophysical Research Letters*, 42, 2805–2813.

McLean, T. (2016) How might Flint's water contamination affect garden soils? Parts 1 and 2. Available at: http://msue.anr.msu.edu/news/how_might_flints_water_contamination_affect_garden_soils_part_1, http://msue.anr.msu.edu/news/how_might_flints_water_contamination_affect_garden_soils_part_2 (accessed Jan. 26, 2016).

Melillo, J. M., Richmond, T. C. & Yohe, G. W. (eds.) (2014) *Climate change impacts in the United States: The Third National Climate Assessment*, U.S. Global Change Research Program, Washinton, DC.

Mohatt, N. V., Thompson, A. B., Thai, N. D. & Tebes, J. K. (2014) Historical trauma as public narrative: A conceptual review of how history impacts present-day health.

Social Science & Medicine, 106, 128–136, DOI: http://dx.doi.org/10.1016/j.socscimed.2014.01.043.

Monastersky, R. (2015) The human age. *Nature*, 157, 144–147.

Morello-Frosch, R. & Jesdale, B. M. (2006) Separate and unequal: Residential segregation and estimated cancer risks associated with ambient air toxics in US metropolitan areas. *Environmental Health Perspectives*, 386–393.

Myers, S. S., Zanobetti, A., Kloog, I., Huybers, P., Leakey, A. D. B., Bloom, A. J., Carlisle, E., Dietterich, L. H., Fitzgerald, G., Hasegawa, T., et al. (2014) Increasing CO_2 threatens human nutrition. *Nature*, 510, 139–142, DOI: 10.1038/nature13179.

Mykyta, L. & Macartney, S. (2012) *Sharing a household: Household composition and economic well-being: 2007–2010*. P60-242, US Census Bureau, Washington, DC.

NASA (National Aeronautics and Space Administration) (2017) NASA, NOAA data show 2016 warmest year on record globally. NASA news release 17-006, available at: https://www.nasa.gov/press-release/nasa-noaa-data-show-2016-warmest-year-on-record-globally (accessed Oct. 5, 2018)

NASA (National Aeronautics and Space Administration) (2018) Long-term warming trend continued in 2017: NASA, NOAA. NASA, available at: https://www.nasa.gov/press-release/long-term-warming-trend-continued-in-2017-nasa-noaa (accessed Oct. 5, 2018)

Nazaryan, A. (2016) In Southeast Los Angeles, your front yard might be a toxic waste site. *Newsweek*, available at: http://www.newsweek.com/2016/04/15/los-angeles-exide-lead-arsenic-pollution-444339.html (accessed April 6, 2016).

NCHS (National Center for Health Statistics) (2014) *Health, United States, 2013: With special feature on prescription drugs*. Hyattsville, MD, available at: http://www.cdc.gov/nchs/data/hus/hus13.pdf - 064 (accessed July 31, 2018).

NGA (National Gardening Association) (2014) Garden to table: A 5-year look at food gardening in America. National Gardening Association, available at: http://www.hagstromreport.com/assets/2014/2014_0402_NGA-Garden-to-Table.pdf (accessed March 3, 2016).

NOAA NCEI (National Oceanic and Atmospheric Administration, National Centers for Environmental Information) (2016) State of the climate: global analysis—annual 2015, available at: https://www.ncdc.noaa.gov/sotc/global/201513 (accessed July 2, 2016).

Okvat, H. & Zautra, A. (2011) Community gardening: A parsimonious path to individual, community, and environmental resilience. *American Journal of Community Psychology*, 47, 374-387, DOI: 10.1007/s10464-010-9404-z.

Oreskes, N., Shrader-Frechette, K. & Belitz, K. (1994) Verification, validation, and confirmation of numerical models in the earth sciences. *Science*, 263, 641–646.

Oreskes, N. & Conway, E. M. (2010) Defeating the merchants of doubt. *Nature*, 465, 686–687.

OSSI (2018) Open source seed initiative. Madison, WI, available at: https://osseeds.org/ (accessed Aug. 17, 2018).

Overland, J. E., Dethloff, K., Francis, J. A., Hall, R. J., Hanna, E., Kim, S.-J., Screen, J. A., Shepherd, T. G. & Vihma, T. (2016) Nonlinear response of mid-latitude weather to the changing Arctic. *Nature Climate Change*, 6, 992.

Parker, L. E. & Abatzoglou, J. T. (2016) Projected changes in cold hardiness zones and suitable overwinter ranges of perennial crops over the United States. *Environmental Research Letters*, 11, 034001.

Pickett, K. E. & Wilkinson, R. G. (2015) Income inequality and health: A causal review. *Social Science & Medicine*, 128, 316–326, DOI: http://dx.doi.org/10.1016/j.socscimed.2014.12.031.

Pulido, L., Kohl, E. & Cotton, N.-M. (2016) State regulation and environmental justice: The need for strategy reassessment. *Capitalism Nature Socialism*, 27, 12–31.

Reich, P. B., Hobbie, S. E., Lee, T. D. & Pastore, M. A. (2018) Unexpected reversal of C_3 versus C_4 grass response to elevated CO_2 during a 20-year field experiment. *Science*, 360, 317–320.

Ruddiman, W. F. (2013) The Anthropocene. In Jeanloz, R. (ed.) *Annual review of Earth and Planetary Sciences*, 41, 45–68. Annual Reviews, Palo Alto, California, DOI: 10.1146/annurev-earth-050212-123944.

Ruddiman, W. F., Ellis, E. C., Kaplan, J. O. & Fuller, D. Q. (2015) Defining the epoch we live in. *Science*, 348, 38–39, DOI: 10.1126/science.aaa7297.

SCENIC (2018) Southwest Climate and Environmental Information Collaborative, available at: https://wrcc.dri.edu/csc/scenic/ (accessed May 24, 2018).

Scott, C. A. (2013) Electricity for groundwater use: Constraints and opportunities for adaptive response to climate change. *Environmental Research Letters*, 8, 035005.

SELC (The Sustainable Economies Law Center) (2018) UrbanAgLaw.org. SELC, Oakland, California, available at: http://www.urbanaglaw.org/ (accessed May 8, 2018).

Siegel, K. R., McKeever Bullard, K., Imperatore, G., Kahn, H. S., Stein, A. D., Ali, M. K. & Narayan, K. M. (2016) Association of higher consumption of foods derived from subsidized commodities with adverse cardiometabolic risk among US adults. *JAMA Internal Medicine*, 176, 1124–1132, DOI: 10.1001/jamainternmed.2016.2410.

Solomon, G. M., Morello-Frosch, R., Zeise, L. & Faust, J. B. (2016) Cumulative environmental impacts: Science and policy to protect communities. *Annual Review of Public Health*, 37, 83–96.

SQS (Subcommission on Quaternary Stratigraphy) (2018) Working group on the "Anthropocene," available at: http://quaternary.stratigraphy.org/workinggroups/anthropocene/ (accessed May 14, 2018).

Stocker, T. F., Dahe, Q. & Plattner, G.-K. (2013) *Climate change 2013: The physical science basis. Working Group I Contribution to the Fifth Assessment Report of the Intergovernmental Panel on Climate Change. Summary for Policymakers*, available at: http://www.climatechange2013.org/images/report/WG1AR5_SPM_FINAL.pdf (accessed Oct. 5, 2018).

Uddling, J., Broberg, M. C., Feng, Z. & Pleijel, H. (2018) Crop quality under rising atmospheric CO2. *Current Opinion in Plant Biology*, DOI: https://doi.org/10.1016/j.pbi.2018.06.001.

UNU-WIDER (2014) World Income Inequality Database (WIID3a), available at: http://www.wider.unu.edu/research/Database/en_GB/wiid/ (accessed July 3, 2018).

USBLS (2018) Employment status of the civilian noninstitutional population, 1940s to date. US Bureau of Labor Statistics, available at: https://www.bls.gov/cps/tables.htm - annual (accessed July 6, 2018).

USCB (2017a) Historical living arrangements of children. US Census Bureau, Washington, DC, available at: https://www.census.gov/data/tables/time-series/demo/families/children.html (accessed July 9, 2018).

USCB (2017b) Current population survey, 2015 and 2016, annual social and economic supplements. US Census Bureau, Washington, DC, available at: https://factfinder.census.gov/faces/nav/jsf/pages/index.xhtml (accessed May 10, 2018).

USCB (2017c) Historical poverty tables: People and families, 1959 to 2016. US Census Bureau, Washington, DC, available at: http://www.census.gov/data/tables/time-series/demo/income-poverty/historical-poverty-people.html (accessed June 28, 2018).

USCB (2017d) 2014 National population projections tables. US Census Bureau, Washington, DC, available at: https://www.census.gov/data/tables/2014/demo/popproj/2014-summary-tables.html (accessed July 8, 2018).

USCB (2018) Urban and rural classification. US Census Bureau, Washington DC, available at: https://www.census.gov/geo/reference/urban-rural.html (accessed May 8, 2018).

Vince, G. (2011) An epoch debate. *Science*, 333, 32–37.

Voicu, I. & Been, V. (2008) The effect of community gardens on neighboring property values. *Real Estate Economics*, 36, 241–283.

von Lützow, M. & Kögel-Knabner, I. (2009) Temperature sensitivity of soil organic matter decomposition—what do we know? *Biology and Fertility of Soils*, 46, 1–15.

Weyant, C., Brandeau, M. L., Burke, M., Lobell, D. B., Bendavid, E. & Basu, S. (2018) Anticipated burden and mitigation of carbon-dioxide-induced nutritional deficiencies and related diseases: A simulation modeling study. *PLOS Medicine*, 15, e1002586, DOI: 10.1371/journal.pmed.1002586.

WID.world (2018) World Inequality Database, available at: http://wid.world/ (accessed July 7, 2018).

Xiao, M., Koppa, A., Mekonnen, Z., Pagán, B. R., Zhan, S., Cao, Q., Aierken, A., Lee, H. & Lettenmaier, D. P. (2017) How much groundwater did California's Central Valley lose during the 2012–2016 drought? *Geophysical Research Letters*, 44, 4872–4879, DOI: doi:10.1002/2017GL073333.

3 Responding to Change as a Food Gardening Strategy

D. SOLERI, S.E. SMITH, D.A. CLEVELAND

Chapter 3 in a nutshell.

- Working with change and the uncertainty it brings is part of gardening, and life.
- A simple model helps us understand change better, and respond more effectively.
- When change is familiar variation, we can cope using existing strategies; when it is a trend toward conditions we've never experienced, new strategies are often needed.
- The impact of change is the result of both exposure to change (how much and how frequently), and sensitivity to change (the extent to which it affects you).
- Our response capacity can reduce the negative impact of change by reducing exposure and, or, sensitivity.
- Vulnerability is the negative impact remaining after our response; it can sometimes be quantified as risk.
- Observation, garden experiments, and working together can all increase our response capacity and decrease our vulnerability.
- Distinguishing between different possible sources of variation makes our observations more useful.
- Informal garden experiments can answer many practical questions about responding to change.
- We can better test hypotheses with more formal experiments, using simple methods to control variation.
- Working together in formal and informal institutions increases our response capacity.
- Equity, transparency, accountability, and autonomy are needed for effective and long-lasting institutions.

When will the rains come this summer, and how intense will they be? Will the aphid population on my broccoli this spring be larger or smaller than usual? As discussed in Chapter 2, gardening always involves working with change and uncertainty, but how we respond to that change may depend on whether it is in the form of familiar variation or long-term trends. Gardeners have lots of experience coping with the uncertainty of short-term, *familiar variation*, like the annual cycle of changing seasons, by adjusting management practices. However, these coping strategies may not be an adequate response to long-term, directional *trends* like anthropogenic climate change (ACC), or increasingly inequitable access to garden resources like land and water, that bring unprecedented changes—changes that could even shut gardens down.

As the trends outlined in Chapter 2 have become increasingly evident, there's been a lot of discussion about how to adapt to them. While adaptation is important, we also need to think longer term about how to *mitigate* negative trends, that is, slow and reverse them, because otherwise the effects of trends like ACC and increasing social inequity will continue to increase in spite of adaptation. Without mitigation, some trends will likely bring changes so extreme that our capacity to adapt will be overwhelmed.

We begin this chapter with an overview and some basic concepts about responding to change, and then describe making observations, doing experiments, and organizing how we work together to respond effectively. More detailed discussion about doing formal experiments can be found in Appendix 3A. As always, we'll use the five key ideas outlined in the Introduction to help understand how to respond to trends in ways consistent with our goals and values.

3.1. From Vulnerability to Resilience

Uncertainty, variation, risk, vulnerability, coping, and adaptation are all words used in everyday conversation, but they are also concepts with specific meanings that can help us understand change in our gardens, and figure out how to respond to it

(Smit and Wandel, 2006; Engle, 2011; Fellmann, 2012). *Conceptual models* can be used to illustrate how these concepts relate to each other and help us manage change effectively. In this chapter we describe a basic model that gardeners can use in responding to change (Fig. 3.1), while keeping in mind that all models evolve as we learn more. Figure 3.2 builds on Fig. 2.1 by adding detail about responding to change that we'll discuss in this chapter. Table 3.1 contains definitions of terms found in the model and this book.

The future, including changes in the environment, climate and society, is uncertain, and the more change there is, the more uncertainty there is. When this change is in the form of familiar variation, probabilities of exposure to the change can be calculated based on past experience. Most of the time, we intuitively estimate probabilities based on our gardening experience and on local knowledge—we learn what to expect.

Familiar, recurring variation in a variable that is normally distributed (not skewed) is quantified in terms of the standard deviation of the mean (Section 2.1), which can be used to calculate the probability of occurrences around the mean. The most common example is the rule of thumb that there is a 68, 95 and over 99% probability that the value of a variable will be expected to lie within one, two or three standard deviations on either side of the mean, respectively (Box 2.3 explains the use of classes of probabilities as likelihood in climate projections). This means that for any normally distributed variable, there is a 95% probability that a new measurement will lie in the range of values between two standard deviations greater than, and less than, that variable's mean. That is, it will be very nearly certain to be within that range.

However, for many environmental variables, familiar variation such as this will be accompanied by trends which will often only be evident by examining data over long periods of time. For example, using temperature records from Davis, California, we can predict when to plant tender garden plants outside in order to avoid late spring frosts (Fig. 3.3). These temperature data show that the trend in last spring frost date is toward earlier dates, which has been observed for many locations in the continental US (McCabe et al., 2015). Thus, in the 1960s if we wanted to plant early with a high probability of avoiding exposure to last frost occurring on average on March 16, we would have planted those tender plants after

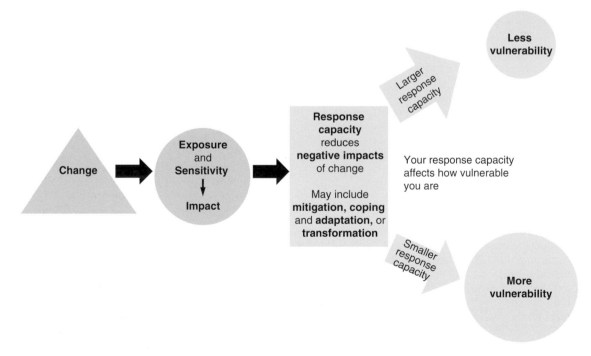

Figure 3.1. A conceptual model about responding to change

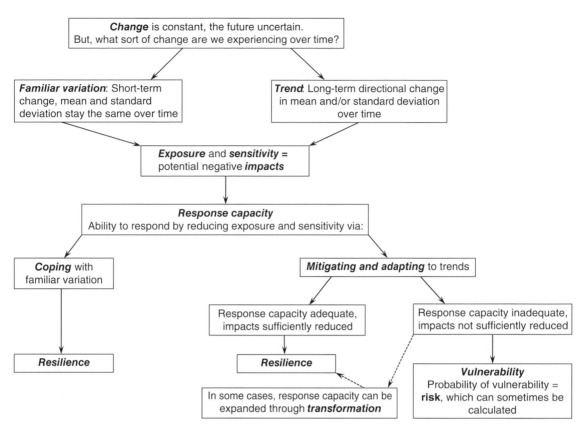

Figure 3.2. Responding to change as familiar variation or trend

May 6, two standard deviations more than that mean. Looking at the 10-year period ending in 2018, spring planting would occur about a month earlier than the 1960s, the result of a trend toward warming earlier in the year. Between 2008–2018, the average last spring frost occurred on February 7. Likewise, using the standard deviation over this period, we can calculate that 95% of the time the final frost would occur sometime before April 4, thus spring plantings after that date are very likely to avoid frost. Still, a small percentage of years will have last frosts later than this. Gardeners' experiences will often result in similar knowledge that will also change through direct experience of familiar variation and trends. Anticipating future trends through long-term projections, as well as using short-term weather prediction tools (Section 8.7.2) can help plan gardening activities such as when to plant frost-tender species outside in the springtime.

The conceptual model (Fig. 3.1) helps in planning for and responding to familiar variation and trends that can have negative consequences. Written out, the model states that the likelihood or probability (Box 2.3) of negative *impact* (*I*) from a change is a function (*f*) of exposure (*E*) and sensitivity (*S*) to the change (Fellmann, 2012). *Exposure* is how frequently and for how long you experience the change; *sensitivity* is the extent to which you will be negatively affected when exposed to it.

$$I = f(E, S)$$

Calculating the probability of an impact is part of quantifying risk (Table 3.1), something gardeners will never need to do, and is often not even possible because there isn't enough information available. However, an understanding of how sensitivity and exposure can interact and compound impact over time can be used in a qualitative way to help us take problems apart and make it easier to figure out effective responses. The model helps us to see

---

that we can minimize exposure to an event with negative impact by escaping it, and reduce sensitivity to it by avoiding the impact or tolerating the damage it causes. For example, squash vine borers (*Melittia cucurbitae*) are common in southern Arizona in spring and summer, and infested plants die quickly (Chapter 9). This means that exposure during the peak squash growing season (June–August) is high if you plant squash. Sensitivity to these pests can also be high because many squash varieties are very susceptible. If there are many vine borer moths laying eggs in a year when you are growing squash (high *E*), and you have a lot of susceptible squash plants (high *S*), you will suffer very high losses (high negative *I*). But through observation and learning gardeners can respond so that their gardening goals are still met.

While we cannot control how many moths are present, we can reduce our exposure by escaping this pest—growing something else, planting fewer squash, or planting very early in the spring or later in the summer, when vine borer moths are less active in southern Arizona. We can reduce our sensitivity by finding and planting only vine borer-resistant varieties, ones that can avoid or tolerate the stress inflicted by borers and still produce a harvest (butternut squash seems to be the most resistant). We can also reduce sensitivity by controlling the borer, for example by wiping the stems of squash plants every few days during the egg laying period if we only have a few plants, or, once they become infested, carefully slitting the stem, removing the larva, and piling soil or compost around the stem at that point, including at least one leaf node, to encourage new roots to grow there.

Our ability to respond to change by reducing exposure and/or sensitivity in ways that reduce negative impacts, and have positive outcomes, is part of our *response capacity*. The greater our response capacity, the larger the range of conditions we can effectively respond to as gardeners, either by coping with familiar variation, or adapting to

Table 3.1. Definitions for understanding and responding to change

Behavioral adaptation	Adjusting behavior to ongoing trends, or in preparation for future conditions being brought by trends
Change	Either familiar recurring variation, or a trend, that is, directional change over time
Coping	The part of our response capacity we use to address familiar variation
Escape	A way to reduce exposure to change
Exposure	Experience of an event—how frequently and for how long; one determinant of vulnerability
Familiar variation	Cyclical, or recurrent change; over time the mean and standard deviation stay the same (Section 2.1)
Impact	A function of exposure and sensitivity to a change
Mitigation	Actions to prevent or decrease change that would have a negative impact. The other part of response capacity to new conditions brought by trends, in addition to behavioral adaptation
Resiliency	The opposite of vulnerability, the result of a successful response to change; can be either stabilizing the current situation in response to familiar variation by coping, or through adaptation, mitigation, or transformation to a positive new state in response to trends
Resist	A way to reduce sensitivity, either by avoiding or tolerating the impact of a change
Response capacity	The ability to respond to change by reducing exposure and/or sensitivity in ways that have positive outcomes; the determinant of resiliency and vulnerability; comprises coping in response to familiar change, and adaptation, mitigation, or transformation in response to trends
Risk	The quantified probability of vulnerability
Sensitivity	The extent to which you could be affected when exposed to a change; one determinant of vulnerability
Transformation	Response by changing to a different, more adapted state or behavior, rather than modifying the existing one
Trend	Unfamiliar, directional change in mean, or in variability, e.g. measured by standard deviation
Uncertainty	A result of lack of knowledge, everything in the future is uncertain; increasing change increases uncertainty
Vulnerability	The extent of negative impact due to change that is beyond our response capacity; inversely related to response capacity

Based on Fellmann (2012); Preston and Stafford-Smith (2009); IPCC (2014b)

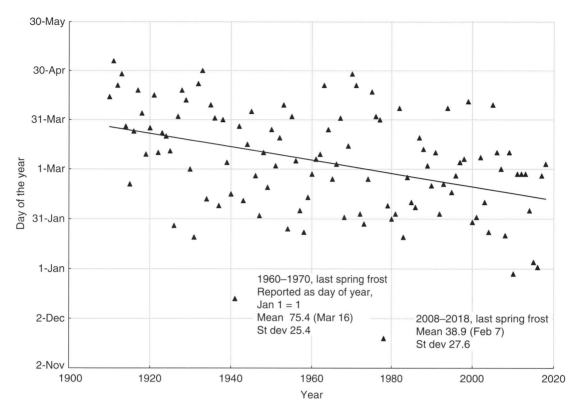

Figure 3.3. Date of last temperature of 0°C (32°F) or below in Davis, California, from November-May for the period 1910–2018

Slope of linear trendline: −0.405 (significantly different from zero, $P < 0.001$, $r = 0.398$). Data source (SCENIC, 2018).

and mitigating trends (Fellmann, 2012). Thinking about our goals for gardening helps to build response capacity that supports those goals. For example, if the primary goal of a community garden is to produce vegetables for members, your response to change might be very different than if the goal is to provide neighborhood youth with opportunities to learn more about gardening.

3.1.1. Coping

Coping is our capacity to respond to the recurring, temporary changes of familiar variation (Fig. 3.2). Coping involves reducing either exposure or sensitivity, or both, as in the squash vine borer example above. All gardeners and farmers have coped with familiar weather variation in the past, such as deciding when to plant based on when the rainy season will start, which is never exactly the same year to year. Farmers in southern Uganda

anticipate the start of rains using a combination of past experience, direct observations of weather indicators (e.g., types and locations of clouds), signs associated with the start of the rainy season (e.g., high night-time temperatures, small whirlwinds), and news of weather in the region (Orlove et al., 2010). Many experienced gardeners in other areas also have indicators of coming rainfall, or high or low temperatures that they have found reliable, although this may be changing due to ACC.

Another example of coping with familiar variation is planning for good fruit production from your tomato plants. Many tomato varieties will not set fruit if maximum temperatures pass a threshold during the two weeks before flowering, because the heat prevents normal pollen growth. An experiment with a common breeding line found about 50% fruit set (50% of flowers produced fruit) when the temperature during pollen growth was

25°C (77°F), but this dropped to about 16% when the temperature was 27.2°C (81°F), and to almost 0% at 28.9°C (84°F) (Peet et al., 1998). If it is very likely that the familiar variation in July temperatures in your area includes temperatures above 27°C, and you plan to grow lots of plants of such a heat-sensitive tomato variety that would be flowering in July, then your exposure and sensitivity will be very high. However, exposure to damaging heat can be reduced by not planting tomatoes, or by having vigorous starts ready to plant outside early in the spring, so that pollen development will occur before such high temperatures. Or, because most tomato varieties in the US are not day-length sensitive (Section 5.8.1), gardeners could use early-flowering varieties that start flowering before high July temperatures. Sensitivity can be reduced by using varieties known to be more heat resistant, or by shading plants to keep them cooler. It's easy to see why having a diversity of practices, and plant varieties, is useful for reducing both exposure and sensitivity.

3.1.2. Adaptation and mitigation

Gardeners have no previous experience with the new conditions that many current trends are bringing to their area, so responding to these trends requires more than coping—it requires *behavioral adaptation* to the changes we are experiencing now, and preparation for the future changes we can expect (Fig. 3.2) (Fellmann, 2012). While coping reduces exposure or sensitivity to familiar variation using established methods, behavioral adaptation to trends often demands new strategies, or new levels of existing strategies, in order to expand response capacity and reduce exposure and sensitivity so that we can garden successfully under new conditions. This relationship between change and our responses is visualized in Fig. 3.4, with the example of maximum average July temperature in Bakersfield, California. Behavioral adaptation can include seeking out biologically adapted crop varieties, and selecting seed to produce plant populations more adapted to the new conditions resulting from trends (Sections 10.3.3, 10.6).

Some current trends are easy to identify, and to make predictions about with certainty (Chapters 1 and 2). We can use this knowledge to begin adapting to those trends before they have a major negative impact on our gardening, and sometimes we can also address a trend's cause and so mitigate it, reducing its magnitude. There are trends affecting gardeners that have been documented or predicted with great certainty, predictions scientists are confident

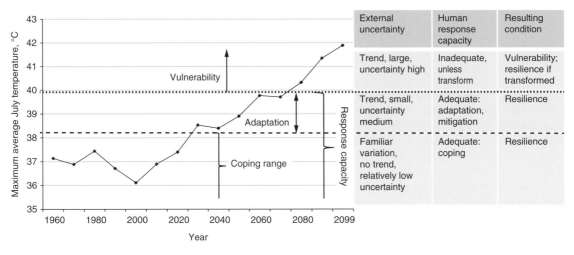

External uncertainty	Human response capacity	Resulting condition
Trend, large, uncertainty high	Inadequate, unless transform	Vulnerability; resilience if transformed
Trend, small, uncertainty medium	Adequate: adaptation, mitigation	Resilience
Familiar variation, no trend, relatively low uncertainty	Adequate: coping	Resilience

Average   1950–99 = 36.9°C   2000–49 = 38.0°C   2050–99 = 40.6°C

Figure 3.4. Coping, adaptation, vulnerability. The example of maximum average July temperature (°C), 1950–2099, Bakersfield, California

Concept based in part on Fellmann (2012) and sources therein. Data (decadal means) from GIF (2018), observed data 1950–2015; 2016–2099 projections based on high emissions scenario [A2] and CNRM CM3 model; station GHCND:USW00023155

about (Box 2.3), and gardeners can adapt to and mitigate some of these.

For example, trends in temperature due to ACC (Section 2.3) mean that some locations that have not had trouble with tomato fruit set in the past will have to plan for it in the future, and will need new strategies to adapt. The practices mentioned earlier as coping strategies used by gardeners who have been working with high summer temperatures as part of familiar variation in their location can also be used as adaptations by gardeners experiencing these high temperatures for the first time as a result of ACC trends. Looking into methods used in areas with climates similar to the one that ACC trends are creating in your area is a great first step in adaptation (Jarvis et al., 2014), as is planning and behavioral adaptation in anticipation of change. Developing infrastructure and institutions that make gardening easier for older city dwellers could reduce sensitivity to the demographic trend of increasing urbanization and aging. Partnerships and policies that prioritize disenfranchised communities' demands for food sovereignty, including gardening services and resources, could mitigate the trend of rising social inequity. Expanding community and school gardens in cities now could mitigate the trend of increasing noncommunicable diseases (NCDs), if the gardens increase fruit and vegetable consumption and physical activity, and reduce consumption of unhealthy food.

Another example of adapting to anticipated changes is responding to the ongoing trend of increasing water scarcity. Some urban community gardens may have a limited amount of municipal water made available free of charge for the whole garden, but any additional water comes at a cost that's prohibitive for members. Trends such as rising temperatures, more severe limits on free water because of decreasing water availability, and more community members wanting food gardens, are ways in which gardeners' exposure to water scarcity is increasing. That exposure can be decreased by developing alternative sources of water for the garden, for example by harvesting rainwater, obtaining greywater from nearby residences for fruit trees, or by requesting a higher free water limit from the city, based on documented benefits the gardens provide. Sensitivity to water scarcity could be reduced by planting more drought-adapted crops, using more efficient watering methods, and conserving water by mulching and other methods (Chapter 8). If reducing exposure and sensitivity don't adequately

respond to water scarcity, community gardeners may need to look for additional sources of fruit and vegetables and other benefits beyond their gardens.

Later in this chapter (Section 3.3) we'll talk about how response capacity can be strengthened not only by changing our individual behavior based on new knowledge, but also by the way we organize our interactions with each other, and share resources.

3.1.3. Vulnerability and resilience

When trends cause negative impact beyond our response capacity, we become *vulnerable* (Fig. 3.4). Vulnerability that can be quantified numerically as a probability is called *risk* (Fig. 3.2). For example, in the IPCC reports, the predicted risks from ACC are described using likelihood classes ranging from virtually certain (99–100% probability) to exceptionally unlikely (0–1% probability) (IPCC, 2014a) (Box 2.3). Often though, biophysical and socioeconomic variables interact in ways that are not well understood, so numerical probabilities for vulnerability can't be calculated, and the risk is classified as "unknown."

Our response capacity can give us resilience, which is the opposite of vulnerability. The greater the response capacity of an organism or a system, such as a community or household garden, in the face of change, the more *resilient* it is, and the less vulnerable (Table 3.1). That is, it can respond to change and stress by remaining in a desirable state through coping, adaptation and mitigation (Engle, 2011) (Fig. 3.4). But resilience does not always mean bouncing back to the way things were before the change.

When a trend results in large changes, resilience can also mean successful *transformation* (Kates et al., 2012), for example to another kind of food garden, perhaps by replacing crops that have been grown in an area for a long time with new ones. For example, walnut varieties that have been grown in California's Central Valley for decades will no longer be able to set fruit in the future due to ACC because there won't be enough *chilling hours* (cumulative hours with temperatures <7.2°C [45°F]) (Luedeling et al., 2009) (Section 5.8.2). This means that gardeners and farmers will need to replant with new varieties, or switch to different fruit trees, and some farmers might even need to give up commercial agriculture if ACC and increasing inequality of access makes water so scarce or costly that farming is impossible for them.

A trend such as the spread of recently introduced pests and diseases can create new vulnerabilities for gardeners and farmers. The Asian citrus psyllid (*Diaphorina citri*) is the vector for *huanglongbing*, or citrus greening disease, caused by a bacterium (*Candidatus* Liberibacter spp.) that infests the phloem tissues (which transport the products of photosynthesis), eventually killing the tree (Section 9.4). The psyllid and the disease were first identified in the US in Florida in 2005, and by 2008 they were in parts of western citrus states including Texas, Arizona, and California. By 2014 the disease was responsible for a 42% drop in orange yields in Florida (Singerman and Useche, 2015). Effective control methods have not been found, so diseased citrus trees die, and gardeners and farmers must adapt by replanting with non-citrus fruit trees, or with annual crops, either one being a significant transformation in their cropping system. Farmers in southern California have started doing this based on predictions of the disease's spread. For some of them the transformation from citrus to vegetables has been made easier by a commitment from the Los Angeles Unified School District to purchase whatever vegetables they can grow (Watanabe, 2013, 2014). The diversity of crops farmers can grow and market, and having supportive institutions, all contribute to farmers' greater response capacity and resilience, including the ability to successfully transform their production when necessary.

Observation and experimentation to figure out effective practices, and working together to connect and organize through networks and institutions, are ways gardeners can improve their response capacity, reduce vulnerability, and support mitigation, and these are what the rest of this chapter is about.

3.2. Observations and Experiments

Through observation and analysis we can understand how our practices affect our gardens and the world beyond, how certain actions are associated with specific outcomes, and what we can do differently to reach our gardening goals. Sometimes cause and effect relationships can even be identified. This process of informal observation, analysis, and action is ongoing and iterative in gardening—as you notice things, you make conclusions about what is going on, and change your practices to reach your goals;

then you observe the results, come to new conclusions, and adjust your practices again, and so on.

3.2.1. Formal science, local science, community science

Science, as the term is commonly used, is a method of systematic empirical observation and experimentation to expand human understanding. Science seeks to minimize bias and may include formal hypothesis testing. Scientific explanations and models, like those based on our observations in the garden, may change as our understanding changes. Formal scientific inquiry, and the knowledge it creates, tends to be very focused, often controlling or accounting for as many variables as possible that are not the direct subject of the study. It may be focused on understanding causal mechanisms, but identifying associations or correlations between phenomena is often a first step.

In the past, formal scientific knowledge was considered to be fundamentally different than the relatively more informal, local, more experience-based knowledge of practitioners. For example, scientists' vs. gardeners' knowledge of garden soils or plants. Obviously scientists' and gardeners' knowledge are different in some ways because they are created in different sociocultural and biophysical places, and use different tools and methods. However, it is increasingly recognized that the local knowledge systems of gardeners, farmers, and others are also similar to scientists' knowledge. Because they are both based on empirical observations over time of the same basic reality, and on testing relationships between variables, their content can be similar—for example both may view the color of maize kernels or bean seeds as the result of a plant's genotype and not its environment. Yet because both are also influenced by their specific social contexts, including the tools available for observing, manipulating, and measuring, their content can be different. For example, gardeners and farmers working with genetically diverse varieties and environments may view crop yield as influenced only by the environment, while scientists working with genetically uniform lines also view genotype as important as a result of observing changes in yield under controlled conditions that gardeners and farmers do not experience (Soleri and Cleveland, 2009) (Section 10.2).

Recognizing the similarities and differences between formal and informal knowledge systems can pave the way for communities and scientists to collaborate, resulting in knowledge and practices that are more useful than either could create independently. That's why recognition of knowledge diversity is one of the key ideas in *FGCW* (Introduction), and one reason for the growth of "citizen science," or more appropriately community or open science (Box 3.1). In *FGCW* we talk about the empirical understanding from these two systems as formal (i.e., formally trained, professional scientists') and local (i.e., experientially based, 'informal,' gardeners' and farmers') knowledge.

Gardeners' and farmers' knowledge of their crops and environments based on years of experience can be more useful for their local situations

---

Box 3.1. Opening up science

There's lots of interest and excitement around "citizen," community, participatory, and open science, broadly defined as some form of partnership between professional scientists and the public (Shirk et al., 2012; Dosemagen and Gehrke, 2017). Realizing the full potential of such partnerships for society requires a conscious effort to open up science so that it can serve the goals of diverse groups of people, be seen as a credible source of knowledge, and support social justice (Soleri et al., 2016). We suggest three ways to support the opening up of science.

First, changing who does science, and who is served by it. "Science" has gained a bad reputation among some people because it has often been used to serve the beliefs and agendas of the most powerful groups and their institutions, ignoring other perspectives, supporting the inequitable status quo, and sometimes inflicting terrible harm (Garrison, 2013). Supporting the capacity to do formal science outside of the dominant institutions of government, business, and formal education can change this. For example, during the twentieth century, plant breeding in industrial countries transitioned from being done by gardeners and farmers, to being mostly done by public institutions, to being dominated by private companies with goals of increasing yield and financial profit. But recently, new public, non-profit and regional groups have begun collaborating with farmers to conduct scientific crop improvement based on different values and goals (Section 10.4).

Second, changing the questions asked. Coalitions of communities and scientists have been formed to address environmental justice questions that were not being asked by scientists, including about air, water, and soil quality (e.g., Macey et al., 2014). Similar coalitions around the community seed sharing movement may be able to contribute to food justice (Soleri, 2017). Continuing with the crop improvement example, instead of asking what variety can provide the highest profits to a seed company, the questions could be ones that gardeners with limited resources would like answered, for example, "How can the diversity of maize be used to improve drought resistance in the maize varieties grown by gardeners in central New Mexico?" One organization asking different questions is the Organic Seed Alliance, which brings plant breeders together with organic gardeners and farmers to develop crop varieties suited to their needs (Sections 3.4, 10.4).

Third, making research assumptions explicit and open to discussion. The belief that formal science is completely objective has been widely discredited, and it is now accepted that, like all knowledge, formal scientific knowledge is limited by its context, and is based in part on values. Sometimes our exposure to different environments, experiences, and ideas is so limited that we simply do not see how context-specific our assumptions are. A plant breeder can be unaware of the social and physical environments gardeners and farmers are working in, leading to assumptions about the way things *are* that she could, and should, test empirically. Doing so can provide surprising and valuable insights (Soleri and Cleveland, 2017).

Value-based assumptions about how we think things *should be* are also important. For example, large, multinational seed corporations and their allies (Section 10.4.4) make many assumptions about how agriculture should be now and in the future, and these need to be made explicit so that the influence of those assumptions on scientific research can be discussed and challenged. We all have value-based assumptions that we are unaware of, but a goal of good science is to minimize their influence on research by making them explicit, and challenging and discussing them among people who may be affected, but may also have different values. Some powerful corporations and institutions have had campaigns defending research favorable to them as "sound science," and calling research with unfavorable findings "junk science" (Oreskes and Conway, 2010; CMD, 2016). Among other things, this is a "misrepresentation of science" (Jasanoff, 2014) as the domain of particular groups in society. As a systematic way of asking questions about empirical reality, science can only be "sound" if it is open to analysis from all people and perspectives, is as objective as possible, and is transparent about its value-based assumptions and motivations.

---

than more generalized formal scientific knowledge. For example, based on many years of collective observations in a neighborhood, gardeners may understand that changes in some plants and the activity of certain insects precede changes in weather and so are important for planning. Similar to the example in Uganda (Section 3.1.1), in South Africa, maize farmers have observed that the blossoming of the native wild aloe, *Aloe ferox*, and red ant activity are indicators of arriving summer rains, and use these as cues for planting (Zuma-Netshiukhwi et al., 2013). Scientists speculate that both of these natural phenomena could reflect changes in soil and air temperatures that precede the rainy season. This local knowledge may be far more useful than relying on formal knowledge from internet or radio weather reports that can arrive too late, and is rarely so locally specific.

3.2.2. Observation

Observation is one of the joys of gardening—there is a lot to see, and there are also smells, textures, tastes, sounds, and more. Our ability to observe can be greatly influenced by our daily habits. Just as eating

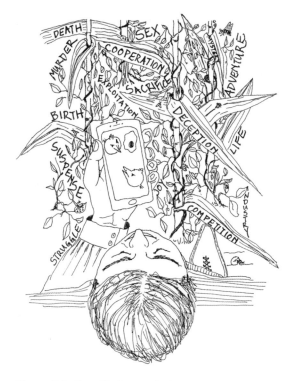

Figure 3.5. Bored in the garden!

lots of salty, sweet, or fatty foods can dull our ability to appreciate the more diverse and delicate tastes of fresh fruit and vegetables (Section 1.1.3), habituation to overt stimulation, such as from the internet and social media, can make it harder to see more subtle stimuli (Grizzard et al., 2015) (Section 1.3), and makes some parts of the garden hard to see at first (Fig. 3.5). Of course, some people are just better observers than others, but our ability to observe can improve quickly with practice. Asking questions in ways that stimulate curiosity, and making the process fun are great ways to get more people interested in making observations—humans love to play!

Some people keep garden journals or diaries of what they observe and do in the garden. These descriptions are data that can be used for asking questions about how things are related, and how to respond to change. Box 3.2 gives some suggestions for making useful garden observations.

Some of the most valuable garden observations are about timing, and are helpful when planning and scheduling garden work. For example, how many days passed between planting my okra seed and the first blossoms, and then the edible fruit or capsules (Fig. 3.6)? Or, how many days from tasseling to having silks in my maize plot? You can also look for changes in relative timing with ACC, as in our example in the Introduction of seeing persimmons ripening early while basil is still thriving. Understanding relative timing can help, for example, in planning when you will plant different crops in an intercropped bed.

Timing of different events in the garden usually varies between seasons and years, and between locations, but as you accumulate observations, you may even create garden calendars to help you plan and keep track of climate trends. Observations like these that monitor *phenology*, the timing of biological processes, are becoming especially important as ACC affects plants worldwide. The US National Phenology Network was established to document this, has active public participation, and now includes observations of some crop plants, as well as many other organisms (Section 3.4.2).

Other important observations to make in the garden are of the amounts of inputs used and outputs produced. Inputs can include seeds, water, compost, and time spent working in the garden and organizing people, and outputs can include the amount of food produced, the garden foods being eaten by whom, neighborhood improvements, and the enjoyment, socializing, and physical activity of gardeners. Environmental, climate, and social changes

(Chapter 2) mean that competition for resources, space, and time are increasing, and so is the need to show evidence that gardens are a worthwhile way to use those resources. Understanding the benefits and costs of our gardening practices helps us adjust them in order to increase the benefit:cost ratio, by modifying or eliminating a crop, practice, or output with a benefit:cost that is so low (much less than 1.0), that it just does not make sense, or increasing a crop, practice, or output that has a high benefit:cost (Section 1.5). Community and school garden organizations can use this information when applying for grants and other types of support.

It's important to measure variables that are good indicators of the goals for the garden. Harvests can be one indicator of garden success. As we describe in Section 1.5.2, recent studies of amounts of garden production have used either total weight per year or season, or harvest weight per plant for a sample of plants of specific crops, extrapolated across gardens. Individual plant production is attractive, especially for gardens that often have multiple species intercropped, making area sown to a crop an invalid indicator. However, variation due to different crop varieties and growing environments could be so large that extrapolation of the data

Question/example	Description	Comparison
How many days after planting does my okra flower, and when are fruit ready to eat?	Descriptive observation Familiarity with garden, e.g., phenology, other characteristics, events	Documenting, comparing before and after passage of time, may continue as long as desired

Start ⟶ time (T) ⟶ End

T_1 ·· T_n

Figure 3.6. Experimenting in the garden: descriptive observation

might be quite misleading, and only some crops, like cucumber, maize, or tomato can be easily measured this way.

An alternative method is to actually weigh everything harvested from your garden, or from a sample of different types of gardens among those you want to document, then use these data to estimate the yield for each type, and total production. Spring-loaded mechanical scales are the easiest and most durable equipment for weighing harvests in the garden; electronic kitchen scales also work but are more delicate and require a battery, solar panel, or other power source.

The amount of food harvested from a garden is the most popular output to measure, but just as food miles is not an appropriate indicator for many of the hopes people have for local food systems (Cleveland et al., 2015), amount harvested may not be the only or best indicator for many gardeners' goals. If the goal is improved nutrition, then other indicators such as types of fruits and vegetables grown, and how much garden produce is eaten and by whom, will need to be measured, in addition to amount harvested, to get a valid idea of the garden's effect. Many other garden outputs are potential benefits, as discussed in Chapter 1, including social and personal ones, that can also be measured. For example an alliance of community gardens and university-based researchers in Minneapolis, Minnesota developed short surveys asking gardeners how many and what type (inter-generational, inter-racial, inter-ethnic, neighbor) of new acquaintances and friendships people developed through the garden (Grewell, 2015).

Inputs also have to be measured in order to estimate benefit:cost ratios, or net benefit. There are

surprisingly few studies of garden inputs and outputs, including one we conducted with friends in our own home gardens in Tucson, Arizona (Cleveland et al., 1985) (Sections 1.5.1, 1.5.2), and quickly learned that input–output documentation is labor-intensive! We measured everything that went into and came out of the garden over a total of five garden years. It was tedious, but without these data, we can only guess about costs and benefits. Droughts and increasing water scarcity make documenting garden outputs per unit of water used especially important (Section 8.5). This can be done for an entire garden, or just one or more garden beds. Water used can be calculated as the time spent irrigating multiplied by rate of water application, measured by using containers of known volume, or more precisely by using a dedicated garden water meter, or an inline meter that can be installed at the faucet or the hose end.

3.2.3. Informal experiments

Informal experiments differ from descriptive observation by focusing on comparing two or more practices or crop varieties, such as comparing a new practice with the one you are already using. They are most useful when you want to get an indication of what might work best in your garden right now, without spending the time to do a formal experiment. For example, does mulching reduce how often I need to water my basil bed? (Fig. 3.7). Small, informal experiments are a good way to answer practical questions about food gardens, and explore ways to reduce exposure or sensitivity to change. For example, to reduce her

Question/example	Description	Comparison
Does mulch (treatment X) reduce how often I need to water my basil bed?	Informal experiment For immediate use, often less than a single season, observations before and after treatment, i.e., days between watering	T_1 (watering frequency) before vs T_2 (watering frequency after) treatment X

Figure 3.7. Experimenting in the garden: informal experiments

community's exposure to food insecurity, a garden organizer might ask: Is it better to involve more households in our community garden, or to encourage a few experienced gardeners to produce as much as possible and distribute some of the harvest to non-gardening households? To decrease their sensitivity, a gardener might ask: Does a chile and garlic spray reduce aphid damage to my chard any better than spraying the aphids off with water?

Informal experiments are small tests driven by gardeners' observations and curiosity, and are the most common type of garden experiment. They can also be valuable for generating hypotheses for formal experiments. However, since you aren't controlling for factors other than the ones you are experimenting with, the results of informal experiments can sometimes be unclear, be very unexpected, and may not be supported by subsequent experience. If this is the case, or if the results lead to questions you want to examine in more depth, you could try a more formal experiment.

Formal experiments test a hypothesis in a more structured way that minimizes confounding variation, making the results of the experiment more certain and generalizable. Formal experiments can tell you more than what worked in one particular place and time, and they are the basis of much agricultural, biological, social, and other scientific research. However, formal experiments require more work, time, and even space than most gardeners want or need to invest, and observation and informal experiments are usually adequate for smaller household gardens. Still, formal experiments can be very useful for larger school and community gardens, or for groups of household gardeners who want to explore more deeply, so we walk through three worked examples in Appendix 3A. Even if you will not be doing a formal experiment, reading that appendix will help improve your observations and informal experiments.

3.3. Working Together

In addition to observation and analysis, the way we organize ourselves and work together is another important strategy for making good use of resources and expanding our capacity to respond to change and uncertainty. There are lots of reasons for gardeners to rely as little as possible on external inputs, but that should not be confused with believing you are autonomous or self-sufficient. In a globalized world with intense pressure on natural and social resources, true self-sufficiency or autonomy is not possible because we are constantly affecting each other's lives. Improving how we work together is key to positive outcomes for the environment and our communities, and for the greatest number of people and other forms of life, now and in the future. Experience has shown that this is true for disasters (Solnit, 2009; IPCC, 2014b), for demanding and enforcing environmental and social justice, and for day-to-day management of shared resources.

Among small-scale farmers worldwide there is a long tradition of organized group work to accomplish tasks requiring lots of labor such as weeding, harvesting, or maintenance of shared infrastructure like irrigation systems (Netting, 1993). Gardeners can work together to protect garden resources and ensure equitable access to them. Trust is a prosocial attitude that is key for creating working relationships with these kinds of benefits. "Trust" may sound vague, subjective, and unscientific, but from individual to global scales, confidence that others represent themselves honestly, do what they say they will, and do not cheat and burden others, is incredibly important. For example, trusting others predicts whether societies are effective at solving shared problems like recycling waste (Mannemar Sønderskov, 2011), and trusting that others will reciprocate your conservation practices was the first requirement for farmers working cooperatively to manage their watershed (Lubell, 2004).

Indeed, one of the reasons that divisive, antisocial political behavior is harmful is because it degrades trust and reduces prosocial behaviors, threatening many of society's basic institutions that we all rely on. Organizing that supports trust and reciprocity through transparent, equitable rule enforcement expands our response capacity, reducing vulnerability (Adger, 2010). The organization needed to build trust and equitably manage our resources occurs through the institutions we create. By *institution* we mean any organized, shared system of interaction that determines participants' constraints and opportunities in relation to a particular process or resource (Box 3.3). Institutions can be informal, such as self-organized groups or the farmer–customer relations described below, but are often formalized, usually through a set of written rules.

Institutions can be valuable and positive, but they can also increase inequity. For example, the USDA's National Organic Program develops and enforces federal regulations for organic agriculture meant to

Box 3.3. Institutions and what can help them be prosocial

Interest in investigating institutions that manage shared resources was stimulated in 1968 by Garret Hardin's now famous paper "The tragedy of the commons" (1968). He argued that open access to shared, finite resources for private benefit would drive self-interested individuals to use those resources to the point of destroying them, leaving both society and the environment worse off—the tragedy. Hardin thought such a tragedy could only be avoided by top-down government controls. Research since then, especially by the late economist and Nobel laureate Elinor Ostrom, and colleagues, has investigated alternative ways of managing shared or *common pool resources* (CPRs) like the ones Hardin referred to. CPRs are resources like irrigation water, land, air, pasture, forests, or fisheries, which are finite, subject to degradation from overuse, but are relatively difficult to prevent others from using. Ostrom and others have now documented many cases where individuals self-organize into *common property management* (CPM) institutions that regulate CPRs over the long-term in ways that prevent degradation and support equitable access (Ostrom, 2008).

Situations such as the one Hardin documented are now seen as tragedies of *open access* to CPRs, that is, situations with a lack of management beyond individual self interest (Cole et al., 2013). CPRs can also be managed as private property by corporations or individuals that have the power to exclude others, or by governments. However, CPM by communities, or other direct use groups, is often the most effective approach to conserving resources in ways that promote social justice, equitably providing benefits to users.

Community food gardens are CPRs, and many garden resources, like water and public urban open space, are becoming CPRs as their scarcity increases, and they become more vulnerable to overuse. For example, many gardeners in the dry areas of the global north have become used to the idea that the amount of piped water they use in their garden is nobody's business but their own. But increasing water scarcity and drought, for example in the western US, means that this is not true—individuals' water use affects everyone, now and in the future. As a result, more and more people are realizing that prosocial water management in support of equity means limiting water use, including in gardens, by agreeing on a system of CPM. Top-down restrictions on water use by water districts and other local and state agencies are reinforcing this transformation of water management. CPM at the community level is a way of making these top-down

measures as fair as possible, preventing a small number of people from disproportionately degrading, or benefiting from, unequal access to shared resources, including limited water supplies, garden plots, or compost.

Below is a list of characteristics of successful CPM institutions (McGinnis, 2011). Many of these characteristics can improve other institutions as well, because they identify features that help ensure the transparency, access, and equity essential for trust and fair, sustained cooperation. After each characteristic we provide a hypothetical example for a community garden (CG). Another set of overarching guidelines for individuals working together in ways that support equity and justice are the Jemez Principles for Democratic Organizing (Section 3.4.3).

Boundaries (biophysical and social) are clearly defined.
> CG: Identify what (land, other resources, tools) and who (individuals, families, households) are part of the community garden.

Congruence between use and a) provision rules (for fairness), and b) fitness to local conditions (for practicality).
> CG: a) The more use, the more responsibility to contribute to the community garden; for example, the more land or water used, and the longer it's used, the higher your dues.
> b) Everyone's use is limited by environmental constraints (amount of land or water available).

Collective choice processes enable most affected individuals to participate in making rules.
> CG: Decisions are made by all members or their recognized representatives in a manner agreed upon, not solely by an executive committee, and not through an unknown decision-making process.

Monitors are accountable to users (or are the users themselves).
> CG: All members can monitor, document, and report rule violations.

Graduated sanctions are applied to rule violators (in increasing levels of intensity according to the extent of the violation).
> CG: Fines and other punishments are commensurate with the rule violation, and also increase with the user's number of violations.

Dispute resolution mechanisms are available to participants at low or no cost.
> CG: A free, agreed upon, transparent resolution process is available to all, and the results of that process are respected by all.

Continued

Recognition by "higher" authorities that users have rights to self-organize and devise their own institutions; institutions not undermined by higher authorities.

    CG: The landowner, community garden umbrella organization, sponsoring government agency, or other authority cannot overrule most decisions by the CG (Section 4.2.2).

Nested enterprises for use, provision, monitoring, enforcement, conflict resolution, and governance.

    CG: All management and problem-solving sub-groups such as membership, or water conservation committees, are part of the community garden, and must be respected by the members, and must themselves recognize whole-group decisions.

---

increase public trust in the organic label, and support large-scale commerce through standardization. Due to the influence of large agrifood corporations looking to take advantage of the growth in organic food sales to maximize their profits, the national standards in the US have been watered down, and many feel that much of the original intent of organic farming and food has been lost (Guthman, 2004; Jaffee and Howard, 2010). As a result of this, and the cost to farmers in money and time for required paperwork, some smaller-scale farmers are establishing their credibility through face-to-face interactions with their customers, relying on personal trust instead of formal "organic" certification, and they have created some alternative institutions to support this. For example, the Participatory Guarantee System is a global, trust-based organic certification alternative especially for small-scale producers, including gardeners who may want to market their produce locally (Kirchner, 2015), with each national or regional program having its own set of rules and norms, and reviewers. Currently over 800 producers are certified in the US, and over 17,000 worldwide.

Similarly, community gardens are institutions that agree on procedures for things like accessing and using garden plots and water, deciding which plants can and cannot be grown, and what inputs can be used. Institutions can change in response to changing needs, and the best ones do, but there is always a shared set of customs and rules that participants follow. At their best, institutions are tools for ensuring that shared resources are protected, that everyone's interest is addressed fairly, and that no one benefits at the expense of others. At their worst, institutions can reinforce great disparities of power and privilege. That's why the prosocial characteristics described in Box 3.3 are important for institutions.

Gardeners often create informal institutions, for example by agreeing to take care of each other's plots when one of them can't do it. Their informal institution is a support network that makes successful gardening more likely, and has shared expectations and rules, even if they aren't written down.

Larger institutions can also be critical for supporting resilient gardens and ensuring that garden benefits will be available when they are needed. Although these institutions may not be self-organized or adhere to all of the standards laid out in Box 3.3, some do respond to grassroots demands. Good examples are the growing number of new laws being passed by regional and city governments that expand response capacity by creating options for households and communities to garden, and for gardeners to sell their garden produce (Section 4.2.2).

3.4. Resources

All the websites listed below were verified on May 13, 2018.

3.4.1. Food gardens as response

Lawson's history of community gardening documents the ways in which community and other gardens have been responses to economic, social, and environmental change in the US since the late nineteenth century (Lawson, 2005).

3.4.2. Experiments

Levitan (1980) is an accessible, basic book about doing experiments in the garden. Experimental design, data collection, and simple analyses are explained with clear illustrations. Still the only book of its kind.

Good statistical practice for natural resources research (Stern et al., 2004) has thorough coverage of how to design, conduct, and analyze experiments with lots of examples from agriculture and natural resources management.

Some plant breeding references in Section 10.7.4 discuss experiments in, or relevant to, the garden.

LibreOffice is an open source, free suite of computer programs developed by an international non-profit collective. LibreOffice closely parallels Microsoft's Office programs, including word processing with Writer (comparable to Word), presentations with Impress (comparable to PowerPoint), and database management with Base (comparable to Access). It is available in many languages and for all operating systems. For gardeners doing experiments, LibreOffice Calc (comparable to Excel) can be used to create spreadsheets, enter data, and perform simple calculations such as random number generation (using the formula =rand()), and descriptive statistics: mean, standard deviation, median, mode, range, minimum, maximum, and others. A few statistical analysis functions also come with Calc: t-tests, analysis of variance (ANOVA), Chi-square, and others. It is what we used to analyze the formal experiments in Appendix 3A.
https://www.libreoffice.org/

A user-friendly, public, online statistical calculator has been created by Vassar College Emeritus Professor of Psychology Richard Lowry. It is also linked to his free online textbook, *Concepts and applications of inferential statistics*.
http://vassarstats.net/index.html

The *Handbook of biological statistics* is an online biostatistics textbook with explanations of different analyses, and often with spreadsheets for doing the analyses in Excel, and sometimes in LibreOffice Calc (McDonald, 2014).
http://www.biostathandbook.com/index.html

The USA National Phenology Network (USNPN) has data sets about plant (including some crops) and animal phenology, much of it collected by the public using their online program and app "Nature's Notebook" (NN). Gardeners can use NN to keep track of garden crop phenology. Many data sets are available on the USNPN website, including tracking and forecasts of seasonal changes such as plant development and pest activity based on ongoing observations.
https://www.usanpn.org/; https://www.usanpn.org/natures_notebook

A website of smartphone data collection apps and much more has been put together by wildlife ecologist Emilio Bruna and his students at the University of Florida.
http://brunalab.org/apps/

The following free smartphone apps include capacity for data form generation, GPS, and photos all uploaded to your project site, and can be shared: Imperial College London, iPhone and Android – http://www.epicollect.net/; University of Washington, Android only – http://opendatakit.org/. California Academy of Sciences' iNaturalist has preset and customizable data collection forms for biological species such as garden crop diversity, for both Android and iPhone.
http://www.inaturalist.org/

The Public Laboratory for Open Technology and Science (Public Lab) is a non-profit organization and open community dedicated to collaborative development of inexpensive, open-source tools for communities to monitor their environmental quality, including air, soil, and water. It sells some ready-made tools, or kits, and directions for building all of the tools are available for free on the website. In addition to the tools, Public Lab actively explores and documents its experiences and best practices for open, more democratic science.
https://publiclab.org/

Considered one of the best free, open source, geographic information system (GIS) software programs, QGIS is a publicly available desktop application in multiple languages for Windows, Mac, Linux, and Android platforms.
http://www.qgis.org/en/site/

3.4.3. Working together

The Jemez Principles for Democratic Organizing are guidelines for working together developed in 1996 in Jemez, New Mexico by a diverse group of people long active in the environmental justice movement in the US. The six principles are simple but profound and very challenging ("be inclusive, emphasis on bottom-up organizing, let people speak for themselves, work together in solidarity and mutuality, build just relationships among ourselves, commitment to self-transformation"), and are a great resource when large and diverse groups try to work together, even if they do not completely adhere to the Principles. Non-profit organizations

in the US and elsewhere are increasingly adopting the Jemez Principles.
http://www.ejnet.org/ej/jemez.pdf

The Organic Seed Alliance has ongoing research on varietal development for organic agriculture, sometimes through farmer or gardener and scientist partnerships, see Section 10.7.4.

The USDA's National Organic Program website includes lists of accepted and prohibited substances for use in certified organic production, and other resources regarding federal certification, all of which are subject to political influence.
https://www.ams.usda.gov/about-ams/programs-offices/national-organic-program

The International Federation of Organic Agriculture Movements (https://www.ifoam.bio/) has been most active in developing and promoting Participatory Guarantee Systems (PGS). The US PGS is Certified Naturally Grown (CNG) (http://www.cngfarming.org/). CNG provides certification through a standardized, online self-evaluation, peer review (farmer to farmer), and involvement by consumer and environmental groups at some sites. The original system was established by farmers in New York state, with input from the extension service, consumers, and the regional Sierra Club (IFOAM, 2008).

3.5. References

Adger, W. N. (2010) Social capital, collective action, and adaptation to climate change. In Voss, M. (ed.) *Der Klimawandel: Sozialwissenschaftliche Perspektiven*, 327–345. VS Verlag für Sozialwissenschaften, Wiesbaden, DOI: 10.1007/978-3-531-92258-4_19.

Cleveland, D. A., Orum, T. V. & Ferguson, N. F. (1985) Economic value of home vegetable gardens in an urban desert environment. *Hortscience*, 20, 694–696.

Cleveland, D. A., Carruth, A. & Mazaroli, D. N. (2015) Operationalizing local food: goals, actions, and indicators for alternative food systems. *Agriculture and Human Values*, 32, 281–297, DOI: 10.1007/s10460-014-9556-9.

CMD (Center for Media and Democracy) (2016) SourceWatch: The advancement of sound science coalition. CMD, available at: http://www.sourcewatch.org/index.php/The_Advancement_of_Sound_Science_Coalition (accessed August 5, 2016).

Cole, D. H., Epstein, G. & McGinnis, M. D. (2013) Digging deeper into Hardin's pasture: the complex institutional structure of "the tragedy of the commons." *Journal of Institutional Economics*, 1–17.

Dosemagen, S. & Gehrke, G. (2017) Civic technology and community science: A new model for public participation in environmental decisions. *Liinc em Revista*, 13, 140–161, DOI: 10.18617/liinc.v13i1.3899.

Engle, N. L. (2011) Adaptive capacity and its assessment. *Global Environmental Change*, 21, 647–656.

Fellmann, T. (2012) The assessment of climate change related vulnerability in the agricultural sector: Reviewing conceptual frameworks. In Meybeck, A., Lankoski, J., Redfern, S., Azzu, N. & Gitz, V. (eds) *Building resilience for adaptation to climate change in the agriculture sector*, 37–61. FAO/OECD, Rome, DOI: 10.13140/2.1.4314.8809.

Garrison, N. A. (2013) Genomic justice for Native Americans: impact of the Havasupai case on genetic research. *Science, Technology & Human Values*, 38, 201–223. DOI: 10.1177/0162243912470009.

GIF (Geospatial Innovation Facility at University of California, Berkeley) (2018) Cal-adapt. California Energy Commission, State of California, Berkeley, CA, available at: http://cal-adapt.org/ (accessed May 13, 2018).

Grewell, R. (2015) *Urban Farm and Garden Alliance Nelson Report UPDATE*, Kris Nelson Community-Based Research Program, Center for Urban and Regional Affairs, University of Minnesota, Minneapolis, MN. 23pp.

Grizzard, M., Tamborini, R., Sherry, J. L., Weber, R., Prabhu, S., Hahn, L. & Idzik, P. (2015) The thrill is gone, but you might not know: Habituation and generalization of biophysiological and self-reported arousal responses to video games. *Communication Monographs*, 82, 64–87.

Guthman, J. (2004) *Agrarian dreams: The paradox of organic farming in California*. University of California Press, Berkeley, CA.

Hardin, G. (1968) The tragedy of the commons. *Science*, 162, 1243–1248.

IFOAM (International Federation of Organic Agriculture Movements) (2008) *Participatory guarantee systems: 5 case studies*, available at: https://www.ifoam.bio/sites/default/files/page/files/studies_book_web.pdf (accessed July 13, 2018).

IPCC (Intergovernmental Panel on Climate Change) (2014a) Summary for policymakers. In *Climate change 2014, mitigation of climate change. Contribution of working group III to the Fifth Assessment Report of the Intergovernmental Panel on Climate Change*. Cambridge University Press, Cambridge, UK and New York, NY.

IPCC (Intergovernmental Panel on Climate Change) (2014b) *Climate change 2014: Impacts, adaptation, and vulnerability. The Fifth Assessment Report (AR5)*. IPCC, Geneva.

Jaffee, D. & Howard, P. H. (2010) Corporate cooptation of organic and fair trade standards. *Agriculture and*

Human Values, 27, 387–399, DOI: 10.1007/s10460-009-9231-8.

Jarvis, A., Ramirez-Villegas, J., Nelson, V., Lamboll, R., Nathaniels, N., Radeny, M., Mungai, C., Bonilla-Findji, O., Arango, D. & Peterson, C. (2014) Farms of the future: An innovative approach for strengthening adaptive capacity. AISA workshop on agricultural innovation systems in Africa (AISA), 29–31 May 2013, Nairobi, Kenya, 119-124, available at: http://www.farmaf.org/images/documents/related_materials/AISA_workshop_proceedings_final__March_2014.pdf (accessed 6 Oct. 2018).

Jasanoff, S. (2014) A mirror for science. *Public Understanding of Science*, 23, 21–26, DOI: 10.1177/0963662513505509.

Kates, R. W., Travis, W. R. & Wilbanks, T. J. (2012) Transformational adaptation when incremental adaptations to climate change are insufficient. *Proceedings of the National Academy of Sciences*, 109, 7156–7161.

Kirchner, C. (2015) Overview of participatory guarantee systems in 2014. In Willer, H. & Lernoud, J. (eds) *The world of organic agriculture. Statistics and emerging trends 2015. FiBL-IFOAM report*, 134–136. Research Institute of Organic Agriculture (FiBL), Frick, and IFOAM, Organics International, Bonn, Germany.

Lawson, L. J. (2005) *City bountiful: A century of community gardening in America*. University of California Press, Berkeley, CA.

Levitan, L. (1980) *Improve your gardening with backyard research*. Rodale Press, Emmaus, PA.

Lubell, M. (2004) Collaborative watershed management: A view from the grassroots. *Policy Studies Journal*, 32, 341–361.

Luedeling, E., Zhang, M. & Girvetz, E. H. (2009) Climatic changes lead to declining winter chill for fruit and nut trees in California during 1950–2099. *PLoS One*, 4, e6166.

Macey, G., Breech, R., Chernaik, M., Cox, C., Larson, D., Thomas, D. & Carpenter, D. (2014) Air concentrations of volatile compounds near oil and gas production: A community-based exploratory study. *Environmental Health*, 13, 82.

Mannemar Sønderskov, K. (2011) Explaining large-N cooperation: generalized social trust and the social exchange heuristic. *Rationality and Society*, 23, 51–74, DOI: 10.1177/1043463110396058.

McCabe, G. J., Betancourt, J. L. & Feng, S. (2015) Variability in the start, end, and length of frost-free periods across the conterminous United States during the past century. *International Journal of Climatology*, 35, 4673–4680, DOI: 10.1002/joc.4315.

McDonald, J. H. (2014) *Handbook of biological statistics*, 3rd edn. Sparky House Publishing, available at: http://www.biostathandbook.com/index.html (accessed Oct. 6, 2018)

McGinnis, M. D. (2011) An introduction to IAD and the language of the Ostrom workshop: A simple guide to a complex framework. *Policy Studies Journal*, 39, 169–183.

Netting, R. M. (1993) *Smallholders, householders: Farm families and the ecology of intensive, sustainable agriculture*. Stanford University Press, Stanford, CA.

Oreskes, N. & Conway, E. M. (2010) *Merchants of doubt: How a handful of scientists obscured the truth on issues from tobacco smoke to global warming*. Bloomsbury Press, New York, NY.

Orlove, B., Roncoli, C., Kabugo, M. & Majugu, A. (2010) Indigenous climate knowledge in southern Uganda: The multiple components of a dynamic regional system. *Climatic Change*, 100, 243–265.

Ostrom, E. (2008) Tragedy of the commons. In Durlauf, S. N. & Blume, L. E. (eds) *The new Palgrave dictionary of economics*, 2nd edn, 360–363. Palgrave Macmillan, DOI: 10.1057/9780230226203.1729.

Peet, M., Sato, S. & Gardner, R. (1998) Comparing heat stress effects on male-fertile and male-sterile tomatoes. *Plant, Cell & Environment*, 21, 225–231.

Preston, B. L. & Stafford-Smith, M. (2009) *Framing vulnerability and adaptive capacity assessment: Discussion paper*. CSIRO Climate Adaptation National Research Flagship, Australia.

SCENIC (2018) Southwest Climate and Environmental Information Collaborative, available at: https://wrcc.dri.edu/csc/scenic/ (accessed May 24, 2018).

Shirk, J. L., Ballard, H. L., Wilderman, C. C., Phillips, T., Wiggins, A., Jordan, R., McCallie, E., Minarchek, M., Lewenstein, B. V. & Krasny, M. E. (2012) Public participation in scientific research: A framework for deliberate design. *Ecology and Society*, 17, 29, DOI: http://dx.doi.org/10.5751/ES-04705-170229.

Singerman, A. & Useche, P. (2015) Impact of citrus greening on citrus operations in Florida. Institute of Food and Agricultural Sciences, University of Florida Gainesville, FL, available at: http://edis.ifas.ufl.edu/pdffiles/FE/FE98300.pdf (accessed July 13, 2018).

Smit, B. & Wandel, J. (2006) Adaptation, adaptive capacity and vulnerability. *Global Environmental Change*, 16, 282–292.

Soleri, D. (2017) Civic seeds: New institutions for seed systems and communities—a 2016 survey of California seed libraries. *Agriculture and Human Values*, DOI: 10.1007/s10460-017-9826-4.

Soleri, D. & Cleveland, D. A. (2009) Breeding for quantitative variables. Part 1: Farmers' and scientists' knowledge and practice in variety choice and plant selection. In Ceccarelli, S., Weltzien, E. & Guimares, E. (eds) *Participatory plant breeding*, 323–366. FAO (United Nations Food and Agriculture Organization), in collaboration with ICARDA (International Center for Agricultural Research in the Dry Areas) and ICRISAT (International Crops Research Institute for the Semi-Arid Tropics), Rome, Italy, Aleppo, Syrian Arab Republic, and Patancheru, India.

Soleri, D. & Cleveland, D. A. (2017) Investigating farmers' knowledge and practice regarding crop seeds: Beware your assumptions! In Sillitoe, P. (ed.) *Indigenous knowledge: Enhancing its contribution to natural resources management*, 158–173. CAB International, Wallingford, UK.

Soleri, D., Long, J., Ramirez-Andreotta, M., Eitemiller, R. & Pandya, R. (2016) Finding pathways to more equitable and meaningful public-scientist partnerships. *Citizen Science: Theory and Practice*, 1, Article 9.

Solnit, R. (2009) *A paradise built in hell: The extraordinary communities that arise in disaster*. Viking/Penguin, London.

Stern, R. D., Coe, R., Allan, E. F. & Dale, I. C. (eds) (2004) *Good statistical practice for natural resources research*. CAB International, Wallingford, UK.

Watanabe, T. (2013) L.A. Unified's local food push is healthy for area economy too. *Los Angeles Times*, November 24, available at: http://articles.latimes.com/2013/nov/24/local/la-me-lausd-food-20131124 (accessed Oct. 6, 2018).

Watanabe, T. (2014) LAUSD food effort makes local farms healthier too. *Los Angeles Times*, February 12, available at: http://www.latimes.com/local/la-me-c1-oranges-lausd-20140212-dto,0,7558479.htmlstory-axzz2wN5R0Igo (accessed Oct. 6, 2018).

Zuma-Netshiukhwi, G., Stigter, K. & Walker, S. (2013) Use of traditional weather/climate knowledge by farmers in the South-western Free State of South Africa: Agrometeorological learning by scientists. *Atmosphere*, 4, 383–410.

Appendix 3A Worked Formal Garden Experiments

S.E. Smith, D. Soleri, D.A. Cleveland

This appendix provides worked examples of formal garden experiments, building on the discussion in Chapter 3, Section 3.2.

3A.1. Formal Garden Experiments to Test Hypotheses

Let's begin with an example of hypothesis testing in a formal garden experiment. A gardener wondered if adding compost to her garden plot, compared with not adding any, would be associated with higher collard yields. If she kept track of last year's production she could do an informal experiment by adding two buckets of compost per plot this year and compare this year's yield with last year's, when she added none, by weighing each harvest and dividing by area planted. But many other variables are probably different between the two years that could affect yield, such as the weather, amount of water applied, or what was grown there the year before the collards. A better informal experiment would therefore be to compare two plots during the same season, one with and one without compost. Yet with only two plots, and no statistical analysis, the results could still be unreliable because the two plots could be different in ways she's not aware of.

To increase the usefulness of the results, she could do a small, formal experiment comparing multiple plots at the same time. The plots should be as similar as possible in every way except that half of them receive no compost, and the other half receive 15 kg (33 lb) of compost m^{-2} (Fig. 3A.1). In this experiment the *dependent variable*, the outcome of interest, is collard yield (g of production per m^2). The amount of compost is the *independent variable*, the one we manipulate to see if there is an effect on the dependent variable. In this example there are two levels, or *treatments* of the independent variable: a "control" with no added compost, and another with 15 kg compost m^{-2} plot. There must be more than one level of treatment for an independent variable, and

there may be more than one independent variable, although analysis becomes more complex and we only discuss examples with one independent variable. The example in Fig. 3A.1 contains six *replications*, or copies, of the same two-treatment experiment, for a total of 12 plots. These plots are arranged randomly in the experimental area, thus the name *completely randomized design* (CRD), and the example in Fig. 3A.1 is just one possible arrangement of those 12 plots. The experiment is designed to see if differences in the independent variable are associated with differences in the dependent variable.

Formal experiments start with a testable hypothesis that clarifies the question, and identifies what treatments to try, and which data to collect. A *hypothesis* is a statement that points directly to both the experiment and the data that can test it. Hypothesis testing focuses on variables we can measure and control, and data that we can observe and analyze. For example, a question such as "What makes my vegetables grow better?" is so broad and vague it is impossible to answer. Among other things, "what" could mean inputs, practices, locations, or seasons; "vegetables" could include a large number of different species and varieties; and "better" could refer to germination vigor, plant size, flavor, color, or amount of harvest. Understanding a garden as a whole system is desirable, as discussed below, but this question is so vague that we wouldn't know what data to collect. It's more useful to identify a specific dependent variable of interest (e.g., collard yield), and an independent variable, like the amount of compost applied, and levels of treatments you have reason to believe may make a difference, and that you can implement. Framed and focused this way, a hypothesis is the basis for planning a formal experiment that you can carry out in your garden.

Hypothesis testing usually starts with a *null hypothesis* (H_0), a statement that the independent variable has no effect on the dependent variable, and a related *alternative hypothesis* (H_A) stating that

---

Question/example	Description	Comparison
Does adding compost increase collard yield?	Formal experiment 2 soil preparation treatments: C, control with no added compost m^{-2} plot A, with 15 kg compost m^{-2} plot	Mean yield (g production area^{-1}) in A vs C plots

Completely randomized design (CRD)

A	A
C	C
C	A
A	C
C	A
C	A

Figure 3A.1. Experimenting in the garden: one example of an experiment with two treatments

there is an effect associated with the independent variable. The alternative hypothesis is a hunch based on our own observations, or what other gardeners or publications have suggested.

For the collard yield and compost example, a H_0 could be, "Mean collard yield between January and June *does not differ* when I add 15 kg of compost m^{-2} plot, compared to no compost." The H_A for that two-treatment experiment (Fig. 3A.1) could be, "Mean collard yield between January and June *is different* when I add 15 kg of compost m^{-2} plot, compared to none." Stated this way the treatments and data are very clear: half of the 1 m^2 plots are prepared without compost, half are prepared with 15 kg of compost added to each; all collard greens

harvested from each plot are weighed and recorded separately. This becomes a three-treatment experiment (Fig. 3A.2) if the H_A is changed to "Mean collard yield between January and June *differs* when 15 or 30 kg of compost are added m^{-2} plot, compared to none," and the experiment is designed accordingly. We discuss design more in Section 3A.2, for now we show this three-treatment experiment with seven replications laid out using a CRD (Fig 3A.2a). We also show a *randomized complete block design* (RCBD), in which the seven replications or blocks (rows in this case) have a complete set of the three treatments randomly arranged within each block (Fig. 3A.2b). A RCBD would be used if you discern or suspect spatial variation at your planting site (see Box 3A.1,

Question/example	Description	Comparison
Does adding compost increase collard yield, and does adding more compost increase yield more?	Formal experiment 3 soil preparation treatments: C, control with no added compost m^{-2} plot A, with 15 kg compost m^{-2} plot B, with 30 kg compost m^{-2} plot	Mean total yield in A vs B vs C plots

[a] Completely randomized design (CRD)

A	C	C
A	B	B
B	C	A
A	B	C
B	C	A
B	C	C
B	A	A

[b] Randomized complete block design (RCBD) with seven blocks (replications)

A	C	B	1
C	A	B	2
B	A	C	3
A	B	C	4
B	C	A	5
C	A	B	6
A	C	B	7

Figure 3A.2. Experimenting in the garden: two examples of an experiment with three treatments

Box 3A.1. Working with variation in garden experiments

Accounting for sources of variation in agricultural experiments can be difficult, but a number of books investigate this in detail (Section 3.4.2). Here we define some components of simple designs for formal garden experiments that take account of variation. These are illustrated in examples given in the remaining part of section 3A.2.

Block: experimental unit that contains a complete set of all the treatments. Blocks are always repeated more than once (replicated) within an experiment.

Check: a familiar crop variety planted alongside new varieties of the same species that are being evaluated, providing a known comparison, a genetic control, grown under the same conditions as the varieties being tested.

Control: a unit (like a tomato plant or a garden bed) as similar as possible to the unit receiving the treatment, but without that treatment (Figs 3A.1, 3A.2). The control provides a comparison for determining whether effects observed in the dependent variable are associated with the treatment, or would occur even without that treatment. Controls allow a gardener to ask: "When everything else is as equal as I can reasonably make it, is there a difference between the treatment(s) and no treatment?" In most cases, experimenters will refer to the control as the "control treatment." It's also possible to conduct an experiment without a control.

Plots: individual entries of a treatment in an experiment, for example, each of the six control plots in the collard experiment in Fig. 3A.1. Plots can include single plants or groups of plants.

Randomization: a chance-based distribution, for example of treatments across an experiment (Figs 3A.1 and 3A.2a), or within a replication (Fig. 3A.2b), done in order to minimize unintentional bias, for example in assigning treatments to different plots.

Replication: one or more copies of the entire experiment (all treatments including the control), to account for any variation that may be unique to an individual copy (Fig. 3A.2b).

Stratification: in the presence of non-random variation, the division of a heterogeneous unit into smaller, relatively homogeneous subunits, so that differences between them can be accounted for in experimental design and analysis. For example, dividing a garden or field according to the major soil types present (Figs 3A.3, 3A.4); identifying zones in a city according to average household income; dividing a year-long experiment according to seasons.

Treatment(s): different levels of the independent variable—the intentional variation you are investigating for its effect on the dependent variable. A control is one form of treatment.

Figs 3A.3, 3A.4). With a clear hypothesis laid out, the next step is to design the experiment to account for sources of variation, other than the independent variable, that may affect the dependent variable.

3A.2. The Challenge of Variation

Food gardens are full of variation because they are complex biophysical and sociocultural systems: there are interactions among different organisms (e.g., plants, people, beneficial insects and microorganisms, other animals, pests, pathogens), materials and objects (e.g., water, minerals and other compounds in soils and water, air, fences, trellises, shades, streets), and processes (e.g., photosynthesis, evapotranspiration, respiration, germination, climate change, irrigation, cultivation, cooperation, rule enforcement). In other words, there are often many independent variables that could affect an outcome (dependent variable), and it is impossible to identify all of them. Therefore, the greatest challenge in making observations and doing experiments that are useful is to recognize, and if possible, account for, the variability present so that it

doesn't obscure the relationship between the independent and dependent variables you are interested in.

Accounting for variation can sometimes be done using simple stratification (Box 3A.1), or by conducting more complicated experiments and analyses that distinguish contributions from different sources of variation and quantify them. Formal, scientific agricultural research often tries to eliminate as many sources of variation that aren't part of the hypothesis as possible, for example, when comparing the yield of different crop varieties, stresses are greatly reduced by using chemical fertilizers, irrigation, and pesticides. This makes it easier to identify relationships between a few key variables such as maximum potential yield of different genotypes under controlled conditions, but it means that the experimental conditions may not represent the real conditions experienced by gardeners or farmers, let alone their practices and preferences. The assumption that needs to be identified and challenged is that results will also apply to real gardens or fields, or that gardeners and farmers would be able, and want, to create similar conditions.

In reality, many gardeners can't or don't want to use the inputs often used in agricultural experiments and conventional industrial agriculture. Also, many scientists acknowledge that food gardens and fields that don't use lots of external inputs are more complex systems, so that outcomes can be different than on researchers' experimental plots. For example, barley breeder Salvatore Ceccarelli found that the barley varieties that did best on experimental stations in Syria were not the same ones that farmers preferred, or the ones that did best in their fields (Ceccarelli, 1996). One reason for this was that the relative performance of different varieties changed between environments, an example of qualitative genotype by environment interaction (Box 10.3). That is, some of the varieties with the best yields in plots on experimental stations had some of the worst yields in farmers' fields, where they were exposed to less fertile soils and drought. In addition, farmers had quality criteria for grain and straw that were different from those of the researchers. Ceccarelli's conclusion was that for low-input systems, screening or selection of crop varieties is usually best done in the environment where they will be grown, and in collaboration with the people who will grow and use them.

Because it is so challenging, until recently scientists have generally neglected whole systems research on the multiple interactions typical of real gardens and fields (Francis, 1986; Vandermeer, 2011), and so we know relatively much less about them scientifically. Yet comparing whole systems can sometimes be most informative for gardeners, for example, comparing the differences in garden yield with different mixtures of crops and varieties, the independent variable, instead of isolating one or two specific crops or varieties as independent variables (see Box 6.2 about the land equivalent ratio).

For most gardeners, reducing variation to the extent done in many conventional, formal experiments would be unrealistic and burdensome, so we need to ask more realistic questions. For example, experiments could be designed to answer a question like: "Given the variation I can expect in my garden, is the independent variable associated with changes in the dependent variable?" A few methods for working with variation make conclusions based on garden observations and experiments more reliable (Box 3A.1).

Effective formal garden experiments are ones that include similar levels of variation in all parts of the experiment, but still allow us to observe the variables we are interested in. For example, if we want to grow an open-pollinated collard variety (Section 10.3), then we know that the plant genotypes in each plot of the experiment outlined earlier (Figs 3A.1, 3A.2) will not be identical, there will be some variation,

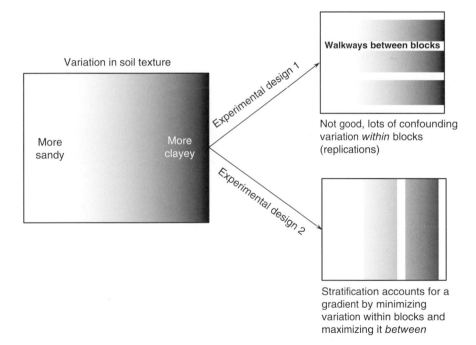

Figure 3A.3. Stratifying to account for a gradient in a garden experiment

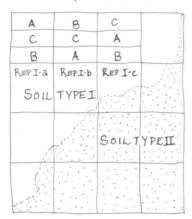

APPROACH 1

WHICH VARIETY YIELDS BEST IN SOIL TYPE I?

THE EXPERIMENT IS CONFINED TO THAT SOIL TYPE, HERE WITH THREE REPLICATIONS—
REPS I-a, I-b, I-c
COMPARE MEAN YIELD AMONG VARIETIES, $\bar{X}A$ vs $\bar{X}B$ vs $\bar{X}C$

A	B	C
C	C	A
B	A	B
REP I-a	REP I-b	REP I-c

SOIL TYPE I

SOIL TYPE II

APPROACH 2

i) WHICH VARIETY YIELDS BEST IN SOIL TYPE I? IN SOIL TYPE II?

OR

ii) WHICH VARIETY HAS THE BEST YIELD OVERALL?
COMPARE THE RESULTS BY VARIETY ACROSS BOTH SOIL TYPES

A	B	C
C	C	A
B	A	B
REP I-a	REP I-b	REP I-c

REP II-a	REP II-b	REP II-c
C	C	A
A	B	C
B	A	B

Figure 3A.4. Two approaches to working with spatial variation in formal garden experiments

A = 15 kg compost m^{-2} added; B = 30 kg compost m^{-2} added; C = control, no compost added

but that's alright. This is because the question being asked is whether changing the amount of compost applied affects production averaged over plants in that genetically diverse variety, and that variation will be taken into account because the genotypes will be distributed randomly across the plots.

Other variation may be realistic but not random, and can bias the results of the experiment, or even make them impossible to interpret. Differences in production between plots may occur because their locations have differences, for example in air and soil temperatures, hours of direct sunlight, plants growing nearby, time spent by the gardener, or amount of gopher damage. These non-random, "confounding" sources of variation can affect the dependent variable as much as, or more than, the independent variables, making the experiment meaningless if we do not address them. The first step in doing this is to design the experiment in a way that allocates random variation similarly *within* all plots, while allocating non-random variation *between* different plots.

Non-random variation in the garden may be structured temporally (e.g., seasonal drought or pest infestations) or spatially (e.g., soil texture or sun exposure), or both. Familiar temporal variation is cyclical, but usually includes greater uncertainty than spatial variation, and unfamiliar variation due to trends like anthropogenic climate change (ACC) (e.g., date of last spring frost, see Fig. 3.3) further increases uncertainty. These sources of variation can be accounted for in formal experiments, for example by doing the experiment in just one season, or by including "year" as a factor in statistics like ANOVA (Section 3A.4). Spatially structured variation is most predictable and easiest to work with. In formal garden experiments, known spatial variation, for example in soil types, can be managed in two ways. If there is a known spatial gradient (Fig. 3A.3), stratification and blocks that are laid out in a manner consistent with that gradient can be used, as described below. In contrast, if there are two areas of clearly demarcated different soil types in the garden, for example clayey and sandy, it can be stratified into those two types, and separate experiments done in each type (Fig. 3A.4, approach 1).

For example, if you have access to many wheat varieties from your local seed library, you may want

Table 3A.1. Worked examples of simple, formal garden experiments

	Dependent variable	Independent variable	Treatments	Design	Analysis
i.	**Fig. 3A.1**				
	H_0: Plots with added compost do not yield any differently than those without added compost				
	Collard yield measured as total g harvested m^{-2}	Compost soil amendment	2: No compost added (control) 15 kg compost added m^{-2}	Completely randomized design (CRD): 12 contiguous plots, 6 randomly distributed replications of each treatment	2-tailed t-test See Fig. 3A.7
ii.	**Fig. 3A.2, example [a]**				
	H_0: Yields of plots with either of two levels of compost added are not different than yields of plots with none added				
	Collard yield measured as total g harvested m^{-2}	Compost soil amendment	3: No compost added (control) 15 kg compost added m^{-2} 30 kg compost added m^{-2}	CRD: 21 contiguous plots, 7 randomly distributed replications of each treatment	1-factor ANOVA (treatment), if ANOVA significant do LSD tests See Fig. 3A.8
iii.	**Fig. 3A.2, example [b], 7 blocks (replications) within, not across, a non-random spatial environmental gradient, such as soil texture (see Figs 3A.3, 3A.4)**				
	H_0: Yields of plots with either of two levels of compost added are not different than yields of plots with none added				
	Collard yield measured as total g harvested m^{-2}	Compost soil amendment	3: No compost added (control) 15 kg compost added m^{-2} 30 kg compost added m^{-2}	Randomized complete block design (RCBD): 21 contiguous plots divided into 7 blocks (replications) with all 3 treatments randomly distributed in each block	2-factor ANOVA (treatment, replication), if ANOVA significant do LSD tests See Fig. 3A.9

to compare the grain yield (dependent variable) of those varieties (variety is the independent variable, the different varieties in your experiment are the treatments) in your garden, which has two types of soil. You might ask, "Which variety has the highest yield in soil (I)?" and plant your experiment only in soil (I). Or, you may want to find a wheat variety best for each soil type, and so plant the same experiment in both soil (I) and (II) (Fig. 3A.4, approach 2). Alternatively, if maintaining multiple wheat varieties is too much trouble, you may want to see which variety is best across both of the soils you cultivate.

Again, the simplest approach in this case is to conduct two experiments, one in each soil type with all varieties, and then compare the results, asking, "Which treatment (wheat variety) has the highest average yield across both soil types?"

This kind of spatial variation can exist within a single household garden, or in different parts of a community or school garden, or in different gardens across a town or city. When such obvious environmental variation is present, these strategies can guide the placement of the replicated experiments outlined in Figs 3A.1 and 3A.2. See Table 3A.1 and figures listed there for examples of the analysis of those experiments.

In the examples above, illustrated in Figs 3A.3 and 3A.4, blocks (replications) are used to account for spatial variation that may be due to the position of any one treatment. Another source of variation is the area bordering the experimental plots, which may have different exposure, including to sunshine, wind, or insects. To control for these edge effects, extra border rows are planted around the plots in the experiment (Fig. 3A.5), and data are not collected from them.

Variation can arise unexpectedly during the course of the experiment. For example, one of us (DAC) taught a university class on small-scale food production in which students did an experiment like the one in Fig. 3A.1, comparing the effects of two treatments: 30 kg compost m^{-2} vs. a control with no compost. Every year, one or more of the plots lost up to a third of the crop to gopher damage. To adjust for this, students calculated the yield on the undamaged portion, and extrapolated it to the damaged portion in order to estimate yield for the entire 1-m^2 plot.

3A.3. Data Collection

After deciding on the data you need to test your hypothesis, but before starting the experiment, try collecting some of those data. It is amazing how quickly it becomes apparent what does and does not work when you actually try it! What data to collect, how to collect them, what instruments to use to measure them, and how to record the measurements, are some of the things that might change after a try at data collection. Remember to make your data collection SMART (Box 3.2): for example, Specific—for each plot, the total grams of collards harvested is divided by the plot area in m^2 to calculate yield; Measurable—for each plot, fresh weight in grams of whole leaves when picked at the leaf base next to the main stem, using a scale, and totaled over the experiment period; Action-oriented—findings will be applied in garden bed preparation the following spring; Realistic—only seven small replications, otherwise it's too much work!; Timed—only from open garden production January through June.

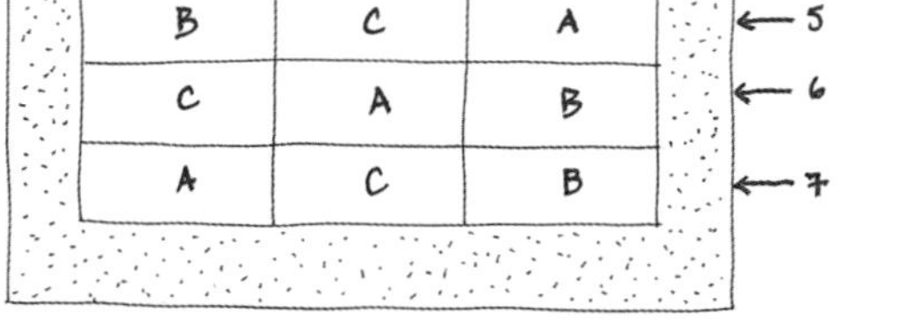

As described in Box 3.2, when multiple people are involved in experiments, or the same experiment is repeated at multiple sites, establishing a protocol is important to ensure that the experiment and data collection are done consistently by everyone and across locations. If more than one person is going to collect data, everyone should meet and practice together so that you agree on what, when, where, and how to do it. Even if only one person is doing the experiment, a protocol improves consistency by laying out exactly how the work will be done, and is something that can be referred to if you forget, if there are questions, or if someone else wants to repeat your experiment.

After testing data collection, make any needed changes, then try the procedures again and keep doing this until everyone responsible for collecting data feels comfortable with the process and the data quality. Once data collection starts, it's a good idea to review the data you collect at the end of each day to make sure they are legible, complete, and make sense. To avoid disappointment, back up all data and keep the backups up to date; keep at least two copies of data sheets in different places, backup digital data onto thumb drives, multiple computers, or in the cloud.

Data collection formats can take many forms including garden diaries of phenology and other processes in the garden; hard-copy spreadsheets and other types of data sheets; digital data entry programs; GPS (global positioning system) data formats for multi-location experiments; drawings and photos (experiment layout, identification of pests, disease, plant nutrient problems, crop species and variety). There are a growing number of publically available data collection applications for smart phones, some including a template for uploading

Figure 3A.5. Border rows minimize edge effects in formal garden experiments

and sharing data (Section 3.4.2). However, digital or online data collection and distribution is not necessary; for most garden experiments simple paper data sheets work well, and data from these can be entered into a digital database if needed. Ultimately it is the data quality and usefulness that are most important.

3A.4. Data Analysis and Interpretation

Hypothesis testing in formal science is usually done by analyzing the data statistically, something that's not necessary for informal experiments. Here we outline a few basic elements of statistical analysis for use with the type of experiments we described earlier. This will also be useful for understanding published research relevant to food gardening.

The tools of statistical analysis, including the commonly used *analysis of variance* (ANOVA) described below, can help to isolate variation associated with treatments from all the other common sources of confounding, non-random variation. Variation that is not consistent across all parts of the experiment, that is non-random, gets in the way of our understanding the relationship of different treatments with the dependent variable. Experimental design and analyzing experimental data statistically make it possible to say whether a treatment effect is "statistically significant," or not. For instance, in Fig. 3A.2b of the collard experiment, statistical analysis showed that both levels of compost amendment yielded significantly more (with mean yields of 3,420 or 3,502 g m^{-2} plot) than no compost (3,304 g m^{-2} plot), so we would reject our null hypothesis (Table 3A.1, iii).

Using the term "significant" when talking about treatment effects implies that the results are most likely not due to random variation. But since we know random variation is always present, these analyses must include statements of probability about that statistical test with these data. These probabilities are shown as "*P*-values" (significance levels) and describe how likely it would be to see the same or a more extreme result than was observed if the effects of the treatments *did not* actually differ. That is, if the null hypothesis is in fact true, how likely is it that you'd see the same or even greater difference between treatments? Formal researchers choose a particular significance level, also called level of error, in such tests, often 5% ($P = 0.05$), or 1%

($P = 0.01$). Interpreting significance levels requires some imagination. For example, you must imagine that you could conduct the compost experiment many, many times under essentially the same conditions. If the statistical analysis of your results has a $P = 0.05$, it means that if those many imagined experiments behaved similarly to yours, in about 5% of those you would conclude that the treatments produce different outcomes, when in fact they don't. You would make an error of this sort 5% of the time. For experiments like the examples we give here, this level of error ($P \leq 0.05$) is considered acceptable.

Conducting formal experiments, and especially handling the variation present in the most appropriate way, is challenging, and rapidly becomes the territory of statisticians. Overlooking or addressing variation incorrectly can make research findings less meaningful, and probably not what you want to base your garden practice or project on. But with an understanding of the basic elements, interested gardeners can conduct simple experiments successfully. Figure 3A.6 shows choices for designing and analyzing the type of garden experiments described in Table 3A.1. Included in the table are two new terms (see Box 3A.1 for other definitions).

- *Experimental design*: organization and layout of your experiment, including how you account for sources of variation other than the independent variable. For example, when variation is unstructured (random), a CRD can be used to randomly distribute individual treatments, with each treatment replicated the same number of times throughout the experiment. When there's reason to believe structured (non-random) spatial variation is present (see Figs 3A.2, 3A.3), use an RCBD with complete blocks ensuring that the entire experiment is replicated in each environmental stratum you have identified.

- *Analysis*: how your data are analyzed, based on treatment number and experimental design. For example, if there are only two treatments, a *t*-test can test if they are different, and if there are three or more treatments, an ANOVA can test if these are different. A 1-factor ANOVA is used when there are treatments only, a 2-factor ANOVA is for experiments with treatments and one other source of variation (aside from random error) identified in the experimental design,

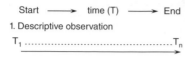

Start ——→ time (T) ——→ End

1. Descriptive observation

T_1 .... T_n

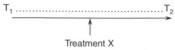

2. Informal experiment

T_1 .... T_2

↑
Treatment X

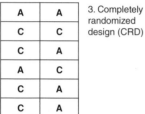

A	A
C	C
C	A
A	C
C	A
C	A

3. Completely randomized design (CRD)

Question/example	Description	Comparison
1. How many days after planting does my okra flower, and when are fruit ready to eat? Fig 3.6	Descriptive observation Familiarity with garden, e.g., phenology, other characteristics, events	Documenting, comparing before and after passage of time, may continue as long as desired
2. Does mulch (treatment X) reduce how often I need to water my basil bed? Fig 3.7	Informal experiment For immediate use, often less than a single season, observations before and after treatment, i.e., days between watering	T_1 (watering frequency before) vs T_2 (watering frequency after) treatment X
3. Does adding compost increase collard yield? Fig 3A.1	Formal experiment 2 soil preparation treatments: C, control with no added compost m^{-2} plot A, with 15 kg compost m^{-2} plot	Mean yield (g production area^{-1}) in A vs C plots
4. Does adding compost increase collard yield, and does adding more compost increase yield more? Fig 3A.2	Formal experiment 3 soil preparation treatments: C, control with no added compost m^{-2} plot A, with 15 kg compost m^{-2} plot B, with 30 kg compost m^{-2} plot	Mean total yield in A vs B vs C plots

4a. CRD

A	C	C
A	B	B
B	C	A
A	B	C
B	C	A
B	C	C
B	A	A

4b. Randomized complete block design (RCBD) with 7 blocks (replicates)

A	C	B
C	A	B
B	A	C
A	B	C
B	C	A
C	A	B
A	C	B

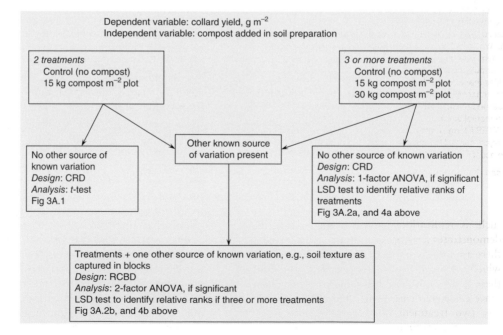

Dependent variable: collard yield, g m^{-2}
Independent variable: compost added in soil preparation

2 treatments
Control (no compost)
15 kg compost m^{-2} plot

3 or more treatments
Control (no compost)
15 kg compost m^{-2} plot
30 kg compost m^{-2} plot

Other known source of variation present

No other source of known variation
Design: CRD
Analysis: t-test
Fig 3A.1

No other source of known variation
Design: CRD
Analysis: 1-factor ANOVA, if significant LSD test to identify relative ranks of treatments
Fig 3A.2a, and 4a above

Treatments + one other source of known variation, e.g., soil texture as captured in blocks
Design: RCBD
Analysis: 2-factor ANOVA, if significant LSD test to identify relative ranks if three or more treatments
Fig 3A.2b, and 4b above

Figure 3A.6. Choices for designing and analyzing formal garden experiments with one independent variable

Formula:
=TTEST(G3:G8,H3:H8,2,2)

J2 $f_x \Sigma =$ =TTEST(G3:G8,H3:H8,2,2)

	Treatment layout in garden	①		Data as recorded (g m⁻² plot)			Data sorted by treatment and yield (g m⁻² plot)			T-test
1	Treatment layout in garden			Data as recorded (g m⁻² plot)			Data sorted by treatment and yield (g m⁻² plot)			T-test
2	None	None		1733	2031		None	Compost	③	0.0297017
3	Compost	Compost		3947	4055		1733	2576		
4	Compost	None		3374	2968		2031	3374		
5	None	Compost		3051	2576		2968	3432		
6	Compost	None		3432	3070		3051	3947		
7	Compost	None		4146	3272		3070	4055		
8							3272	4146		
9										
10	Formula: =AVERAGE(G3:G8)					②	Means	2,687.50	3,588.33	
11										

Figure 3A.7. Analyzing a formal garden experiment, see Fig. 3A.1, CRD, t-test

Null hypothesis: Plots with added compost do not yield any differently than those without compost.

This is a simple experiment comparing two treatments using a Completely Randomized Design (CRD) and a t-test.
Treatments:

None = control, no compost added
Compost =15 kg compost added m⁻²

1. Two treatments: None (a control) and Compost, placed randomly in 12 1–m² plots. This is a very simple design where an equal number of treatments are simply placed randomly within the garden, known as a Completely Randomized Design (CRD).
2. Sort data by treatment, calculate means. For example, for the control treatment, **G3:G8**, generated in Calc by = AVERAGE(G3:G8) and then hit return.
While the yield of the Compost treatment seems higher because of the variation among the plots we can't be sure. In this simple two-treatment example we can do a t-test to evaluate the different sources of variation and their effects on yield.
3. Conducting a t-test
Because there are two treatments a t-test is best here for determining significance of differences in treatment effects. Using a formula is ideal. **While in Cell J2 enter the following formula**: =TTEST(G3:G8,H3:H8,2,2)
and then hit return.
This formula includes data for the two treatments and two additional values. For the first of these ("tails"), a "2" should be entered. This signifies that a "two-tailed test" will be done. In such a test you will be evaluating whether one treatment is significantly different than the other, either significantly less **or** significantly more. This is a conservative sort of test and is the type typically done in most analyses. For the second (final) value in the formula you should also enter "2" if the treatments are laid out as they are here (randomly in the experimental area).
The TTEST formula provides a probability (P) value as output. The P value here (cell **J2**) is **less** than 0.05 (the "alpha" value) and so we can reject our null hypothesis that adding compost made no difference in yield. If the TTEST P value was greater than 0.05 we would have accepted our null hypothesis that adding compost did not affect yield.

We can therefore conclude that adding compost *did* result in a significantly different yield - in this case it was higher.

such as replications. In cases where ANOVA demonstrates a significant treatment effect, then there are specific statistical tests that can indicate which treatments differ from each other. One of these is the least significant difference (LSD). Once calculated based on values from ANOVA, any two treatment means that differ by an amount *greater* than the LSD are considered significantly different.

In Table 3A.1 we outline the same simple garden experiments discussed above (Figs 3A.1, 3A.2), and illustrate how changes in treatment number and experimental design can change the method of analysis. We also walk through the screenshots (Figs 3A.7 – 9) from analyzing these experiments using the free, open source software LibreOffice Calc (Section 3.4.2). Table 3A.1 gives the figure numbers for screenshots of analyses of each example. If you

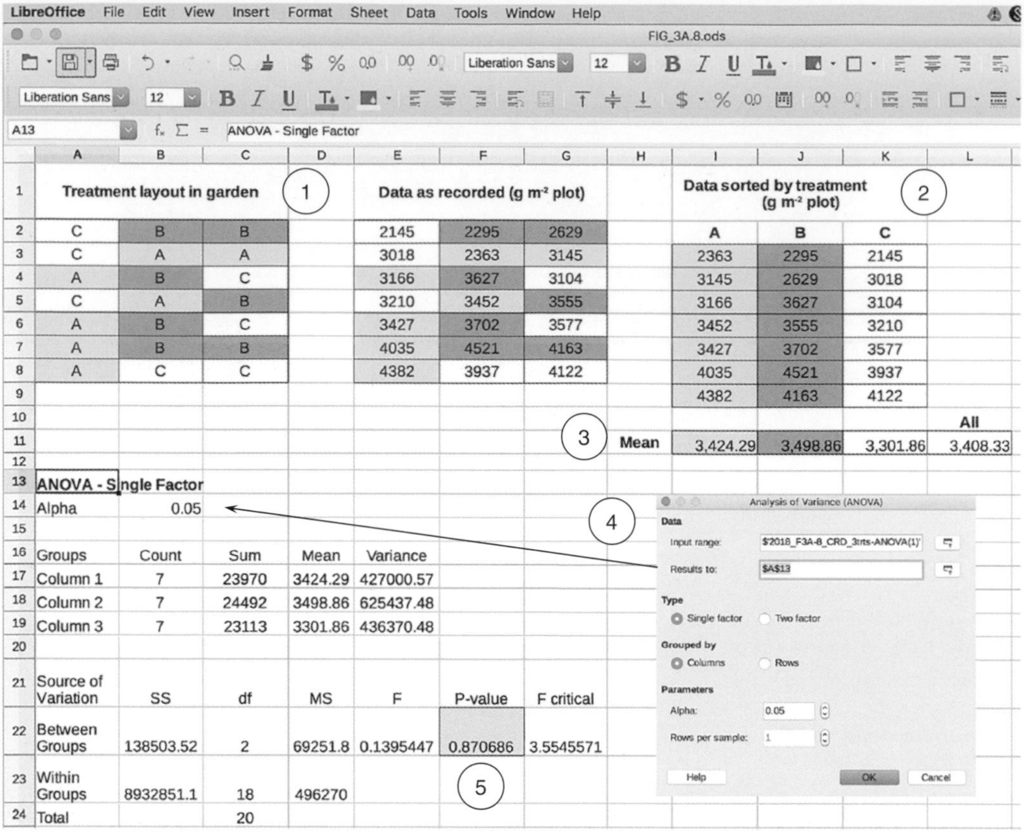

FIG_3A.8.ods

A13 f_x Σ =   ANOVA - Single Factor

Treatment layout in garden (1)

A	B	C
C	B	B
C	A	A
A	B	C
C	A	B
A	B	C
A	B	B
A	C	C

Data as recorded (g m⁻² plot)

E	F	G
2145	2295	2629
3018	2363	3145
3166	3627	3104
3210	3452	3555
3427	3702	3577
4035	4521	4163
4382	3937	4122

Data sorted by treatment (g m⁻² plot) (2)

A	B	C
2363	2295	2145
3145	2629	3018
3166	3627	3104
3452	3555	3210
3427	3702	3577
4035	4521	3937
4382	4163	4122

(3) **Mean**

	A	B	C	All
Mean	3,424.29	3,498.86	3,301.86	3,408.33

ANOVA - Single Factor

Alpha   0.05   (4)

Groups	Count	Sum	Mean	Variance
Column 1	7	23970	3424.29	427000.57
Column 2	7	24492	3498.86	625437.48
Column 3	7	23113	3301.86	436370.48

Source of Variation	SS	df	MS	F	P-value	F critical
Between Groups	138503.52	2	69251.8	0.1395447	0.870686	3.5545571
Within Groups	8932851.1	18	496270			
Total		20				

(5)

Analysis of Variance (ANOVA) dialog box

Data
Input range: $2018_F3A-8_CRD_3trts-ANOVA(1)
Results to: A13
Type: ⦿ Single factor   ○ Two factor
Grouped by: ⦿ Columns   ○ Rows
Parameters
Alpha: 0.05
Rows per sample: 1
Help   OK   Cancel

Figure 3A.8. Analyzing a formal garden experiment, see Fig. 3A.2a, CRD, 1-factor ANOVA

Null hypothesis: Yields of plots with either of two levels of compost added are not different than yields of plots with none added.

A simple experiment comparing three treatments (control and two levels of compost) using a Completely Randomized Design (CRD) and a 1-factor analysis of variance (ANOVA).
Treatments:

A = 15 kg compost added m⁻²
B = 30 kg compost added m⁻²
C = control, no compost added

1. Three treatments: two levels of compost and none (a control) placed randomly in 21 contiguous 1–m2 plots. This is a very simple design (layout) - CRD - where an equal number of treatments are simply placed randomly within the garden. We have used this design because we have no reason to think environmental variation is consistently organized in our garden. For example, no reason to think that nearby plots are more similar to each other than far away plots. Because there are more than two treatments, we will evaluate treatment effects using ANOVA.
2. In order to do ANOVA in Calc, organize the data so that treatments are in columns. This is a 1-factor (or 1-way) ANOVA because the treatment represents the only known source of variation in the experiment.
3. Calculate treatment means. For example, for treatment A, **I3:I9**, generated in Calc by =AVERAGE(I3:I9) and then hit return.
Means for the three treatments appear to be quite similar. ANOVA will allow us to evaluate the different sources of variation - including the treatments - and their effects on yield.
4. Conducting a 1-factor ANOVA
Highlight the block of data (**I3:K9**) and then click on Data → Statistics → Analysis of variance (ANOVA). See the screen capture to the right to see what the resulting box should look like. Once you indicate where you want the results table to go (**A13**), click OK and it will appear, shown on the left.
5. ANOVA tables such as this one include *P* values. Here this value (cell **F22**) is **greater** than 0.05 (our chosen level of error or "alpha" value) and so we can **accept** our null hypothesis that adding compost made no difference in yield.

We can therefore conclude that adding compost *did not* result in significant change in yield.

The table shows that ANOVA partitions total variation in the experiment into that between groups (Columns, representing treatments) and within those groups. The extent to which these sources of variation differ relative to each other will determine whether a significant treatment effect exists. A relatively large amount of variation between groups compared to that within groups would suggest that treatment effects are significant.

Notice though that the data as recorded seem to indicate a second source of variation may be influencing results. Plots on one side of the experiment (top) seem to yield much less than those at the other side (bottom). A second experimental design, called a Randomized Complete Block Design, might have been a better option given these experimental conditions. Figure 3A.9 presents an example of how this experiment might have looked if this design had been used instead of the Completely Randomized Design presented here.

A36 $f_x \Sigma =$ =L43*SQRT((2*D31)/7)

Treatment layout in garden (1) — **Data as recorded (g m⁻² plot)** — **Data sorted by treatment (g m⁻² plot)** (2)

A	B	C		F	G	H		K	L	M	N
C	B	A	Rep 1	2145	2295	2234	Rep 1	**A**	**B**	**C**	
B	C	A	Rep 2	3193	3023	3145	Rep 2	2234	2295	2145	Rep 1
A	C	B	Rep 3	3166	2921	3210	Rep 3	3145	3193	3023	Rep 2
C	A	B	Rep 4	3210	3452	3555	Rep 4	3166	3210	2921	Rep 3
A	B	C	Rep 5	3427	3702	3577	Rep 5	3452	3555	3210	Rep 4
B	C	A	Rep 6	4039	3926	4021	Rep 6	3427	3702	3577	Rep 5
C	B	A	Rep 7	4327	4518	4495	Rep 7	4021	4039	3926	Rep 6
								4495	4518	4327	Rep 7
											All
							Mean	3,420.00	3,501.71	3,304.14	3,408.62

ANOVA - Two Factor

Alpha   0.05

Completely Randomized Design, Figure 3A.8

	A	B	C	All
Mean	3424.29	3498.86	3301.86	3408.33

Groups	Count	Sum	Mean	Variance
Column 1	7	23940	3420	510773
Column 2	7	24512	3501.7	500171
Column 3	7	23129	3304.1	514285

					P-value	F critical
Row 1	3	6674	2224.7	5690.3		
Row 2	3	9361	3120.3	7681.3	0	1.03464948
Row 3	3	9297	3099	24247		
Row 4	3	10217	3405.7	31366		
Row 5	3	10706	3568.7	18958		
Row 6	3	11986	3995.3	3686.3		
Row 7	3	13340	4446.7	10872		

Source of Var	SS	df	MS	F	P-value	F critical
Rows	9084349	6	2E+06	271.08	4.267E-12	2.99612038
Columns	137980.67	2	68990	12.352	0.0012212	3.88529383
Error	67023.333	12	5585.3			
Total	9289353	20				

87.04535416

Treatments	Difference	Significant?
A – C	115.857	Yes
B – C	197.571	Yes
B – A	81.714	No

Analysis of Variance (ANOVA) dialog box:
Data — Input range: $2018_F3A-9_RCBD_Trts_ANOVA! ; Results to: A12
Type: Single factor / ● Two factor
Grouped by: ● Columns / Rows
Parameters: Alpha 0.05 ; Rows per sample: 1
Help   OK   Cancel

df	t 0.05	df	t 0.05
1	12.706	21	2.08
2	4.303	22	2.074
3	3.182	23	2.069
4	2.776	24	2.064
5	2.571	25	2.06
6	2.447	26	2.056
7	2.365	27	2.052
8	2.306	28	2.048
9	2.262	29	2.045
10	2.228	30	2.042
11	2.201	35	2.03
12	**2.179**	40	2.021
13	2.16	45	2.014
14	2.145	50	2.008
15	2.131	55	2.004
16	2.12	60	2
17	2.11	70	1.994
18	2.101	80	1.989
19	2.093	90	1.986
20	2.086	100	1.982

Figure 3A.9. Analyzing a formal garden experiment, see Fig. 3A.2b, RCBD, 2-factor ANOVA

Null hypothesis: Yields of plots with either of two levels of compost added are not different than yields of plots with none added.

A simple experiment comparing three treatments (control and two levels of compost) using a Randomized Complete Block Design (RCBD) and a 2-factor analysis of variance (ANOVA).
Treatments:

A = 15 kg compost added m⁻²
B = 30 kg compost added m⁻²
C = control, no compost added

1. Three treatments: two levels of compost and none (a control) in 1-m² plots placed randomly within 7 replicates (blocks) that share a similar environment. Note that each treatment is randomly placed within each replicate. This is known as a Randomized Complete Block Design. We have used this design because we believe that environmental variation is consistently organized in our garden and that nearby plots are more similar to each other than far away plots. Because there are more than two treatments, we will evaluate treatment effects using ANOVA. In addition to treatments, the analysis will also include replicates.

Continued

Figure 3A.9. Continued.

2. In order to do ANOVA in Calc, organize the data so that treatments are in columns and replicates are in rows. This is a 2-factor (or 2-way) ANOVA because there are two known sources of variation in the experiment, treatments and replicates.

3. Calculate mean values. For example, for treatment A, **K3:K9**, generated in Calc by
=AVERAGE(K3:K9)
and then hit return.

The means are not exactly the same as those in the previous experiment (Fig 3A.8), but they are quite similar. See the box below (cells **K16:N16**) for those values from Fig 3A.8. Compare these to examine the two experimental designs and how their use may affect your final conclusions.

4. Conducting a 2-factor ANOVA

Highlight cells **K3:M9**, then go to the menu, Data → Statistics → Analysis of variance (ANOVA). See the screen capture to the right to see what should be entered in the box. Once you indicate where you want the results table to go (**A12**), click OK and it will appear, shown on the left.

5. ANOVA tables such as this one include *P* values. Here this value for Columns (treatments, cell **F30**) is *less* than 0.05 (our chosen level of error or "alpha" value) and so we can *reject* our null hypothesis that adding compost made no difference in yield.

We can therefore conclude that adding compost *did* result in a significant change in yield.

The table shows that ANOVA partitions total variation in the experiment into that between two groups, treatments (Columns) and replications (Rows). It also estimates variation that cannot be attributed to *either* treatments or replications, which is known as error. The extent to which these sources of variation differ relative to the error will determine whether a significant treatment (or replication) effect exists. A relatively large amount of variation between groups compared to that within groups would suggest that treatment effects are significant.

The analysis shows that treatments within a replication are more similar to each other and considering all the treatments as a group, the replicates differ significantly from each other. This is demonstrated by the very small *P* value associated with rows (replications) in Cell **F29**. Here, a substantial amount of the total variation can be attributed to replicates. Accounting for this through the analysis reduces the error and makes it more likely to identify a significant treatment effect if one is present. Since there is a significant treatment effect in this case (Cell **F30** is < 0.05), data from this table can be used to determine which treatment means are significantly different from each other.

6. One way to determine whether two particular treatment means are significantly different is to calculate a "least significant difference" (LSD) value. If the difference (ignoring sign) between any two means is greater than this value, then these means can be considered significantly different.

$$\text{LSD} = t_{\alpha/2} \times \text{SQRT}[(2 \times \text{MS error})/\,r]$$

With: $t_{\alpha/2} = t$ value from the table shown ($P = 0.05$) with df = the Error degrees of freedom from the ANOVA table (6a), SQRT = square root (calculate this using "=SQRT" function in Calc), MS (mean square) error is taken from the ANOVA table (6b), and r = the number of replications in the experiment (rows in this analysis, 7).

7. For this analysis, the LSD would be:

$$\text{LSD} = 2.179 \times \text{SQRT}[(2 \times 5585.3)/7] = 87.04535$$

The same equation as entered in Calc on the spreadsheet in Figure 3A.9 is:
=L43*SQRT((2*D31)/7)
and then hit return.

8. Taking the difference between each of the treatment means, we can use the $P = 0.05$ LSD value calculated in step (7) above (87.04535) to identify any absolute value differences greater than the LSD.

For this experiment we can conclude that the yield of treatment C (Control) differs significantly from that of both treatments A and B, and also that yields of treatments A and B do not differ from each other.

want to conduct a formal garden experiment, but are unsure about the basic elements given in Table 3A.1, or your experiment does not fit these examples, we strongly suggest consulting the local extension service, college, or university for assistance in designing and conducting an experiment that will have reliable, useful results.

References

Ceccarelli, S. (1996) Adaptation to low/high input cultivation. *Euphytica*, 92, 203–214.

Francis, C. A. (1986) *Multiple cropping systems.* Macmillan Publishing Company, New York, NY.

Vandermeer, J. (2011) *The ecology of agroecosystems.* Jones & Bartlett Publishers, Sudbury, MA.

PART II STARTING THE GARDEN

4 Garden Placement

D. Soleri, S.E. Smith, D.A. Cleveland

Chapter 4 in a nutshell.

- Physical, historical, and social factors are important when choosing a new garden site, or working with an existing one.
- Orienting a garden to the sun can make management easier and reduce the inputs needed.
- Knowing the sun exposure of a garden over the course of a day, and a year, helps to plan garden crops and planting patterns, and optimize growing environments.
- Planting patterns can make the best use of garden space, and make it easier for plants to grow.
- Air and water movement affect temperatures, water loss, and other processes in the garden.
- Contaminants reaching gardens through the soil, air, or water can harm gardeners directly, or when they eat contaminated garden produce.
- Simple strategies can sometimes minimize the movement of contaminants into gardens; at other times contamination requires larger social and political responses.
- There are various ways to gain access to land, which is often the biggest challenge for gardeners, especially in urban neighborhoods.
- Long-term garden improvements and gardener involvement require secure tenure.
- Community involvement and support of gardens and gardeners is important for a safe and welcoming garden environment.

A thriving, pleasant garden pulls you in; the plants seem happy, the space is enticing, it's the kind of place people want to spend time in, and leave hoping to return to, or dreaming about creating a similar space themselves. What we call "garden placement" is an important part of creating a garden like that. This is because where a garden is placed affects not only the light, temperature, soil, and water that plants need to grow, but also how easy it is to have thriving plants, and how attractive the garden is to people. In this chapter we focus on how garden placement affects these variables and can be used to respond to change and optimize garden benefits. We use *placement* as an umbrella term for location, directional orientation (north-, south-, east-, west-facing), and layout of gardens, as well as the patterns of gardens or plants in the landscape. Placement of the garden, and plants within the garden, can create microenvironments that are better places for plants to grow, and for people to be. Garden placement also considers how a location was used in the past, and its social context today. Good placement creates spaces that people enjoy, and so are more likely to spend time in, and reap the benefits of food gardening. Of course, you have to work with what's available, and sometimes the only positive aspect of a garden site is people's commitment to it, but this commitment can be the key for a successful garden.

Unless you know what to look for, you may see one place in the landscape as pretty much the same as another. While simple distinctions, such as flat vs. sloped, paved vs. unpaved, or wet vs. dry, are easy to notice, every place has many other characteristics. Understanding and making the most of these can make a big difference in how change impacts your garden, what you are able grow, how many months of the year the garden can be productive, the kinds and amounts of inputs needed, how much time people want to spend gardening, how safe the garden produce is to eat, and how long you can continue to garden there.

The best location for a food garden is where it's easy to fit gardening into daily schedules, and to spend short amounts of time observing and caring for the garden. For example, newly planted seeds or transplants (Chapter 6) do best if checked daily for soil moisture, shade and pest problems. Proximity makes harvesting something for meals quick and easy, and can encourage more fresh fruit and vegetable consumption (Kamphuis et al., 2006). However, establishing or accessing a garden close to home or work is not always possible or practical, and many of us don't live close to the community or school gardens we are part of.

In this chapter we outline how to manage physical, historical, and social factors when choosing a new garden site, or working with an existing one. This is especially important in situations that are problematic in terms of sunlight, contamination, or tenure, or where large investments, for example in land, improvements, perennial plants, or terracing, are being considered. First, we discuss physical space, including changing sunlight and shadow during the day and year, temperature, air and water flow, and planting patterns. Then we look at historical and social contexts, including contamination, access, noise, and security.

4.1. The Physical Space

In rural areas there are usually more choices of where to put a garden than in cities, but in either setting placement can have a big impact on garden success because it affects the frequency and intensity of wind, frost, erosion, runoff, contamination, noise, and especially sunlight, in the garden.

4.1.1. Location and orientation

Understanding the sun's path at your garden site, and anticipating how it will change through the days and seasons, is key to optimizing garden and plant placement. *Orientation* is placement in relationship to the path of the sun, and is determined by *latitude* (measured in degrees of distance north or south of the equator, which is 0° latitude), depicted as lines parallel to the equator distributed around the globe between the two poles, and *longitude* (measured in degrees of distance east or west from the primary meridian 0° longitude, which runs through Greenwich, England), typically depicted on the globe as vertical lines at the equator, converging at the north and south poles (Fig. 4.1). In *FGCW* we give latitude and longitude in decimal degrees north (N) and west (W), respectively, which together identify a precise geographic location on the sphere of the Earth, that defines much of our experience of the sun and weather over the day and year, and how we can expect anthropogenic climate change (ACC) to affect us.

The Earth's axial tilt (~23.45°—we round it to 23.5°—away from perpendicular to its orbiting path, Fig. 4.1), together with where our planet is in

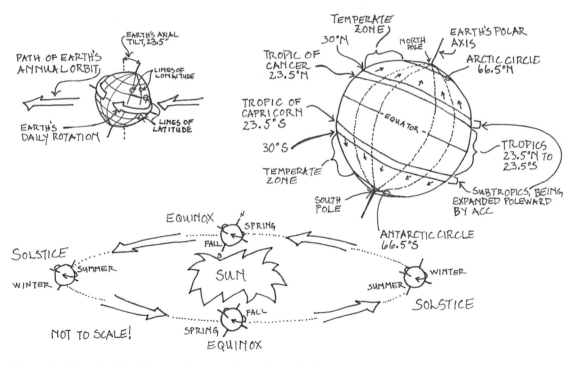

Figure 4.1. Terms for identifying location and orientation on the Earth

its annual orbit around the sun, result in the seasons and daylengths we experience at a specific geographic location. Those living in North America, the Mediterranean, West Asia and other locations in the northern hemisphere will be closer to the sun from late March to late September, with days longer than nights, and warmer temperatures, and farther from the sun from late September to late March, with days shorter than nights, and cooler temperatures (Fig. 4.2). Spring and fall are the transitions in between a location's maximum inclination toward (summer solstice) and away (winter solstice) from the sun. Locations in the southern hemisphere such as Australia and southern Africa and South America experience the same shifts as the northern hemisphere, but they occur at opposite times of year, due to the Earth's axial tilt. The latitudes 23.5° north and south of the equator (0°) are called the Tropics of Cancer and Capricorn, respectively, and are the highest latitudes where the sun appears directly overhead in the summer. The area between them is the *tropics*. The *subtropics* are the areas lying between the Tropic of Cancer and 30.0° N and between the Tropic of Capricorn and 30.0° S (Fig. 4.1). Defining these zones is still a matter of discussion, especially if climate is a factor, because that can change with altitude, and is changing with ACC (Box 4.1), but for simplicity we use this geographic definition (Corlett, 2013).

An easy way to understand the sunlight reaching a location at particular times is to visualize it using a simple version of an astronomical tool called the celestial sphere, which is used in planetariums. The *celestial sphere* is a projection of the path of objects we see in the sky like our sun and moon, stars, and planets, as experienced at a particular location and time (Fig. 4.3). As we use it here, a location, at a specific date, and even time of day, is the center point of a horizontal, circular plate that is the horizon extending out in all directions as far as the eye can see. This horizontal plate is the base of the celestial sphere's upper half, with the sky as seen from there rising up from the edges of the plate. You can divide the plate like a pie into quarters, using a compass to locate the four cardinal directions—north, south, east, west—on that plate. This creates a template to trace the sun's path as experienced in your garden over the seasons, or even over a day. Two measurements are used to locate positions on the celestial sphere. First, the *azimuth* is the distance in degrees around the perimeter of the circular plate moving clockwise from 0° at north,

with east, south, and west being 90°, 180°, and 270°, respectively. Second, *altitude* (or elevation, not the same as altitude or elevation above sea level) is the angle in degrees from the horizon to the object in the sky being measured. Figure 4.3 shows celestial spheres with the sun's path on June 21 and December 21 2017, the summer and winter solstices (the sun's highest and lowest altitude paths in the northern hemisphere, respectively), for Las Cruces, New Mexico and Sacramento, California, using the azimuth and altitude information from Suncalc.org (Section 4.3.1).

The examples in Fig. 4.3 illustrate the seasonal changes in the sun's path as experienced at specific locations and times; in Table 4.1 the resulting changes in sunlight are given for those locations and dates. During summer in subtropical and temperate zones (from latitudes of approximately 23.5°–30.0° and 30.0°–66.5° north [or south] of the equator, respectively—Box 4.1), the sun rises and sets farther to the north (or south in the southern hemisphere), and moves in a higher arc (altitude) above the horizontal plate of the celestial sphere. During the winter, the sun's path traces an arc lower on the horizon, that is shorter and closer to the plate of our celestial sphere. Together these seasonal differences in the Earth's position relative to the sun result in fewer hours of sunlight and therefore less energy gain in the winter compared with summer. Despite the differences in latitude between these two locations, the change in the sun's altitude between June and December 21 is the same, which makes sense because no matter the location, this change is always the sum of the angles of the Earth's maximum axial tilt towards and away from the sun ($23.45° + 23.45° = 46.90°$) (see Table 4.1).

Using the celestial sphere, you can stand at a location and visualize the sun's path across the sky, the sunlight reaching you, and how that changes over the day, and year. Of course, this is a starting place that is modified by local topography, including mountains, buildings, and trees that also affect the sunlight reaching a location. The fastest and easiest way to estimate the sun's path, and the sunlight and shadows in the garden, is to use the free public website Suncalc.org (Section 4.3.1). Height information for some buildings and other topographic features that may shade your garden can be found in some publicly available maps (e.g., Google Earth Pro). But if buildings surrounding your garden are not included in those maps, you can make

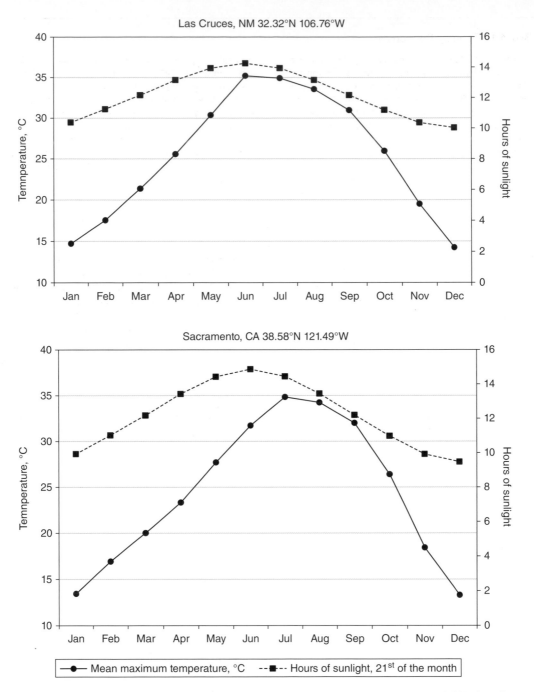

Figure 4.2. Hours of sunlight on 21st of each month, and mean maximum monthly temperature (°C) in Las Cruces, New Mexico and Sacramento, California, 1981–2015

Hours of sunlight, 2015, http://aa.usno.navy.mil/data/docs/Dur_OneYear.php; mean maximum monthly temperature, http://ncdc.noaa.gov/cdo-web/, 1981–2015

Box 4.1. Defining the tropics

It's easiest to define the tropics as we do in the text, based on physical location, and that definition has always included the weather patterns present in those zones. But ACC is changing the climate and circulation patterns, like the Hadley tropical circulation, that have traditionally characterized the tropics and subtropics. The Hadley circulation creates a dry, subtropical zone beyond its poleward edge, for example in the Sonoran, Chihuahuan, and Mojave Deserts north of the Tropic of Cancer in the northern hemisphere. As a result of ACC, conservative estimates are that the width of this tropical circulation and related patterns increased by a total of 2° latitude between 1979–2005 (Seidel et al., 2008), meaning a poleward move in each of the northern and southern hemispheres of about 111 km (69 mi), pushing drier subtropical conditions into new environments at higher latitudes. This is one of the complex and profound climate changes that are underway, and another example of how changes in the composition and functioning of the atmospheric envelope around the Earth are altering traditional associations between location and weather, with many implications, including for our food gardens.

your own estimates using a simple form of triangulation, the calculation of height using basic trigonometry (Section 4.3.1). If for some reason Suncalc.org is not accessible, Box 4.2 outlines the same basic calculation for estimating shadows at a location, although we've found the results can differ by a very small amount due to small discrepancies in altitude estimates.

Estimating shadow length in the garden is fun and informative, and good estimates are important when large investments of work or money are being made. For example, in the northern hemisphere

winter, when the sun's path is low on the horizon, tall landscape features to the south of your garden will likely shade it some or all of the time, but in summer may not be a problem, or in late summer may even provide beneficial relief from afternoon sunlight. In the northern hemisphere, planting on the south side of features like walls in the winter may provide direct sunlight in the day as well as the benefit of night-time thermal radiation (Section 4.1.3). This understanding can help you save time and resources, for instance, recognizing that some locations, or places within them, are simply better

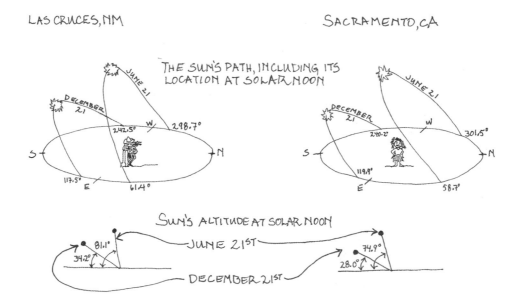

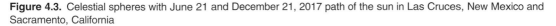

Figure 4.3. Celestial spheres with June 21 and December 21, 2017 path of the sun in Las Cruces, New Mexico and Sacramento, California

---

Garden Placement

Table 4.1. Changes in hours of sunlight and the sun's path on June and December 21, 2017 in Las Cruces, New Mexico and Sacramento, California

Location	Jun 21	Dec 21	Jun 21 azimuth, degrees		Dec 21 azimuth, degrees		Altitude of sun at solar noon, degrees	
	Hours of sunlight		Sunrise	Sunset	Sunrise	Sunset	Jun 21	Dec 21
Las Cruces, NM 32.32 N 106.76 W	14.27	10.03	61.4	298.7	117.5	242.5	81.1	34.2
	Hours of sunlight difference, December compared to June	−4.24	Total degrees of sun's path in June	237.3	Total degrees of sun's path in December	125.0	Change in sun's altitude, December to June	46.90
Sacramento, CA 38.58 N 121.49 W	14.87	9.47	58.7	301.5	119.9	240.2	74.9	28.0
	Hours of sunlight difference, December compared to June	−5.40	Total degrees of sun's path in June	242.8	Total degrees of sun's path in December	120.3	Change in sun's altitude, December to June	46.90

Data source, Suncalc.org

left fallow during the winter because of lack of adequate light and heat.

The seasonal changes described above are the result of how the Earth's axial tilt affects the angle at which the sunlight strikes the planet, and a location's exposure to the sun over the course of the Earth's daily spin around its axis, and annual revolution around the sun. These, and the landscape immediately around the garden, affect the sunlight and heat reaching the garden.

4.1.2. Light

Sunlight is the energy source for all plant growth via photosynthesis (Section 5.4) and the amount of sunlight, its timing during the day and year, and its quality (direct vs. indirect), are major determinants of what garden harvests are possible. So the sunlight a location receives is a big focus in garden placement. The celestial sphere visualizes hours of sunlight possible and its angle at different times in a location and date. But what does this mean for the light and temperatures experienced there, and how can we optimize those in our gardens?

The Sun's energy comes in the form of radiation of different wavelengths: relatively short wavelength invisible ultraviolet light (<~400 nanometers [nm]); intermediate wavelength visible light (~380–780 nm) that green plants use for photosynthesis; and invisible, longer wavelength infrared light (~700 nm–1mm). About 48% of the solar radiation reaching the Earth's atmosphere makes it directly to the surface (Pidwirny, 2013). The rest is either reflected back into space (29%), or absorbed or scattered by molecules and particles in the atmosphere (23%), reaching us as indirect, diffused light.

Location in space and time has a big effect on how strong sunlight is. *Insolation* is the amount of incoming solar radiation received per unit surface area, commonly measured in watts (a unit of power) per square meter (Wm^{-2}), with a global average of 340 Wm^{-2}. The sun is so far away that we imagine sunlight striking the earth as parallel rays, and all other things being equal, insolation is highest when incoming solar radiation is perpendicular to a surface, such as the soil surface or plant leaves (Khavrus and Shelevytsky, 2010). When the receiving surface is not perpendicular to the rays, the same amount of light energy is spread over a larger surface area, meaning there is less energy per area, that is, insolation is lower.

Box 4.2. Manually estimating shadows on your garden

Any object of height (H) in direct sunlight will cast a shadow whose length (L) can be found with this equation:

L = H/TAN ∠A

(shadow length = height of object divided by the tangent of the sun's altitude in degrees)

The sun's altitude (A) depends on your location and the date, changes over the course of a day (Section 4.1.1 and Fig. 4.1), and is measured in degrees, as is its tangent. Many apps and online calculators can provide tangents and other trigonometric functions, and they can be calculated with Calc in LibreOffice (Section 3.4.2), or see the NASA site in Section 4.3.1 for tables. It's important to note that this equation is for the shadow cast by a pole, and that the shadow lengths being calculated are the length of the pole's shadow extending in a straight line opposite (180° from) the sun (Fig. 4.4). Say for example, we want to know "What is the length of the shadow from a 90 cm (35 in) high east-to-west fence in Los Angeles, California on October 31, 2019 at 3:30 p.m.?" The sun's altitude at that time and date is 26.53° (find this from one of many sources online, Section 4.3.1). The tangent of 26.53° = 0.4992, therefore:

L = 90/0.4992 = 180.3 cm

This means that when measuring in a direct line with the sun's azimuth, a 90 cm (35 in) tall pole in the fence will cast a shadow about 180 cm (70 in) long in Los Angeles on October 31, 2019 at 3:30 p.m.; at noon the same day the shadow of the same pole would be 103 cm (40 in) long. But if the object the

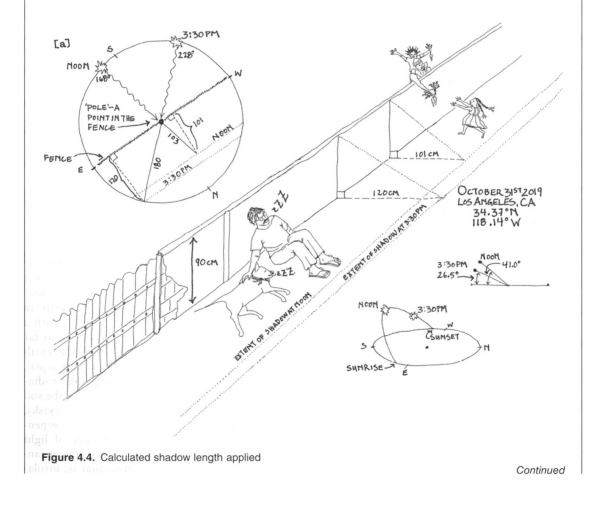

Figure 4.4. Calculated shadow length applied

Continued

sun is striking is more than a single vertical pole—in this example a fence running east to west—it is the relationship between the azimuth of the sun and that object that will determine the *width* of the shadow cast by that object. To estimate the width of this shadow, you can use trigonometry as shown by [a] in Figure 4.4, finding the adjacent side of the right triangle, or you can measure the line from the calculated length that is perpendicular to the fence. In this example, the shadow will extend north, perpendicular from the fence about 120 cm (47 in) at 3:30 p.m., and about 101 cm (39 in) at noon on that same date.

Approximate differences in proportion of total possible insolation due to seasonal or daily changes at one location or between different locations in the sun's path can be calculated using a simple equation: as the altitude or angle of the sun above the horizon decreases from 90° (perpendicular, directly overhead, the *zenith*) so does the insolation at a rate represented by the sine of that angle, indicated as "SIN" in the equations here. For example, when the sun's rays strike a surface at 90° the insolation is maximal or 100% of possible, sine = 1.0, and decreases as the sun's altitude decreases, and it gets closer to the horizon.

$$\text{SIN} \angle 90 = 1.0$$

$$\left[\begin{array}{l} \text{sine of a } 90° \text{ angle equals} \\ \quad 1.0, \text{ or } 100\% \text{ insolation} \end{array} \right]$$

$$\text{SIN} \angle 45 = 0.71$$

$$\left[\begin{array}{l} \text{sine of a } 45° \text{ angle equals} \\ \quad 0.71, \text{ or } 71\% \text{ insolation} \end{array} \right]$$

That is, when the sun's altitude above the horizon is 45°, the solar radiation per unit^2 of surface is 71% relative to when its altitude is 90°. Another way to think of it is that the same amount of solar radiation is being spread over a larger area compared to when the sun's altitude is 90°. Comparing maximum summer and winter insolation potential shows the range present at a given location. For example, in Sacramento, California on June 21 at its highest point (solar noon, when the sun is closest to the zenith) the sun's altitude is 74.9°, while on December 21 it is 28.0°, shown here rounded to the nearest degree (Fig. 4.3):

$$\text{SIN} \angle 75 = 0.97$$

$$\text{SIN} \angle 28 = 0.47$$

That means midday insolation in December is 48% (0.47/0.97) of what it is in June. That's why midday sunlight in your winter garden is much less warming than it is in the summer. For gardens in very high latitudes, grown in the winter or shade, low levels of insolation may limit photosynthesis and growth, but often other limitations, such as insufficient water or low temperatures, also contribute.

The angle of the Earth's axis and surface relative to the sun's rays also affects the amount of atmosphere the rays must pass through. The lower the sun's altitude the more the sun's rays are diffused by the atmosphere. This is another reason why sunlight is more intense at lower latitudes than higher ones, and higher elevations—at high latitudes the sun's rays strike the Earth's surface at smaller angles and travel through more of the atmosphere. For example, compare the altitude of the sun at solar noon on June 21 in Sacramento, California (74.9°) and in lower latitude Las Cruces, New Mexico (81.1°) (Fig. 4.5).

At solar noon on June 21, Sacramento has an insolation of 965.3 Wm^{-2}, while for Las Cruces it is 988.0 Wm^{-2}. The effect of latitude becomes even more obvious when you compare insolation at solar noon on the same date in higher latitude Redding, California (40.6° N, 122.4° W), insolation 956.3 Wm^{-2}; and much lower latitude Oaxaca de Juárez, Oaxaca, Mexico (17.1° N, 96.8° W), insolation 994.5 Wm^{-2}. These are approximations that vary due to atmospheric conditions and exact location and time, but they give you a feeling for how the latitude of your garden's location affects insolation.

We've seen how insolation changes over the seasons (Fig. 4.5), but insolation also changes over the course of a day, increasing from sunrise up until local solar noon, and then decreasing toward sunset as the angle of the sun's rays decrease. Light intensity at any location is greater in summer than winter, and at midday than morning or evening since this is the season and time of day when the Earth's surface is closest to perpendicular to the sun's rays. There is less intense light earlier as well

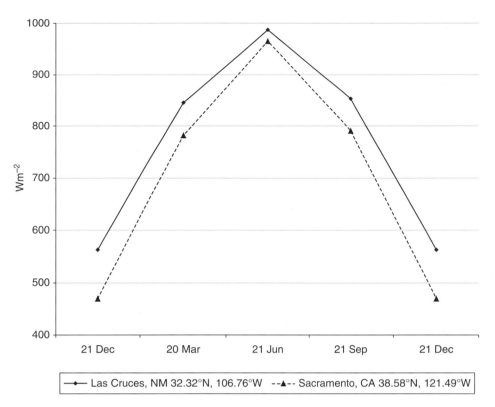

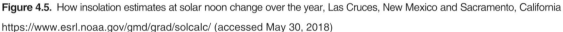

Figure 4.5. How insolation estimates at solar noon change over the year, Las Cruces, New Mexico and Sacramento, California
https://www.esrl.noaa.gov/gmd/grad/solcalc/ (accessed May 30, 2018)

as later in the day, but it is the cumulative effect of light energy on temperature that also matters (Section 4.1.3).

Insolation is also greater when it is direct, that is, the sun's rays are unimpeded. It's easy to feel this when comparing a sunny day to a cloudy one, or to a shaded spot; light that has been diffused by clouds or the atmosphere, or in the shade where only indirect light is present has much less energy than direct sunlight. Some plants can survive on only indirect light, usually because they evolved to live as an understory shaded by taller vegetation, but most garden plants require at least some direct light, having evolved to grow in areas people have cleared, leaving plants exposed to lots of direct sunlight.

4.1.3. Temperature

In addition to the energy from sunlight used in photosynthesis, sunlight also transmits energy to the air and the objects whose surface it strikes, and this energy can accumulate, changing the temperature of these objects, sometimes a lot.

Objects on the Earth including plants, soil, walls, windows, rocks, and water exposed to sunlight absorb more energy from short-wave solar radiation and reach higher temperatures than the surrounding air, because these objects have greater *mass* (material substance, quantified by weight on Earth) and *density* (mass unit^{-1} volume) than air. The increase in temperature of an object when exposed to sunlight is its *solar* or *thermal gain*. When an object is warmer than the air around it, it emits long-wave thermal (also called infrared) radiation, a form of heat, until the object and the air around it reach the same temperature, that is, until they come into equilibrium.

For the same amount of insolation, solar gain depends on an object's mass, density, and its *albedo*, or the proportion of the insolation that is reflected off the object and back into the atmosphere. Light-colored surfaces like sand and snow are the most reflective, with albedos close to 1; new asphalt, deep water, and other dark surfaces have albedos close to 0, because they absorb so much of the solar radiation they receive. For example, when industrial soot lands on polar ice surfaces, and ACC raises temperatures, the polar caps start melting. The change from high albedo ice and snow to lower albedo sooty surfaces and dark water (IPCC, 2013, Chs 4 and 9) means that more radiation is being absorbed, warming the polar region, adding further to the warming and melting caused by ACC (Riihelä et al., 2013).

Objects that absorb solar insolation transfer energy to cooler surrounding air. When the temperature of a brick wall near a garden differs from the temperature of the air around it, the energy in the form of heat is transferred from the wall into the surrounding cooler air and warms it. This transfer will continue until the wall and the air around it are the same temperature (equilibrium). This process can moderate temperatures in your garden. For example, a thick, high-density wall can store a relatively large amount of energy when warmed by the sun during the day, when the sun also warms the air. During the night, the air in the garden will become cooler as warm air rises. As this occurs a temperature differential develops between the relatively warm wall and the cooler air around it, and the heat in the wall then radiates out, warming the air surrounding it, which can help keep nearby plants warm.

Indirect light has lower insolation than direct light, so shading that reduces the amount of direct sunlight striking the garden will reduce solar gain and subsequent thermal radiation of walls, soil, and plants. That's why shading summer crops like tomatoes and peppers just before and during flowering keeps them and the soil cooler and can protect heat-sensitive pollen. Solar gain can be substantial, even on leaves. For example, in an Australian vineyard, grape leaves in full sunlight were found to be as much as 15°C (27°F) hotter than adjacent leaves in the shade (Jones et al., 2009). In spite of this solar gain, leaves that are transpiring water through their stomata may be able to stay cooler than the surrounding atmosphere because of the cooling effect of evaporation (Section 5.5), but will require more water.

These basic temperature dynamics can be used in the garden, for example by avoiding planting close to objects with substantial solar gain in the hotter months of the year, but intentionally planting near them in the cooler months when the thermal radiation can moderate cold night-time and early morning temperatures. Containers filled with water placed near tender plants will help keep them warm during the night by radiating heat they absorbed during the day, until a temperature equilibrium is reached. Similarly, dark-colored objects with low albedo can be used to increase solar gain in the winter, and light-colored, high albedo objects used to decrease gain in the summer (Sections 6.5, 8.9.2).

Being aware of the changing path of the sun over the year and its effect on temperatures helps in planning where to place a garden, or which areas to plant with what crops during different seasons. Generally, under hot, sunny conditions morning sunlight exposure and afternoon shade is preferable. This way, plants' light requirements can be met before air temperatures get very high and before plants, soil, and other exposed objects are heated by solar gain. In these conditions, afternoon sun exposure can be a problem because upper soil layers have dried out due to evaporation, and solar gain of exposed features in the garden would be greatest, eventually radiating heat to garden plants, and can extend the period during which plants experience heat stress.

In warm regions or seasons, cities can become *urban heat islands* due to low albedo (lots of asphalt, dark rooftops), accumulated solar gain in high-mass buildings, less cooling evapotranspiration (few plants or bodies of water), heat release from energy use (air conditioning, vehicles, lighting), and reduced removal of heat due to urban topography (tall, densely-packed buildings block air flow), among other reasons (Voogt, 2004). This effect tends to be more pronounced in poorer urban neighborhoods, because they are often more crowded and have fewer public amenities like street trees and parks that could provide evapotranspiration and shading (Buyantuyev and Wu, 2010). Planting heat- and drought-adapted deciduous trees and vines such as jujube, grape, fig, and mesquite can help shade and cool the garden in the summer, but still let in winter sunlight.

In the cooler months, low temperatures may stress sensitive garden plants. Just as some plants are not adapted to temperatures above a certain threshold (Section 5.8.2), some tropical or subtropical plants not adapted to low temperatures suffer *chilling injury* such as failure to grow, blossom drop, or fruit damage, even if temperatures are above 0°C (32°F) but below about 10°C (50°F) (Wang, 2004, rev. 2016). Many crops usually grown in summer are vulnerable to chilling damage, including squash, chayote (though it can return from its roots), melon, tomato, eggplant, *papaloquelite*, basil, and others. Chilling damage should not be confused with the *chilling requirement* of some crop varieties (Section 5.8.2).

In the same process described earlier, on cloudless nights when air temperatures drop to or near freezing (0°C, 32°F), the solar gain from the day's sunlight radiates from objects in the landscape into the air and is not trapped or radiated back by cloud cover (Snyder and de Melo-Abreu, 2005). Low mass objects such as a wooden fence or the leaves of a plant radiate heat into the air, rapidly depleting their daytime solar gain, until their temperature equals that of the air. When a surface reaches the temperature at which water in the air condenses (the *dew point*), and if that temperature drops to freezing, condensation on the surface takes the form of ice crystals, and sometimes crystals form inside plant tissues. This is called a (*radiative*) *frost* because it usually occurs when there is no wind that could mix layers of colder and warmer air, and when there is an *inversion* of the typical, overall pattern of decreasing temperature with increasing height above the Earth's surface in the lower atmosphere (*troposphere*). Instead, in an inversion heat energy rises rapidly from the surface and objects there, including plants, leaving colder air at the soil surface, especially in low areas (from cold air drainage, Section 4.1.4), with warmer air above it.

Frost damage, which is typically caused by ice crystal formation within plant cells, can be prevented by interrupting heat loss and trapping some heat around the plants by covering them with a cloth before night-time cooling begins. Another effective strategy is to irrigate the soil surface just before a cold period, because water in the soil will store and radiate more heat, warming the air around the plants (Snyder and de Melo-Abreu, 2005). Using containers of water that provide night-time thermal radiation to a small passive greenhouse or *cold frame* (a growing area covered with glass or translucent plastic that holds the warmth, Section 4.3.2), is based on the same principle.

Freezing damage also occurs when ice crystals form in plant tissues, desiccating and damaging the cells. The low temperatures and damage caused can be the same with frost and freeze, but the reason for the cold temperatures and effective responses differs. There is freezing damage when air temperatures are at or below freezing because freezing air moves into the area, removing the heat being given off by soil, structures, plants, and other objects at night. The more rapid the drop in temperature below freezing, the greater the damage because there is less time for protective processes inside the plant like the movement of solutes into cells or intercellular spaces that can slow freezing. Preventing freeze damage to garden plants is generally more difficult than preventing simple frost damage and usually involves an energy-intensive heating method. Sometimes, as with frost, water in the soil or even on the surface of plants will provide some freezing protection. This is because as water cools or freezes it releases energy, which can temporarily *increase* the air temperature, and may reduce freezing damage in plants (Snyder and de Melo-Abreu, 2005). The simplest and most efficient protection from freezing is to learn about local weather, including projections for how it will change, and to plant species and varieties tolerant of the low temperatures that are very likely; use orientation, solar gain, and perhaps a cold frame to extend the growing season; or don't grow susceptible crops during periods when freezes are likely.

4.1.4. Flow: Air circulation and drainage

Air movement is only visible when a breeze moves something like dust, steam, leaves, or your hair, but actually the air around us is constantly moving. Colder air is denser (and therefore heavier since density = mass/volume) than warm air, that is, the molecules are closer together, because they are moving more slowly, hitting and rebounding off of each other less than in warmer air. Immediately next to the Earth's surface (the bottom of the troposphere) heavy, colder air sinks and relatively light, warmer air rises, so gardens in low areas, or low spots within a garden, are likely to be most affected by *cold air drainage*: the downward movement of colder air into low areas as warmer air rises, an inversion of the larger-scale temperature gradient (Section 4.1.3). To minimize damage, gardens can be placed at mid elevations to avoid cold air drainage in low areas in the cold season, as well as the colder temperatures occurring at high elevations due to lower atmospheric pressure. As with water, the downward flow of cold air into the garden can be blocked, for example by walls and buildings.

Looking at the topography around your garden, or even using public geographical tools like OpenTopoMap and Google Earth (Section 4.3.1) can help you understand cold air drainage in the area.

Higher air temperatures lead to more evapotranspiration from plant and soil surfaces (Section 5.5 and Chapter 8), which typically will increase over the course of a warm day as air temperatures rise and solar gain heats structures, plants, water, and soil. In addition, when stomata are open, water is transpired from the plant to the immediately surrounding air according to the vapor gradient; from high vapor content inside leaves to lower vapor content in the surrounding air. As a result, the air immediately in contact with the plant, the *boundary layer*, is moister than the air farther away and cooler because of evaporation. If a breeze replaces this moist air with drier air, transpiration will increase. If the rate of transpiration exceeds the rate at which water moves through the plant to the leaves, the plant will wilt. Prolonged wilting can kill plant cells and tissues. This is one reason why hot, dry, windy weather is so stressful for plants. Setting up windbreaks (Section 8.3.2) to protect plants from drying winds will help to reduce evapotranspiration, plant stress, and the amount of water needed.

In addition to air flows, water movement is an important consideration for garden placement. For example, low areas and high water tables can result in saturated soils that suffocate garden plants, but providing better drainage and using raised beds can be effective responses (Chapter 8). Information about flooding potential at garden sites will be increasingly important as many locations experience more short-duration, high-intensity rainfall due to ACC (Section 2.3.2). In flood-prone locations, exploring options for storing or diverting heavy rainfall needs to be part of planning for more resilient gardens in the Anthropocene.

4.1.5. Planting patterns to complement garden placement

Optimizing sun exposure through planting patterns of different crops that are growing together in the garden is based on the same principles that apply to placement of the garden itself: orientation, and changing exposure to light and temperature over the season and day. Knowing the mature size and form of the crops you are planting is the place to start, and roughly mapping the garden can help (Section 6.2.3).

The plants themselves can be used to create microenvironments in the garden that make it more hospitable for other plants, or at least do not make it less hospitable. For example, in the northern hemisphere during the shorter days of winter, a tall trellis for climbing runner beans is best placed north of shorter plants such as chard, cilantro, or lettuce so that those plants can receive as much sunlight as possible. However, in the summer, tall maize plants grown to the south and west of more tender plants like tomatillo or zucchini can provide those with some relief from midday and afternoon sun. Some hardy, aggressively climbing summer annuals such as long bean or deciduous perennials like chayote (often grown as an annual), grape, and kiwi can be trained into shading trellises. A row of very drought- and heat-resistant nopal cactus can be great for optimizing sunlight and wind exposure for more tender garden plants, while providing delicious pads and fruit. Garden beds and annual crops can also be rotated to different places relative to garden perennials at different times of year to take advantage of changing shading and exposure.

4.2. Historical and Social Contexts– Environmental Justice and Garden Placement

While physical location is important, the history and larger social environment of a location also need to be considered when deciding about the placement of food gardens. Discrimination and inequity in the past and present leave their mark on people and the environments where we live, work, play, and tend our food gardens. Unfortunately, social and environmental injustice has a very long history and ongoing presence around the world.

Environmental injustice is social injustice manifest in biased patterns of access to and use of high quality natural resources, environments, and opportunities, and patterns of exposure to environmental hazards. Environmental injustice mostly affects the least powerful, based on markers such as "race," socioeconomic group, ethnicity, legal status, religion, gender, age, and sexual orientation. The powerful have more opportunities and resources to adapt to negative changes, and so are less vulnerable in the face of change (Section 3.1). That's why, even though ACC is negatively affecting everyone on Earth, it is not always a great social leveler, and if current social structures persist, will likely increase injustice, not only in space—from one place to another—but also in time between generations (Hansen et al., 2017).

As described in Section 2.4, environments can be thought of as sources or sinks, or both. Source environments are often controlled or exploited by the economic and political elites; the environments of the less powerful are often sinks, dumps for the externalized wastes and other negative consequences of those elites. *Externalized* costs are those borne by society in general, and often powerless groups in particular, and not by those who generate wastes, or consume the products that create that waste.

Environmental injustice in the US, for example, is the result of structural injustices in the legal, economic, educational, and other systems (Williams and Mohammed, 2013). Among many other things, lack of power also affects the options available for garden placement. Busy streets and older buildings and other structures are present in many neighborhoods, but lack of power means that the information and resources needed to remove and remediate the resulting contamination are disproportionately lacking in neighborhoods where the least powerful live. Similarly, lack of green space and denser urban canyons in disenfranchised areas create more unpleasant and unhealthy environments, with contamination that is more concentrated, and difficult to avoid.

Individual actions to remedy some of the personal consequences of environmental injustice may be effective, but usually require resources, for example, replacing a fence around your yard painted with lead-based paint, or acquiring containers and clean garden soil. Even if resources are available, environmental injustice usually affects entire communities, for example because contaminants in the air, soil, and water move freely regardless of property lines or borders. This is one of the reasons why the most effective solutions are ones achieved by community groups and their allies demanding accountability and policy change.

The kinds of problems experienced by less powerful communities may not even be considered when environmental dangers are assessed. For example, studies found that poor neighborhoods in Brooklyn, New York are exposed to multiple sources of contamination: waste dumps, manufacturing, and small-scale sources like dry cleaners and car shops in residential areas, that state and federal monitoring are not designed to detect (Corburn, 2005). This creates yet another burden of environmental injustice—community members themselves must document problems and then demand changes.

This is the case in New Orleans, Louisiana, where residents have been working with the non-profit Louisiana Bucket Brigade to sample air quality in "fenceline" communities, those directly adjacent to sources of contaminants (Ottinger, 2010). Luckily, inexpensive, easy to use tools and support are becoming more available for community groups concerned about contamination in their environment (Section 4.3.4).

Ultimately, all activities, including food gardens, will be more beneficial and enduring if they recognize environmental injustices, develop strategies that address power imbalances, and seek solutions based on equity. A first step is to know what problems to look for.

4.2.1. Contamination

A *contaminant* is any substance introduced into the environment that can harm plants, people, other animals, or overall ecological function. Harm can be not only physical and biological, but also social and cultural, for example transgenes moving into traditional maize varieties (Oaxaca, Mexico), contamination of local rivers sacred to Navajo farmers (northern Arizona), or oil spills and sea-level rise in coastal environments home to Cajun fisherfolk (Louisiana), all of which many local people see as violating their cultural and economic rights. Contaminants can affect different people differently, and children, women of reproductive age, the elderly, ill, and stressed people are especially sensitive to many contaminants. Avoiding or minimizing contamination is essential for creating gardens that are enjoyable and safe for everyone, especially when they are located in schools, clinics, or senior and refugee centers.

The most common contaminants, especially in urban areas, include heavy metals or semi-metals (e.g., lead, barium, cadmium, chromium, arsenic, nickel, copper), synthetic organic compounds (e.g., PCBs [polychlorinated biphenyls], organophosphates, PAHs [polycyclic aromatic hydrocarbons]), fibers of the mineral asbestos, and microorganisms that cause illness (e.g., *Escherichia coli*, *Salmonella* spp., *Giardia intestinalis* [aka: *G. duodenalis*, *G. lamblia*]). Dust and other airborne particulates are contaminants that are unhealthy to inhale for extended periods.

Contaminants reach garden plants and the people working and playing in the garden through three pathways: a) the soil, either naturally occurring as with asbestos or arsenic in some soils, or introduced substances that settle in the soil, such as pesticides, asbestos from deteriorating roof tiles, lead from old paint (Säumel et al., 2012); b) air, including vehicle exhaust, blowing dust, pesticide drift, volatile organic compounds; and c) water, as rain, runoff or irrigation containing substances from fossil fuel burning, fertilizers and pesticides (Fig. 4.6).

The history of a site can quickly provide a lot of information about possible soil contamination, for example the past use, storage, or disposal of solvents, fuels, paints, lubricants, herbicides, insecticides, fungicides, fertilizers, pressure-treated wood, asbestos pipes, automobile brake pads, or medicines and medical waste. In addition, some disease-causing microorganisms can be passed from the feces of wild or domestic animals to humans, such as the intestinal parasite *Giardia intestinalis* that

can be acquired when people eat produce watered with *Giardia*-contaminated water, or drink the water directly. Animal manure that has not been well composted can introduce pathogens to the garden including the bacteria *Listeria*, *Salmonella* and *Escherichia coli* O157:H7, all of which can be ingested by gardeners or those eating garden produce. Recommendations as to how hot (often >55°C, 131°F), and for how long (often >5 days) composting should be differ and will depend on the type of manure and the risk, and we strongly suggest investigating your manure source, and consulting with local extension agents or university researchers for appropriate methods. See the resources (Section 4.3.4) at the end of this chapter for suggestions of who to contact for information about possible known contaminants in your area.

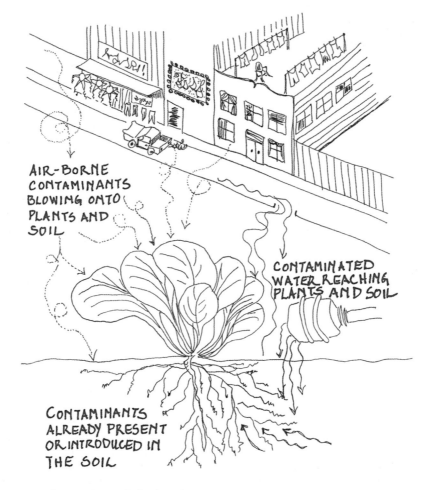

Figure 4.6. Pathways of contamination in food gardens

In addition to the history of the site itself, its location is also an indication of possible contamination, for example gardens next to busy streets, old buildings, or industrial sites such as for lead battery recycling (Section 2.4). Lead (Pb) is the most common and dangerous garden soil contaminant (Wortman and Lovell, 2013). By the time leaded fuels were banned for road use by most western industrialized governments in the mid 1990s, 5–6 million MT (metric tons) of Pb had been used in fuel and its processing in the US, and 75% of this went into the air and has remained in our environment ever since (Mielke et al., 2011). Similarly, another approximately 5–6 million MT of Pb was present in lead-based paints used until that was stopped in 1978 in the US. So now the primary source of Pb in gardens is the soil where airborne particles or paint chips have settled. Soil testing for Pb is important if there is any concern, and especially if the site is near a major street or a structure built before 1978.

Lead in the soil binds easily with organic matter and clays, so soils rich in organic matter will reduce the amount of Pb taken up by plants. Acidic soils (pH ≤5) with little organic matter make Pb more available to garden plants, and so does incompletely decomposed organic material (Section 7.5.1). Less than 1% of Pb in soil is water soluble, so deep watering to try to wash it down below the root zone will not help (EPA, 2005).

The US EPA has established upper limits for Pb of 400 ppm (parts per million) in soils where children play, and 1,200 ppm in other residential soils (EPA, 2001). Again, Pb testing is important if there is reason to suspect contamination, and especially if children and women of childbearing age are present, because the nervous systems of children and fetuses are particularly vulnerable to Pb damage. The state of California recommends a limit of 80 ppm of Pb in residential soils, but suggests that where produce is to be grown gardeners should consult their Human and Ecological Risk Office if there is a possibility of Pb contamination (DTSC, 2015). This is certainly worth checking for urban gardens; a recent study of Los Angeles, California soils included 27 samples from community food gardens with Pb levels exceeding 80 ppm (Clarke et al., 2015), with some samples in garden areas having levels as high as 1,720 ppm.

Any garden site near a street will have some airborne deposition of contaminants from past and present vehicle exhaust, brake linings, and other particulates; the heavier the traffic, and the more stopping and starting that occurs, the greater the contaminant concentration. Although fuels no longer contain lead, other contaminants such as carcinogenic PAHs are still used in gasoline (Srogi, 2007). In addition to proximity, presence of airborne contaminants is also affected by topography and wind. The amounts of most airborne contaminants including particles, drop noticeably with distance from the source (Zhu et al., 2002), so the farther away from a road or other source of airborne contamination the better. When the garden and street are in an area surrounded by buildings, this urban canyon inhibits dispersal of airborne contaminants and so concentrations can be high even away from the street. To some extent this same phenomenon can be used to protect the garden. When there is a clear source of airborne contamination like a street, dry cleaner, or a furniture refinishing or car repair shop that uses sprayed paints and solvents, any sort of vertical barrier between the source of pollution and the garden will help block some of the airborne particles, reducing the amount reaching the garden. Fences, walls, buildings, hedges, and trees are all physical barriers that can help keep some of these contaminants from landing on garden plants and soils. The taller and more solid the barrier, the more effectively it will block air and contaminant flow (Hagler et al., 2011).

Water can also contain contaminants such as nitrates in groundwater polluted by fertilizer run-off, lead from old pipes and soldered pipe joints entering municipal and home water systems, and arsenic from soils and mining activity. For example, the town of Dewey-Humboldt, Arizona is in an area with naturally high levels of arsenic in the soil and groundwater, but nearby mining has raised concentrations further (Ramirez-Andreotta et al., 2013). Research with home gardeners there found the water (77%), soil (16%), and garden vegetables (7%) to be sources of gardeners' daily arsenic intake, but cooking vegetables in water would increase their arsenic content. Analyses allowed researchers to identify limits to soil levels of arsenic for safely growing produce, which differed by crop. For example, lettuce requires the lowest concentration (1.56 mg arsenic kg^{-1} soil), while chard and spinach can be safely grown at the highest tolerable soil concentrations (12.4 mg kg^{-1}).

Harvested rainwater from streets may contain pollutants from fuels and asphalt, which can be minimized by not using the "first flush" of street runoff for irrigation (Section 8.7.3). Similarly, any materials that have been treated with preservatives or pesticides such as pressure-treated or creosoted wood, or salvaged lumber that may have been treated, will likely leach those compounds into the soil and water in the garden, and it is better to avoid using them entirely. Instead, use concrete blocks or lumber from redwood or cedar (red cedar, *Juniperus virginiana* or western red cedar, *Thuja plicata*), ideally recycled. Section 8.9 describes types of garden beds, with suggestions of some materials to use.

No matter how they reach the garden, contaminants such as heavy metals that settle in the soil then enter roots and move through the plant's vascular system, crossing a series of plant tissue 'barriers' that capture some of the contaminants and so reduce their concentration in the next tissues (Fig. 4.7). These barriers, in the order that they occur for contaminants in the soil are: soil–root; root–shoot;

shoot–flower/fruit, so contaminants from the garden soil and/or water taken up by the roots are most concentrated in the roots, then shoots and leaves, and least concentrated in flowers, fruits and seeds (Finster et al., 2004). If garden soils or water are contaminated, generally the safest crops to grow are those you eat the fruit or seeds of —such as tomato, chile, pepper, eggplant, pea, bean, and squash. Leafy greens are best grown in containers with clean soil and water, away from fumes and dust.

If soil tests show high levels of Pb or other contaminants, uncontaminated (clean) soil needs to be brought in for planting areas. Planting in raised containers with clean soil, and building up a 0.3–0.6 m (1–2 ft) buffer of wood chips or other material between the ground and the new garden soil (Pfeiffer et al., 2014) to prevent mixing with contaminated soil are effective strategies to reduce exposure to lead (Mitchell et al., 2014). You can

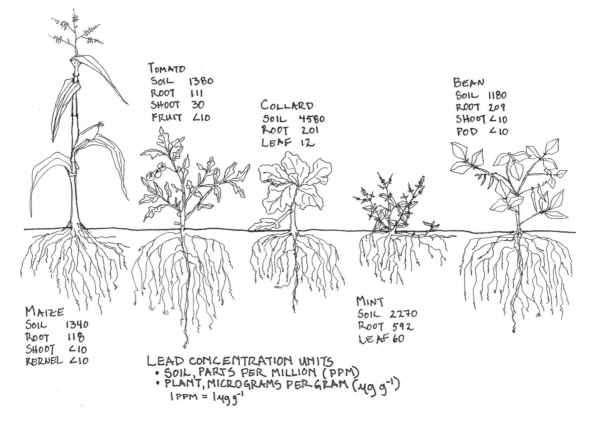

Figure 4.7. Examples of lead levels in different parts of garden vegetables

Lead in soil—parts per million, lead in plant tissues—micrograms Pb g⁻¹ dry plant material.
Source: Finster et al. (2004)

also minimize exposure to soil contaminants by covering contaminated soil with thick mulch or non-edible vegetation to prevent blowing dust. The greatest source of Pb and many other contaminants in garden produce is from swallowing particles on unwashed roots, leaves, pods, and fruits (Wortman and Lovell, 2013). Washing produce is a good idea in general, but if you think contamination might be present be sure to wash all produce vigorously with clean water, peel washed root crops, and then rewash them with fresh water. But if contamination is suspected, root crops shouldn't be grown unless the soil is tested and found to be safe.

Soil tests for contaminants can be expensive, costing much more than soil tests for nutrients, but there are some less expensive options. One study of garden sites in New York City found significant positive correlations between high levels of zinc (Zn) and the heavy metals lead, cadmium, and barium in those soils (Mitchell et al., 2014), possibly because Zn is present in substances that also contain the heavy metals, such as paint. Because Zn levels can be measured with inexpensive soil nutrient tests, those scientists suggest that high Zn levels could be used as an indicator for the need to test for heavy metal contamination.

When thinking about sources of contamination also consider how nearby gardens and other landscapes are managed. Sometimes potent agrochemicals are used by gardeners, farmers, and landscapers. This includes chemicals that are no longer legal, are legal but used inappropriately, or are homemade poisons that can be just as dangerous as purchased ones. For example, traditional homemade pesticides include ingredients such as arsenic, mercury, and nicotine, all now known health and environmental hazards. Discussions and education focused on neighborhood health can be a way to increase awareness and stimulate change for everyone's benefit, and can be especially effective if the shared dangers are explained in a non-accusatory way. Often people just do not know the toxicity of what they are using. Negotiation and agreement on the use of safer alternatives would be the best outcome in such situations. If negotiation and change are not possible then take steps to protect your garden area such as placing barriers between it and the source of contamination to prevent air and waterborne movement into your garden, testing and/or replacing your soil that may already be contaminated. Ultimately, action as a community will be most effective to change practices inside or outside the community that are harming it.

As mentioned above, government monitoring of contaminants is often inadequate, especially in communities experiencing environmental injustice, and so sometimes the only way change occurs is when community members monitor pollution themselves (Section 2.4). For example, in the US communities have collaborated with the government or non-governmental organizations (NGOs) to collect data on water quality, ongoing industrial pollution, pesticide drift, and other contaminants in their neighborhoods and homes. Whether open, community, participatory, or "citizen" science (Section 3.2.1), this sort of research is invaluable, especially for documentation and change in the face of environmental injustice. In Section 4.3.4 we give information on groups that can help with tools for documentation, measurement and organizing, and their number is increasing.

4.2.2. Land access and tenure

Unsurprisingly, the land for gardening, like housing, reflects social history and environmental injustice. For example, in Oakland, California, low-income, mostly black and brown populations have been pushed into the flatlands of that city where natural weathering, topography, and social policies have resulted in the highest levels of lead contamination, including of garden soils (McClintock, 2015).

A major consideration for garden placement, particularly in cities, is land access and tenure. In the US many people garden in their own yards, or on balconies and roofs. Yet, in 2013 an estimated 12% of gardeners were growing their gardens in other situations, like community gardens or in other people's yards (NGA, 2014), especially in urban clusters (>2,500 people) where over 71% of the estimated 42 million food-gardening households were located in 2013. As urban populations and urban food gardening both increase, this proportion will too. In most large US cities (≥620,000 inhabitants), 20–50% of residents live in rental units, and so even if they grow container gardens at their residence, any more extensive gardening must be done on land that is probably not theirs. Urban gardeners face several major challenges for gaining access to land for gardens: a) scarcity of appropriate sites; b) uncertain or very short tenure (often just one season—Thibert, 2012), which is a disincentive to longer-term investments such as planting perennials and improving soil; and c) lack of policies for addressing these issues, including appropriate zoning.

The difficulty of finding land for gardens that is not in private yards is exacerbated by the increasing need for housing, the rise in urban property values, and past planning that did not include gardens or other green spaces, or failed to make their presence known to the public. This means that most urban gardens have been located opportunistically, where there happens to be free space and/or a consenting landlord, but with no or very limited tenure rights, as well as other challenges. However, increasing recognition of the multiple benefits of green space for urban residents and communities (Chapter 1) means that allocation of land to gardens is receiving much more support now from municipal authorities and planners than in the past.

Public lands and right of ways, municipal complexes, schools, libraries, clinics, hospitals, and non-profit organizations including land trusts (Birky and Strom, 2013) are all potential gardening sites. Rooftop gardens using a lightweight soil mixture and waterproof barrier between beds and the roof create new space for gardening, often with better sunlight and temperatures than gardens at ground level, and provide environmental and aesthetic benefits (Sanyé-Mengual et al., 2017). Access, water availability, safety, and building engineering and codes, are among the new challenges with rooftop gardens. The species and varieties of plants best suited to rooftops may also be different than at ground level, due to different weather and shallower soil depth. An example of a resource where you can learn more about these is given in Section 4.3.3.

Public or privately owned vacant lots or existing green spaces can also be used for food gardens. Still, studies in New York City found that community gardens established in such lots are assumed to be temporary, even by gardeners, and that unless land is legally designated specifically for food gardens, these may simply be used as a way to eliminate urban blight until market-driven alternatives take over (Drake and Lawson, 2014) (Section 1.6.1). Even designating land for "urban agriculture" is insufficient if the goal is community gardens, especially for low-income people, because that land can be appropriated for commercial urban farming (Mees and Stone, 2012).

Private and government landowners often have concerns about added costs including liability insurance, water, and access. Tenure agreements that address these issues may be reached informally, but then gardeners have no legal recourse when landowners change their minds, or have different interpretations of the agreement. There is a growing number of organizations investigating ways to address these problems and developing templates for formal agreements, including between private landowners and gardeners or gardening groups (Section 4.3.3). Such formal agreements can give landowners assurance that costs will not be passed on to them, removing this worry about making their land available for food gardens, and offer more explicit, secure tenure conditions to gardeners.

Aside from concerns about potential costs, some private landowners may not see any direct financial benefit to themselves in making their property available, and so not be interested. Responding to public pressure, some cities and states now have legislation offering owners clear incentives for making land available for growing food. For example, in 2013 the California Urban Agriculture Incentive Zones Act (AB 551) changed the property tax law so that if a landowner allows her empty urban lot to be used for agriculture for at least five consecutive years, she will pay a lower, agricultural tax rate. A similar law was passed in Utah in 2012 (the Urban Farming Assessment Act, SB122), and in Maryland in April 2014 (Property Tax Credit Urban Agricultural Property, HB223). Also in California, the 2014 Personal Agriculture Act, (AB 2561) makes it possible for people to garden for home consumption in many locations including rental properties where gardening may have been prohibited in the past, and the 2013 California Cottage Food Operations Act (AB 1616) allows certain products including herbs, herbal teas, vegetable chips, and other foods made from garden produce to be processed on a small scale in the home and sold to the public.

Having access to reliable, affordable, good quality water for gardens can also be an issue when looking for garden space. Water access is becoming more difficult with rising temperatures and regional droughts due to ACC. If water needs to be hauled from another source by truck or even from nearby with buckets, it can discourage gardening (Pfeiffer et al., 2014), and needs to be considered in garden placement.

In some areas water for gardening has been virtually free but this is changing. For example, until recently many cities in the California Central Valley had no residential water meters, instead a flat monthly fee was paid no matter how much was used, creating no incentive for careful water management. Now meters are being installed, leading to different

Table 4.2. An example of residential and agricultural water rates, City of Santa Barbara, California, May 2018

		$ per 100 ft³ month⁻¹		
Date		Residential, single family tiered rate, based on water use		
	ft³ month⁻¹	0–400	401–1200	>1200
July 2014		$3.28	$6.39	$13.44
July 2018		$4.44	$12.96	$23.98
		Agricultural irrigation rate		
		100% calculated irrigation requirement[1]		Use above 100% calculated irrigation requirement[1]
July 2014		$1.56		$13.44
July 2018		$3.01		$23.98

[1]As calculated by water company based on area, type of plants, and evapotranspiration.
www.santabarbaraca.gov/gov/depts/pw/resources/rates/wtrsewer/changes.asp (accessed May 29, 2018)

attitudes toward water use, which will probably affect gardens. More and more regions are also introducing or increasing tiered water rates that rise with the amount of water used (Table 4.2). But, even those who are ready to pay for high water use are under public pressure, including "drought shaming," to stop them from using a disproportionate amount of that shared, finite resource. Greywater and rainwater runoff (Sections 8.7.3–8.7.4) are important supplements or replacements for piped water, but access can be logistically and legally challenging in urban areas. The urban agriculture legal support websites in Section 4.3.3 include information and links to resources about water access for city food production.

4.2.3. Noise, praedial larceny and other social issues

Making food gardens inviting places is important, otherwise people will not want to be there and the gardens will not thrive. However, a welcoming, well-equipped, productive garden can also attract unwanted attention. Excessive noise, crime including theft of garden produce (*praedial larceny*) and tools, social aggression, or other factors can quickly discourage gardeners. Many of these problems are the legacy of poverty, neglect, structural racism, and other inequities that continue today (Section 2.4).

If gardens are supported or promoted as an act of charity, then gardeners are relegated to the role of victims and recipients of largess. When community gardens are seen as another manifestation of the inequity and exclusion already experienced by some

members of the community, damaging the garden can be an expression of frustration and resentment (Section 1.6). No amount of fencing, locks, alarms, and other security paraphernalia will be able to prevent this, and spending lots of time and money turning the garden into a fortress is undesirable, usually escalating instead of resolving the problem. A garden can create divisions, and it can reflect or exacerbate pre-existing ones. Another reason community gardens may not be supported is that they may not be a priority, and it is better to invest energy in what is really wanted by the community. As described in Chapter 3, this is when having a clear idea of the larger goal is critical and can make the difference between a resilient, useful community project and one that is not. For example, if the issue is a lack of activities for youth, then a soccer field or basketball court may be more meaningful than a garden, at least as a first response.

Gardens can also be part of the process of problem resolution, an opportunity to start changing existing unjust relationships, which may not mean treating everyone equally because this simply perpetuates existing inequities. Making this change will only occur if the gardeners, and eventually their community, see the garden as something positive and genuinely theirs, where their voices count. This is most likely to happen when justice and equity are explicit focuses of the garden.

4.3. Resources

All the websites listed below were verified on May 31, 2018.

4.3.1. Orientation, light and temperature

The best and most user-friendly site for solar path and shadow information that we have found is Suncalc.org. It can search by address or city name globally (no need for geocoordinates), can be queried for past or future dates and times, and height of landscape objects can be entered to estimate shadows at specific times; apps for all systems are also available. Email response to questions is quick and helpful.
http://www.suncalc.org

If Suncalc.org cannot be accessed, the University of Oregon's online sun chart program generates custom charts based on coordinates or zip codes with various options for output data and format. It has not been updated since 2007, but still works.
http://solardat.uoregon.edu/SunChartProgram.html

An online solar calculator has been created by the US government's Earth Systems Research Laboratory, Global Monitoring Program, a part of the National Oceanic and Atmospheric Administration (NOAA). Locations worldwide can be identified using geographic coordinates or by clicking them on an interactive map of the world. Data are given for specified date and time; outputs include azimuth and altitude of sun, solar noon, sunrise and sunset, and there's a useful glossary of terms and some diagrams.
http://www.esrl.noaa.gov/gmd/grad/solcalc/

At this site you enter a location name or geocoordinates, date and time, then you can draw an object and its orientation and the website will calculate and visualize shadow cast then, or over the course of a day.
http://www.findmyshadow.com/index.php

Triangulation for estimating the height of objects that may cast shadows across a garden can be found on a number of websites, such as this one from the University of Tennessee, Knoxville, Tennessee.
http://www.tiem.utk.edu/~gross/bioed/bealsmodules/triangle.html

The Community Collaborative Rain, Hail and Snow Network (CoCoRaHS) at Colorado State University, Fort Collins, Colorado is a public-participation project documenting precipitation at thousands of sites in the US and Canada. It has a 2016 resource guide for master gardeners that has informative explanations about weather relevant to gardening, although it is disappointing that only 4 of its 108 slides are about climate change, and those are mostly devoted to photos!
https://www.cocorahs.org/

OpenTopoMap is a sister project to OpenStreetMap; both are crowd-sourced, open online maps.
https://opentopomap.org/
https://www.openstreetmap.org/

Google Earth Pro is now free, and you can estimate height information for buildings in major cities and other features using the 3D path option of the ruler tool.
https://www.google.com/earth/download/gep/agree.html

NASA (US National Aeronautics and Space Administration) has an educational website with explanations and tables of sine, cosine, and tangent values of angles from 1 to 90°. In addition, these may all be calculated within Calc in LibreOffice (see Section 3.4.2).
https://www.grc.nasa.gov/WWW/k-12/airplane/sincos.html

NOAA has a glossary of weather terms:
http://w1.weather.gov/glossary/index.php

A number of university extension services have websites or documents available online that give easy-to-understand discussions of frosts, freezes, and plants. See for example Perry (2001) and
http://biomet.ucdavis.edu/frostprotection/Principles%20of%20Frost%20Protection/FP005.html

4.3.2. Cold frames

There are many websites and other resources available with directions for building and using cold frames, such as this one by the University of Missouri's Extension Service.
http://extension.missouri.edu/p/g6965

Building cold frames with found or used materials (old windows, recycled lumber, cinderblocks, bricks) is usually quite easy. Be sure to check for possible contaminants such as lead paint or creosoted or pressure-treated wood though. For example, this one uses straw bales and old windows.
https://www.highmowingseeds.com/blog/a-cold-frame-for-many-occasions/

4.3.3. Placement issues for urban food growing

The Sustainable Economies Legal Center (SELC) has created a website with legal advice and resources relevant to urban agriculture, especially in the western US. There are templates for different formal contractual agreements, garden-relevant zoning and health laws, details of insurance coverage depending on the type of garden, where it is located and how the produce is used, and lots of other valuable information.
http://www.urbanaglaw.org

The Community Law Center in Baltimore, Maryland has some of the same resources with an eastern US focus.
http://communitylaw.org/urbanagriculturelaw project/

RUAF, a "global partnership on sustainable Urban Agriculture Food Systems" based in the Netherlands is active in many projects worldwide and its useful website has lots of training material, publications, and other resources, including about rooftop gardening, and many urban agriculture policy issues.
http://www.ruaf.org/

4.3.4. Contamination

The California State Office of Environmental Health Hazard Assessment is an example of a state-level agency that has set its own toxicity standards for currently monitored contaminants such as lead, arsenic, cadmium, and a number of synthetic chemical compounds such as PCBs, DDT, 2,4-D, chlordane and others.
http://www.oehha.ca.gov/risk/chhsltable.html

For information regarding possible contaminants see health department environmental toxicology websites, for example this one for Los Angeles County:
http://publichealth.lacounty.gov/eh/TEA/ToxicEpi/index_ToxicsEpi.htm

A nonprofit organization in Texas has a short video about making raised beds (Section 8.9.2) from pallets that is one way to elevate gardens away from contaminated soil.
http://foodisfreeproject.org/resources/

For nonprofit organizations that provide support to community groups wanting to monitor their environments, and training and tools for doing so, see Public Lab, Section 3.4.2; Drift Catchers; and Louisiana Bucket Brigade, Section 9.9.6.

Based at a college and focused on Pennsylvania, the nonprofit Alliance for Aquatic Resource Monitoring (ALLARM) has a well-developed approach to community-based water monitoring and is a resource for anyone wanting to do that work.
http://www.dickinson.edu/allarm

4.4. References

Birky, J. & Strom, E. (2013) Urban perennials: How diversification has created a sustainable community garden movement in the United States. *Urban Geography*, 34, 1193–1216.

Buyantuyev, A. & Wu, J. (2010) Urban heat islands and landscape heterogeneity: Linking spatiotemporal variations in surface temperatures to land-cover and socioeconomic patterns. *Landscape Ecology*, 25, 17–33.

Clarke, L. W., Jenerette, G. D. & Bain, D. J. (2015) Urban legacies and soil management affect the concentration and speciation of trace metals in Los Angeles community garden soils. *Environmental Pollution*, 197, 1–12, DOI: http://dx.doi.org/10.1016/j.envpol.2014.11.015.

Corburn, J. (2005) *Street science: Community knowledge and environmental health justice*. MIT Press, Cambridge, MA.

Corlett, R. T. (2013) Where are the subtropics? *Biotropica*, 45, 273–275.

Drake, L. & Lawson, L. J. (2014) Validating verdancy or vacancy? The relationship of community gardens and vacant lands in the U.S. *Cities,* 40, Part B, 133–142, DOI: http://dx.doi.org/10.1016/j.cities.2013.07.008.

DTSC (California Department Of Toxic Substances Control) (2015) *HERO HHRA Note Number: 4*. California Department Of Toxic Substances Control (DTSC) Office Of Human And Ecological Risk (HERO), Sacramento, CA, available at: https://www.dtsc.ca.gov/AssessingRisk/upload/HERO-HHRA-Number-4-October-6-2015.pdf (accessed Oct. 6, 2018).

EPA (2001) *EPA fact sheet: Identifying lead hazards in residential properties*. US Environmental Protection Agency, Washington, DC.

EPA (2005) *Ecological soil screening levels for lead*. US EPA, Washington, DC.

Finster, M. E., Gray, K. A. & Binns, H. J. (2004) Lead levels of edibles grown in contaminated residential soils: a field survey. *Science of The Total Environment*, 320, 245–257.

Hagler, G. S. W., Tang, W., Freeman, M. J., Heist, D. K., Perry, S. G. & Vette, A. F. (2011) Model evaluation of roadside barrier impact on near-road air pollution. *Atmospheric Environment*, 45, 2522–2530.

Hansen, J., Sato, M., Kharecha, P., von Schuckmann, K., Beerling, D. J., Cao, J. J., Marcott, S., Masson-Delmotte, V., Prather, M. J., Rohling, E. J., et al. (2017) Young people's burden: Requirement of negative CO_2 emissions. *Earth System Dynamics*, 8, 577–616, DOI: 10.5194/esd-8-577-2017.

IPCC (Intergovernmental Panel on Climate Change) (2013) Climate change 2013: The physical science basis. Working Group I contribution to the Fifth Assessment Report of the Intergovernmental Panel On Climate Change. IPCC, Geneva, available at: http://www.climatechange2013.org/images/report/WG1AR5_SPM_FINAL.pdf (accessed Oct. 6, 2018).

Jones, H. G., Serraj, R., Loveys, B. R., Xiong, L., Wheaton, A. & Price, A. H. (2009) Thermal infrared imaging of crop canopies for the remote diagnosis and quantification of plant responses to water stress in the field. *Functional Plant Biology*, 36, 978–989, DOI: http://dx.doi.org/10.1071/FP09123.

Kamphuis, C. B. M., Giskes, K., de Bruijn, G. J., Wendel-Vos, W., Brug, J. & van Lenthe, F. J. (2006) Environmental determinants of fruit and vegetable consumption among adults: A systematic review. *British Journal of Nutrition*, 96, 620–635, DOI: 10.1079/bjn20061896.

Khavrus, V. & Shelevytsky, I. (2010) Introduction to solar motion geometry on the basis of a simple model. *Physics Education*, 45, 641.

McClintock, N. (2015) A critical physical geography of urban soil contamination. *Geoforum*, 65, 69–85, DOI: http://dx.doi.org/10.1016/j.geoforum.2015.07.010.

Mees, C. & Stone, E. (2012) Zoned out: the potential of urban agriculture planning to turn against its roots. *Cities and the Environment (CATE)*, 5, 7.

Mielke, H. W., Laidlaw, M. A. S. & Gonzales, C. R. (2011) Estimation of leaded (Pb) gasoline's continuing material and health impacts on 90 US urbanized areas. *Environment International*, 37, 248–257.

Mitchell, R. G., Spliethoff, H. M., Ribaudo, L. N., Lopp, D. M., Shayler, H. A., Marquez-Bravo, L. G., Lambert, V. T., Ferenz, G. S., Russell-Anelli, J. M., Stone, E. B., et al. (2014) Lead (Pb) and other metals in New York City community garden soils: factors influencing contaminant distributions. *Environmental Pollution*, 187, 162–169, DOI: http://dx.doi.org/10.1016/j.envpol.2014.01.007.

NGA (2014) *Garden to table: A 5 year look at food gardening in America*. National Gardening Association, available at: http://www.hagstromreport.com/assets/2014/2014_0402_NGA-Garden-to-Table.pdf (accessed Oct. 6, 2018).

Ottinger, G. (2010) Buckets of resistance: Standards and the effectiveness of citizen science. *Science, Technology & Human Values*, 35, 244–270, DOI: 10.1177/0162243909337121.

Perry, K. B. (2001) Frost/freeze protection for horticultural crops. Horticulture information leaflet 705. NCSU Cooperative Extension, available at: https://content.ces.ncsu.edu/frostfreeze-protection-for-horticultural-crops (accessed Oct. 11, 2018)

Pfeiffer, A., Silva, E. & Colquhoun, J. (2014) Innovation in urban agricultural practices: responding to diverse production environments. *Renewable Agriculture and Food Systems*, FirstView, 1–13, DOI: doi:10.1017/S1742170513000537.

Pidwirny, M. (2013) Energy balance of Earth, available at: https://editors.eol.org/eoearth/wiki/Energy_balance_of_Earth (accessed May 28, 2018).

Ramirez-Andreotta, M. D., Brusseau, M. L., Beamer, P. & Maier, R. M. (2013) Home gardening near a mining site in an arsenic-endemic region of Arizona: Assessing arsenic exposure dose and risk via ingestion of home garden vegetables, soils, and water. *Science of The Total Environment*, 373–382, 454–455, DOI: http://dx.doi.org/10.1016/j.scitotenv.2013.02.063.

Riihelä, A., Manninen, T. & Laine, V. (2013) Observed changes in the albedo of the Arctic sea-ice zone for the period 1982–2009. *Nature Climate Change*, 3, 895–898.

Sanyé-Mengual, E., Oliver-Solà, J., Montero, J. I. & Rieradevall, J. (2017) The role of interdisciplinariety in the evaluation of the sustainability of urban rooftop agriculture. *Future of Food: Journal on Food, Agriculture and Society*, 5, 46–58.

Säumel, I., Kotsyuk, I., Hölscher, M., Lenkereit, C., Weber, F. & Kowarik, I. (2012) How healthy is urban horticulture in high traffic areas? Trace metal concentrations in vegetable crops from plantings within inner city neighbourhoods in Berlin, Germany. *Environmental Pollution*, 165, 124–132.

Seidel, D. J., Fu, Q., Randel, W. J. & Reichler, T. J. (2008) Widening of the tropical belt in a changing climate. *Nature Geoscience*, 1, 21–24.

Snyder, R. L. & de Melo-Abreu, J. P. (2005) Frost protection: Fundamentals, practice, and economics, Vol. 1. In Natural Resources Management and Environment Department (ed.) *Environment and natural resources series 10*. FAO, Rome.

Srogi, K. (2007) Monitoring of environmental exposure to polycyclic aromatic hydrocarbons: A review. *Environmental Chemistry Letters*, 5, 169–195.

Thibert, J. (2012) Making local planning work for urban agriculture in the North American context: A view from the ground. *Journal of Planning Education and Research*, 32, 349–357, DOI: 10.1177/0739456x11431692.

Voogt, J. A. (2004) Urban heat islands: hotter cities. *Actionbioscience*, available at: http://www.actionbioscience.org/environment/voogt.html – primer(accessed Oct. 22, 1018).

Wang, C. Y. (2004, rev. 2016) Chilling and freezing injury. In Gross, K. C., Wang, C. Y. & Saltveit, M. (eds) *The commercial storage of fruits, vegetables, and florist and nursery stocks*, 62–67. Agricultural Research Service, USDA.

Williams, D. R. & Mohammed, S. A. (2013) Racism and health I: Pathways and scientific evidence. *American Behavioral Scientist*, DOI: 10.1177/00027642134 87340.

Wortman, S. E. & Lovell, S. T. (2013) Environmental challenges threatening the growth of urban agriculture in the United States. *Journal of Environmental Quality*, 42, 1283–1294.

Zhu, Y., Hinds, W. C., Kim, S. & Sioutas, C. (2002) Concentration and size distribution of ultrafine particles near a major highway. *Journal of the Air & Waste Management Association*, 52, 1032–1042.

5 How Plants Live and Grow

D. Soleri, S.E. Smith, D.A. Cleveland

Chapter 5 in a nutshell.

- Anthropogenic climate change (ACC) affects garden crops in many ways that vary from place to place.
- Potential benefits of ACC include increased photosynthesis and expanded growing zones or seasons for some species.
- Potential costs of ACC include reduced quantity and quality of harvest due to rising atmospheric carbon dioxide (CO_2) concentration, high temperatures and precipitation extremes, and mismatches for existing crops between a changing climate and local soils, water, and daylengths.
- A basic understanding of how plants live and grow can help gardeners respond to change and solve problems.
- The vascular system connects all parts of the plant and circulates water and nutrients from the roots, and carbohydrates and hormones from the shoots.
- Photosynthesis in plants combines CO_2 from the air and water from the soil, capturing the sun's energy in the chemical bonds of carbohydrates.
- Respiration releases the chemical energy in carbohydrates for use by plant tissues, and produces CO_2 and water.
- Most of the water used by plants is taken up by roots, moves through the plants, and evaporates through pores (stomata) on leaves (transpiration).
- Plants obtain CO_2 for photosynthesis from the air through open stomata.
- Photosynthesis, respiration, and transpiration are affected by water availability and temperature.
- Plants respond to water and heat stress using three strategies: escaping stress by completing the reproductive cycle before the stress occurs in their environment, avoiding stress by changing their growth pattern when stress is present, and tolerating stress by adjusting internally to maintain growth.

- Excess salts in the soil or water interfere with water uptake and growth, but some plants can tolerate salt better than others.
- Some plants' reproductive cycles have temperature or daylength requirements, so they will not flower and produce seed and fruit if these are not met, even if other conditions are favorable.

Air, water, soil, and light. We know these are the basic resources essential for plants in our food gardens. But how do plants use each of these to live and grow to produce the benefits we want, either through their harvest, their appearance, aroma, or other characteristics? In this chapter we explore answers to that question, answers that will help us garden more successfully and respond to trends that are affecting our gardens, communities, and the world.

Knowing the basics of how plants live and grow, including their *morphology*, or physical form (Fig. 5.1), increases gardeners' capacity for responding to these trends.

All three main ACC trends—higher concentrations of CO_2 in the atmosphere; warmer temperatures; more variable precipitation—will have an impact on our gardens (Section 2.3). In some cases, the impacts will be predictable and the responses fairly straightforward, such as the need to plant more drought- and heat-adapted crops in response to rising air temperatures and increasing water scarcity. Others will be less predictable, such as exactly how the growth of some crops and weeds will change under higher atmospheric CO_2. Let's start with the plants themselves.

5.1. The Vascular System in Plants

Water, minerals, food, and other vital substances circulate as liquid solutions through plants in a network of cells called the *vascular system*. Plant roots take up water in the soil as a result of differences in water potential or energy (Section 8.2). Water uptake and vascular transport is driven by the evaporation of water from pores on the plant surface (*stomata*),

---

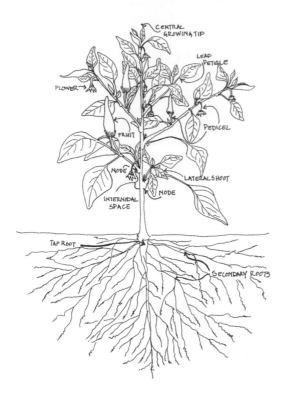

Figure 5.1. Basic plant morphology, the example of Solterito chile, a eudicot

a process known as transpiration (Section 5.5), which increases solute concentration, lowering the water potential. Water in cells with a higher concentration of solutes has lower water potential because the water molecules are attracted to the solute particles and, as water is lost from cells, the remaining water is that which is held most tightly.

As water potential decreases in a cell, water moves across water-permeable cell membranes into it from adjacent cells that have lower solute concentration and so higher water potential, a process called *osmosis*. In other words, if two liquids are separated by such a membrane, water will move out of the more dilute solution, the liquid with a lower concentration of solutes like salt or sugars, and into the more concentrated solution that has relatively lower water potential. This movement will continue until both liquids have the same concentration of solutes, and so the same water potential. Water loss from plant surfaces into the air increases solute concentration in leaf cells, which drives the process that brings water up from the roots through the vascular system. This process also creates a

pulling force (*tension*) on adjacent water molecules because of the *cohesiveness* of water molecules—their attraction to one another, and their *adhesion*—attraction to other substances such as cell walls, as they are drawn up the plant through the vascular system to replace water lost in transpiration. When the solute concentration is greater in the root cells than in the soil, water will move into the roots. However, under drought conditions with low soil moisture, the remaining soil water is held too tightly to soil particles (Section 8.2.1)—and so has very low water potential—for these mechanisms to pull water out of the soil and into the plant's roots, and the plant experiences water stress (Section 5.6), and can eventually die.

The vascular system is easier to understand with an overview of the two basic types of flowering plants, eudicots and monocots. Members of these two groups contrast in the arrangement of their vascular systems, in their germination, growth and structure, and often in their susceptibility or resistance to certain pests, diseases, or chemicals (Chapter 9). *Eudicots*, including many broadleaf garden crops such as amaranth, bean, chile, cucurbits, tomato, and most fruit trees, have two *cotyledons*, or seed leaves, in their seeds, and branched or net-like leaf veins. *Monocots*, including date palm, garlic, lemongrass, onion, maize, and other cereal grains, have only one cotyledon, and usually have major leaf veins that are parallel to each other, running the length of the leaf. In larger seeds the difference between monocots and eudicots is obvious. For example, a bean or squash seed can be easily split into its two halves, the cotyledons, while a maize kernel or onion seed, with only one cotyledon, cannot.

A very few common food garden plants are neither eudicot nor monocot, including avocado (*Persea americana*), cherimoya and related species (*Annona* spp.), and the popular culinary herb from southern Mexico, *hoja santa* (*Piper auritum*), all originating in the Americas and members of a separate and ancient group of flowering plants known as "magnoliids."

The vascular system is a vital part of both the aboveground shoot system of a plant, and the underground root system (Fig. 5.2). The *xylem* carries xylem sap, containing water and nutrients, from the soil, up to the leaves; the *phloem* cells carry phloem sap, containing the carbohydrates produced by photosynthesis and growth-regulating hormones produced in the growing tips, to the rest of the plant. In monocots, vascular bundles that

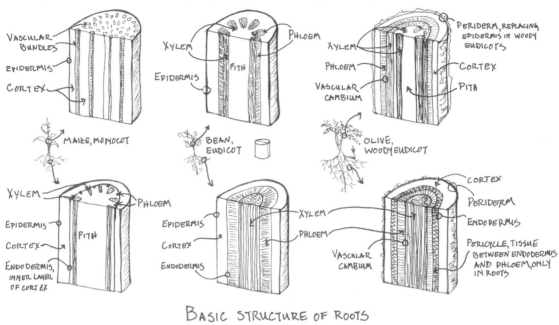

BASIC STRUCTURE OF SHOOTS

VASCULAR BUNDLES
EPIDERMIS
CORTEX
MAIZE, MONOCOT

XYLEM
PITH
EPIDERMIS
PHLOEM
BEAN, EUDICOT

XYLEM
PHLOEM
VASCULAR CAMBIUM
PERIDERM, REPLACING EPIDERMIS IN WOODY EUDICOTS
CORTEX
PITH
OLIVE, WOODY EUDICOT

XYLEM
EPIDERMIS
CORTEX
ENDODERMIS, INNER LAYER OF CORTEX
PITH
PHLOEM

EPIDERMIS
CORTEX
ENDODERMIS
XYLEM
PHLOEM
VASCULAR CAMBIUM

CORTEX
PERIDERM
ENDODERMIS
PERICYCLE, TISSUE BETWEEN ENDODERMIS AND PHLOEM, ONLY IN ROOTS

BASIC STRUCTURE OF ROOTS

Figure 5.2. Monocot and eudicot vascular tissue arrangement

include xylem and phloem tissues run throughout the leaves, stems, and roots. In contrast, in most eudicots vascular tissues form a continuous layer or ring under the *epidermis*, the outer surface of green plant parts.

In woody eudicots, the phloem is separated from the xylem by the *vascular cambium*, a thin layer of growing tissue that develops into either xylem or phloem. These rings of phloem, vascular cambium, and xylem tissue make grafting and layering possible in eudicots (Section 6.7), whereas they are not possible in monocots. The *periderm*, which includes the *bark* (*cork*), is the outer layer of the trunk, and older branches and roots in woody eudicots. It is a more rigid, woody tissue that includes a layer of dead cells on the outer surface that replaces the epidermis. Underneath the epidermis in green shoots and stems, and under the outer layers of bark in woody plant parts, lies the *cortex*, the tissue layer surrounding the vascular system.

The movement of water under tension, facilitated by cohesion and adhesion, and by osmosis, is a passive transport process that does not require energy from the plant. In contrast, *active transport* across membranes from areas of lower to those of higher solute concentration does require chemical energy from the plant. Movement of phloem sap is an active transport process that involves *phloem loading* and *unloading*, to move carbohydrates into and out of nearby phloem cells and into other cells where they are needed, making rapid plant growth possible (Rennie and Turgeon, 2009; Turgeon, 2010).

Because most photosynthesis occurs in the outer and upper layers of the plant exposed to sunlight, they are the *source* of carbohydrates, and the movement of phloem sap is primarily inward toward the main stem and downward toward the roots, the primary *sink* or destination for carbohydrates. However, sap in the phloem may also move up and out towards other sinks, for example early in the growing season when carbohydrates stored in the roots and tubers of perennials flow up to the stems and leaves.

5.2. Plant Growth and Flowering

Plant growth is the result of division (*mitosis*) of the *somatic* (non-reproductive) cells in the growing tips of shoots and roots, followed by enlargement of the newly produced cells. There are two major *patterns* of aboveground growth in most garden plants. *Determinate growth* occurs when the central and secondary shoots all end in reproductive growth, that is, in flowers. Examples include many garden crops like the brassicas, bush-type bean, eggplant, lettuce, Jerusalem artichoke, herbs like basil and cilantro, and maize, with its tassel on top of the central stalk and ears on the ends of secondary shoots below. Climbing bean, vining-type tomato, tomatillo, winter squash, chayote, and other garden plants have *indeterminate growth*, where each new shoot growing laterally from the previous one can produce vegetative and reproductive growth, depending on the growing conditions. Indeterminate garden plants grown for fruit or seed can produce a harvest over a long period, whereas with determinate plants all the fruit and seed tend to ripen at the same time. The capacity to keep growing and producing a harvest, even after a period of no production due to heat or drought, is why indeterminate plants may be better adapted to the warmer temperatures and variable precipitation of ACC (Olesen and Bindi, 2002).

Most flowering plants have two distinct *phases* of growth. During *vegetative* growth, roots, stems, and leaves grow, and during *reproductive* growth flowers, seeds, and fruit grow. A plant's *reproductive cycle* is the time it takes from initiation of flower buds to produce mature seeds. The total length of time an individual plant lives is its *lifespan*. In some plants, reproductive cycle and lifespan are nearly the same length of time, in others they are not.

Annual plants, which include many common garden crops like amaranth, maize, melon, squash, and *verdolaga*, take one year or less to complete their reproductive cycle and lifespan, after which they die. *Biennial* plants like carrot and onion have a two-year lifespan. In most biennials there is only vegetative growth for the first and part of the second year, and in the remainder of the second year reproductive growth occurs. This leads to flowering and fruit and seed maturation, and then the parent plant dies. *Perennials*, like many fruit trees, nopal cactus, and some herbs, have a lifespan longer than two years, usually going through their reproductive cycle every year after an initial period (one or more years) of only vegetative growth. Some chiles and tomatoes are short-lived perennials in frost-free areas. Sometimes lifespan is relatively long, as in the case of olive trees, which can live for hundreds of years. Many agaves, whose swollen leaf bases, roots, and flower stalks are eaten, and leaf fibers used for weaving, have a lifespan of about 20 years. During this time they go through only one reproductive cycle, producing a flower stalk once and then dying.

Most annual and biennial garden crops are *herbaceous*, with tender, green, pliable, non-woody aboveground growth. Some perennials such as banana and yam are herbaceous as well, but the trunks and branches of most perennial fruit trees and vines become woody with age—hard, rigid, and covered with bark. Some perennials such as fig, grape, jujube, persimmon, pomegranate, and pistachio are *deciduous*; they have a repeating seasonal cycle of losing their leaves, and becoming dormant, followed by a period of growth, leafing out, and flower and fruit production (Fig. 5.3). Non-deciduous *evergreen* perennials, like avocado, citrus, guava, and loquat, retain their leaves and grow throughout the year. Some deciduous (e.g., fig) and evergreen (e.g., loquat) trees are *drought deciduous*, loosing their leaves during extreme drought, which reduces transpiration and helps to avoid water stress.

Flowers contain *gametes* or sex cells: male gametes (sperm cells) are produced in pollen grains that originate in *anthers*, and female gametes (egg cells) are produced in *ovules*, contained in *ovaries* (Fig. 5.4). Seed and fruit production normally occurs when pollination is followed by fertilization (Section 10.5). Some plants like okra have *perfect* flowers that contain both male and female structures (Table 5.1). *Monoecious* plants have separate male

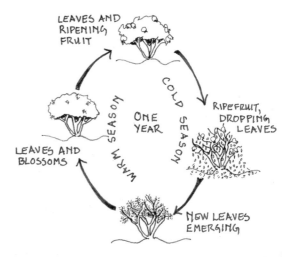

Figure 5.3. Pomegranate is a deciduous perennial with a yearly reproductive cycle

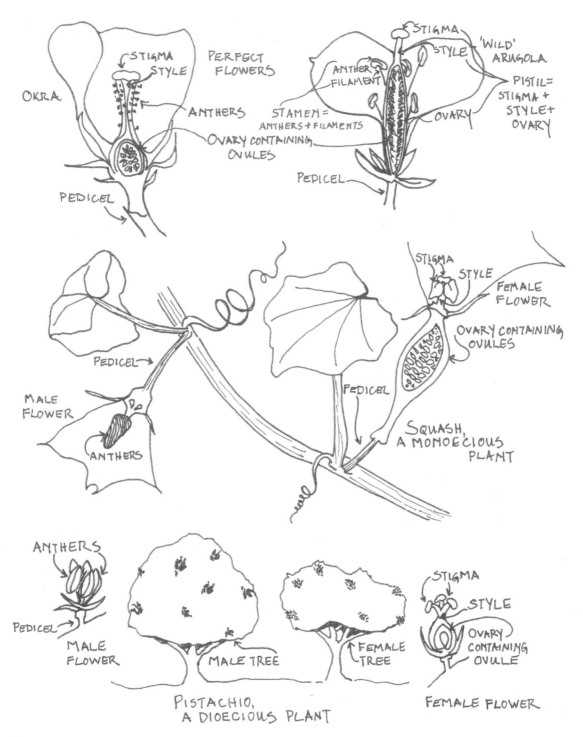

Figure 5.4. Flower and pollination types in some garden crops

Table 5.1. Flowering and pollination types in some garden crops

Flowering type	Definition/description	Garden examples
Perfect	Both male (staminate) and female (pistilate) parts present in individual flowers, mostly self-pollinating	Bean and pea, okra, pepper, tomato; stone fruit; some carob, olive, persimmon, pomegranate
Imperfect (male and female parts occur in different flowers)		
Monoecious	Separate male and female flowers on same plant, but some varieties with perfect flowers	Cucurbits, maize; some mulberry, olive, persimmon, pomegranate, walnut
Dioecious	Male and female flowers on separate plants	Asparagus, spinach; carob, some varieties of fig, mulberry, persimmon, pistachio
Pollen movement primarily via …	Typical flower characteristics	Garden examples
Wind, gravity, rain	Many inconspicuous, small flowers not attractive to pollinators, lacking color or fragrance; light, dry pollen	Asparagus, beet, chard, date palm, maize, olive, spinach
Insects including honey and native bees, other animals	Attractive to pollinators; showy, fragrant, white, or brightly colored; heavy, moist, sticky pollen; some with nectar rewards	Cilantro, cucurbits, fava bean, melon, scarlet runner bean

Some material based on McGregor (1976)

and female flowers on the same plant as in most cucurbits, and maize. *Dioecious* plants such as date palm and some varieties of carob, fig, mulberry, persimmon, and pistachio bear female flowers on one plant, and male flowers on another.

Many plants relying on pollinators have undergone natural selection for showy, colorful flowers, and growing flowering crops like fava bean and garbanzo with both nectar and pollen rewards for pollinators adds beauty to the garden. Encouraging a wide range of pollinators, including native bees, results in higher rates of fertilization, better seed set, and larger fruit (Garibaldi et al., 2013). Still, to be certain you may want to hand-pollinate the flowers of herbaceous garden plants that only last a short time. For example, squash blossoms wither and drop off after one day, and hot, dry conditions can shorten the flowers' viability even further.

5.3. Roots

Roots provide structural support by anchoring plants in the soil. Roots also take up water and minerals and transport them to the shoots, and absorb the oxygen needed for respiration in root cells (Section 5.4). *Root hairs* are elongated extensions of single epidermal cells concentrated just above the actively growing part of the root tip, and they absorb most of the water and soil nutrients that enter the plant (Muller and Schmidt, 2004; Datta et al., 2011). Root hair area varies with variety, species, and growing environment. To appreciate the importance of root hairs, consider the results of a famous 1937 study that estimated the number and area of root hairs of one container-grown rye (*Secale cereale*) plant just before flowering (Dittmer, 1937), based on extensive sample counts. The researchers estimated the plant had more than 14.3 billion root hairs with a total length of over 10,600 km (>6600 mi)—farther than from Los Angeles, California, to Beijing, China—and a surface area of nearly 400 m² (4320 ft²), 84 times more than the area of the plant's total aboveground parts, and 1.7 times more area than the rest of its root structure.

Growing root tips excrete *mucigel*, a gelatinous coating that makes it easier for them to move through the soil, helps with the uptake of soil nutrients (Section 7.3), may repel competing roots, and probably has other qualities we don't yet understand. Many flowering garden crops also obtain nutrients through mutualistic symbiotic relationships between their roots and soil microorganisms (Box 7.1).

Some plants have large, fleshy tuberous roots that store carbohydrates and water, including commonly eaten garden crops such as beet, carrot, cassava, radish, and sweet potato. Other garden "root" crops are actually not roots at all, but tuberous stems, like Jerusalem artichoke and potato, or extremely short stems topped by fleshy modified leaves, as with garlic, onion and bunching onion (Section 6.7). In *FGCW* we use "root" to refer to true roots.

There are two, easy-to-identify types of roots, classified by their origins and how they grow: tap and fibrous roots (Fig. 5.5). Eudicot garden crops, for example amaranth, carrot, chile, okra, and sweet pepper, and most fruit trees, usually have a *tap root*, a dominant vertical root that is an extension of the *radicle* in the seed embryo. As the tap root grows, secondary *lateral roots* grow out from it, and smaller tertiary roots from those (Fig. 5.1). In contrast, monocots like maize and onion commonly have *fibrous roots*, none of which dominate, and which spread out laterally and downward in a fine mass. In plants with fibrous roots the radicle dies early in seedling growth and roots grow from buds low down on the shoot or stem, and then many secondary and tertiary roots grow from

those (Esau, 1977). These are called *adventitious roots* because they do not grow from the radicle. Tap roots can make use of nutrients and water deep below the surface. In comparison, fibrous root systems are better at taking up water closer to the soil surface, and from a wider radius around the plant, a trait adapted to limited, intense rainfall. Hopi farmers plant their maize about 1.8 m (6 ft) on center, and below the thick layer of sandy surface soil (Section 6.2.4), so the fibrous roots have a large area from which to collect short bursts of summer rain that moves down quickly through the sandy soil.

Plants with tap roots may not transplant well, especially when they are mature. If their tap root is cut off or damaged, some can recover by developing new roots in a pattern similar to a fibrous root system, as happens with tomato transplants that are capable of growing adventitious roots from nodes on their stems. But usually this happens if the plant is young and vigorous, and its shoot system is relatively small. An exception to this is olive, which when grown from seed has a deep tap root

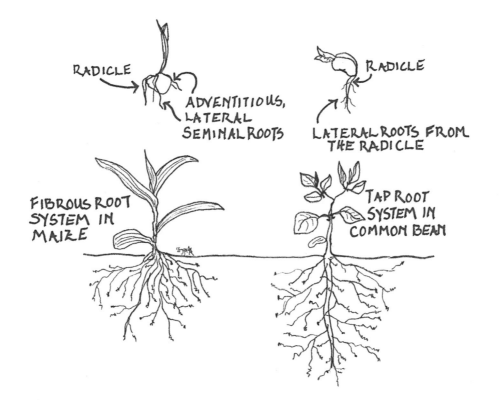

Figure 5.5. Fibrous and tap roots

but is amazingly hardy and can be transplanted with its tap root pruned back and a small root ball, because an adventitious, fibrous root system takes over, and the tree can recover to produce fruit.

Root systems are also affected by a plant's growing environment, including the density and types of neighboring plants, the soil, distribution of nutrients, and irrigation patterns. Plant roots only grow where water is available, not in dry soil. *Hydrotropism* is the preferential growth of plant roots toward areas with more water (Cassab et al., 2013). Although still not well understood, it seems that sensors in root tip cells detect relative water availability in the soil they are touching; these signal hormones and other substances that stop growth on the side of the root closest to the most moisture, while those on the opposite side of the root continue growing. The end result is growth curving toward that moisture.

Plant roots also have *compensatory growth*, meaning that roots will proliferate where soil conditions are favorable, making up for slower root growth in less favorable areas (Bengough et al., 2006). Hydrotropism and compensatory growth are good to keep in mind when irrigating plants because their root systems will develop most strongly where the soil is consistently watered and has good structure and nutrient content (Sections 7.3.2, 8.4). Frequent shallow irrigations, for example 5–8 cm (2–3 in) deep, result in shallow root systems. Under hot, dry conditions, moisture in this surface layer is lost quickly by evaporation, leaving less for the plants, which have to be watered frequently. For best water use efficiency beginning at the seedling stage, try for deeper and less frequent irrigations that wet the soil to below the current root zone. This encourages plants to develop a deep root system due to hydrotropism and compensatory growth, regardless of whether they are fibrous- or tap-rooted species. If plants have been watered shallowly for a while, don't switch abruptly to less frequent deep irrigations—make the transition gradually, observing how the plants react.

Soil texture and structure also affect root growth (Section 7.2). Roots will proliferate where soils are loamy, and not compacted, and there is adequate pore space with a structure that allows roots, air, and water to move easily through the soil. Very compact clayey soils have little soil structure, and are difficult for roots to grow through. The result can be roots that are thick and deformed from trying to grow through the soil, and are less able to absorb water or anchor the plant.

Poor soil drainage and too much water can also cause shallow rootedness as deeper roots die back in the waterlogged soil. Both under- and over-watering when there is poor drainage can result in salt accumulation because naturally occurring salts are not leached or washed below the root zone. High salt content can slow growth and even kill the plant (Section 8.8.5).

5.4. Photosynthesis and Respiration

Photosynthesis is the process by which green plants capture solar energy and store it in carbohydrates, and it is the most important direct link between the living and non-living matter on Earth (Beerling, 2012). Through this connection the evolution of life on Earth has been affected by, and in turn has affected, Earth's geophysical characteristics, including the soils and atmosphere (Box 5.1). Photosynthesis converts the sun's energy to chemical energy in the molecular bonds of carbohydrates that are the energy source that plants use for maintenance, growth and reproduction, and are the source of energy for essentially all life on Earth.

Chloroplasts are the structures in plant cells where photosynthesis occurs, and individual cells near a plant's surface—mostly in leaves—may contain from a few dozen to more than 100 chloroplasts, depending on the cell and its phase of development (Okazaki et al., 2010). Both the number of chloroplasts per cell and the rate of photosynthesis decline as leaves and other plant parts age (Olesinski et al., 1989), a reason to prune off older leaves, to make more resources available to younger, more productive ones. Inside chloroplasts is *chlorophyll*, a green pigment molecule that captures sunlight to fuel the photosynthetic reaction between CO_2 in the air, and H_2O in the plant. The CO_2 and H_2O molecules contribute the C, H and O atoms that are transformed during photosynthesis into oxygen gas (O_2) and the carbohydrate glucose (Fig. 5.6).

Carbon dioxide from the air + water from the soil + energy in sunlight (via photosynthesis in chloroplasts) → a carbohydrate (glucose) stored in plant tissues + oxygen released into the air

$$6\ CO_2 + 6\ H_2O + \text{sun's energy} \rightarrow C_6H_{12}O_6 + 6\ O_2$$

Box 5.1. Evolution and forms of photosynthesis in food garden plants

Plant photosynthesis has profoundly affected the Earth's environment and climate in the past, and continues to do so today. Billions of years ago the Earth's atmosphere was dominated by water vapor and then nitrogen (N); CO_2 was present but O_2 was not. The evolution of photosynthesis, about 3.5 billion years ago, first in single-celled organisms, and later in multicellular organisms, changed this by removing CO_2 from the atmosphere and releasing O_2.

To grow and maintain themselves, the cells of organisms respire, taking O_2 out of the atmosphere and emitting CO_2. However, some of the C captured in photosynthesis remains in the tissues of living, dead, and decaying plants, and the soil microbes that consume them. Some of this C remains out of the atmosphere for very long times in partly decomposed or fossilized forms (Section 2.3.1). As a result of photosynthesis and long-term C storage or *sequestration* in living and dead organisms, and other processes, the CO_2 composition of the Earth's atmosphere changed over time (Table 5.2). Atmospheric O_2 content rose to today's level of 21% about 850 million years ago (Hohmann-Marriott and Blankenship, 2011). While photosynthesis and other factors resulted in changes in atmospheric concentrations of O_2 and CO_2, these changes themselves, along with further environmental changes, also drove the evolution of different forms of photosynthesis.

The first type of photosynthesis to evolve was probably what we now refer to as C_3, originally in marine algae, from which land plants evolved. C_3 photosynthesis gets its name from the three-carbon carbohydrate molecule that the C in CO_2 is initially

captured in. As plants spread across the Earth, the total contribution of atmospheric O_2 from photosynthesis increased, and combined with ongoing climate changes slowly increased the ratio of O_2 to CO_2 in the atmosphere (Sage et al., 2011).

As the Earth became drier and cooler and the atmosphere became increasingly oxygen-rich, CO_2 was limiting for photosynthesis, especially in areas with high temperatures and light intensity, where water might also be limiting (Sage et al., 2011). Water scarcity triggers stomatal closure (Section 5.6), which in turn decreases access to CO_2 and therefore photosynthesis. We can see an evolutionary response to these changing conditions in the fossil record, where for each 100 ppm decrease in atmospheric CO_2 and accompanying increase in O_2, there was a 2–4% increase in the density of stomata on leaves, and a decrease in stomata size (Franks et al., 2013). This is because many smaller stomata are more efficient at CO_2 uptake with minimal water loss—they can open and close more quickly, an important adaptation to decreasing CO_2 and limited water (Section 5.5).

Another adaptation to these atmospheric changes and to decreasing water availability was the independent evolution in multiple locations and plant lineages of two other forms of photosynthesis—CAM and C_4. With high temperatures and light intensity, both of these are more efficient than C_3 in terms of *physiological water use efficiency* (WUE), which is (units of CO_2 captured)/(units of water transpired) (Condon et al., 2004; Silvera et al., 2010). Physiological WUE affects *agronomic WUE*, which is usually measured as (amount of crop harvested)/(total amount of water, including precipitation and irrigation) calculated for a defined area, and what gardeners are most interested in (Section 8.3). Many gardening practices we talk about work to reduce excess transpiration and evaporation, which moves agronomic WUE closer to physiological WUE.

In Crassulacean acid metabolism (CAM) plants, CO_2 intake and transpiration (water loss) occur at night when temperatures and evaporation rates are generally lower, while the light capture and C fixation parts of photosynthesis occur during the day with stomata shut. Separating these processes in time increases physiological WUE. While CAM plants use relatively less water, they usually grow very slowly compared to C_3 plants, because not much CO_2 can be stored in their cells at night and this is quickly used in photosynthesis during the day. Many lineages of CAM land plants are thought to have evolved about 20 million years ago (Brenchley et al., 2012; Edwards

Table 5.2. Estimates of when major changes occurred in the CO_2 concentration in the Earth's atmosphere

Time period	Estimated atmospheric CO_2 concentration, parts per million (ppm)
~550 million years ago[a]	2,800–4,700
~250 million years ago[a]	1,500
800,000 years ago[a]	200
<280 years ago in 1750, before the Industrial Revolution[b]	280
Today	>400, and rising

[a]Franks et al. (2013)
[b]IPCC (2013)

Continued

Box 5.1. Continued.

and Ogburn, 2012). CAM plants that you might grow include agave, nopal cactus, and pineapple.

About 30–20 million years ago, C_4 photosynthesis evolved as another response to the changing conditions on Earth, also in multiple locations and plant groups (Silvera et al., 2010; Sage et al., 2011; Edwards and Ogburn, 2012). In C_4 plants, named for the four-carbon molecule they produce during photosynthesis, CO_2 intake (in mesophyll cells) and C fixation (in bundle-sheath cells) are spatially separated, allowing C_4 plants to concentrate CO_2 in bundle sheath cells during the day. Carbon fixation is more efficient in the bundle-sheath cells because photorespiration (see below) is reduced due to the higher concentration of CO_2 there than in the mesophyll cells. Therefore, C_4 plants do not need to have stomata open as much as C_3 plants for a given amount of C fixation. But C_4 photosynthesis requires more energy for each glucose molecule produced and will only have an advantage compared to C_3 photosynthesis in environments with intense light and high temperatures. Examples of C_4 crops are the leaf and grain amaranths, maize, millet, sorghum, and sugarcane. *Verdolaga* is a C_4 garden green that under drought conditions can switch to CAM-like photosynthesis (Lara et al., 2003).

To get a feeling for their difference in water use, you can compare the ratio of physiological WUE in CAM, C_4, and C_3 plants which on average is 6:2:1— in CAM plants it is three times greater than in C_4, and six times greater than in C_3 plants when each are grown under ideal conditions for their type (Sage and Kubien, 2003; Yamori et al., 2014). The fact that both CAM and C_4 modifications of C_3 photosynthesis evolved multiple times and places suggests that they involve the simplest, most direct structural and physiological changes to photosynthesis that provide a selective advantage under specific environmental conditions, and these changes also mean they may respond differently to some of the trends of ACC (Section 2.3.1, Table 2.4).

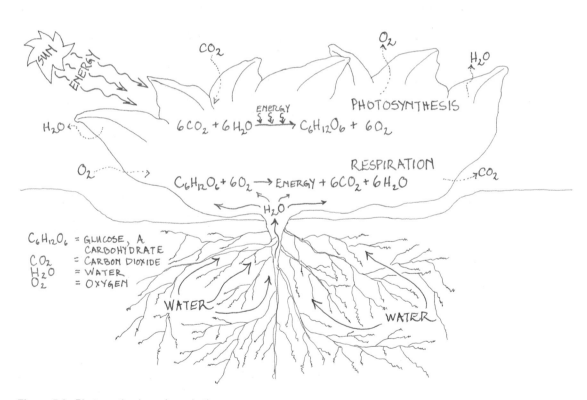

Figure 5.6. Photosynthesis and respiration

Most plants grown in food gardens initially capture the C in CO_2 in a three-carbon molecule, and so are called C_3 plants, although there are important exceptions (Box 5.1).

CO_2 is only 0.04% of our atmosphere, amounting to about 848 petagrams or 848 billion MT (9.35 x 10^{11} US tons) of C. Photosynthesis by plants on land results in the net capture of about 54 petagrams of C each year (Haberl et al., 2007; Smith et al., 2012), called *net primary production*—the C removed from the atmosphere through photosynthesis (*gross primary production*) minus that lost back to the atmosphere during respiration annually. Most C captured by photosynthesis in a given year will return to the atmosphere via respiration in plants and over the next few years as plant material is used for food, for example by humans, or by microorganisms in the soil or a compost pile (Sections 7.4–7.6). This CO_2 is classified as *biogenic CO_2*, because the C it contains was removed from the atmosphere by photosynthesis a short time ago, and returns soon after, it has little or no effect on ACC. A very small amount of C removed from the atmosphere by photosynthesis is sequestered for long periods in stable forms of humus in the soil, in peat (Box 6.4), in dead phytoplankton on the seabed, or through fossilization into oil, natural gas, and coal.

Respiration is the series of chemical reactions in living cells through which plants and animals access the energy captured by photosynthesis, typically in the simple sugar glucose. That sugar is combined with oxygen—from the atmosphere or from photosynthesis—to produce CO_2, water, and usable energy. During respiration in the presence of oxygen, which takes place in *mitochondria* (organelles found in most cells of multicellular organisms), glucose is *oxidized*, that is, it loses electrons to oxygen, releasing chemical energy, and producing CO_2 and H_2O (Fig. 5.6).

Glucose + oxygen (via respiration in mitochondria) → energy available for cellular processes + CO_2 released into air + H_2O released into the cell and plant vascular system

$$C_6H_{12}O_6 + 6O_2 \rightarrow energy + 6CO_2 + 6H_2O$$

The amount of C from the air that is captured in photosynthesis and stored in plant tissue—about 43% of the weight of dry plant material (Latshaw and Miu, 1924; Atwell et al., 1999)—is directly affected by the availability of water, CO_2, and solar energy, so a shortage of any one of these will limit photosynthesis, slowing or stopping plant growth and productivity. Other factors that affect the amount of C captured include the plant species, nutrients, soil and air temperatures, wind, and diseases.

If nothing else was changing, higher concentrations of atmospheric CO_2 might encourage increased plant growth in cases where CO_2 limits photosynthesis, especially C_3 photosynthesis. When CO_2 is limited, including when stomata close under hot conditions or as a response to water loss, photorespiration rather than photosynthesis can occur. *Photorespiration* is a chemical reaction pathway in plant cells that instead of capturing energy in chemical bonds during CO_2 fixation, uses O_2 and chemical energy in order to produce a molecule that was thought to make photosynthesis overall less efficient at capturing solar energy. Because of the increased supply of atmospheric CO_2 due to ACC, it is thought that rates of photorespiration will be reduced in C_3 plants, whose photosynthesis is more CO_2-limited compared to C_4 plants. If so, this would give C_3 plants an advantage as atmospheric CO_2 concentration rises.

However, there is accumulating evidence that photorespiration plays a role in nitrogen (N) uptake from the soil by plants, with new research investigating the hypothesis that photorespiration is not inefficient, but rather provides the conditions and substances that make nitrate (NO_3^-) assimilation possible in plants (Bloom and Lancaster, 2018). Decreased photorespiration is one reason hypothesized for reduced nutrient content of food harvested under raised atmospheric CO_2 (Box 2.4), and a C_3 advantage under higher atmospheric CO_2 cannot be assumed (Reich et al., 2018). The relationship between rising atmospheric CO_2 concentration, photorespiration, soil N, and plant yield is an active area of research. For now, finding varieties with lower N requirements or more efficient N use, and replenishing organic matter with available forms of N, could be the most effective responses.

Greater CO_2 concentration might also be a benefit under dry conditions as more time with stomata closed means less water loss from the plant via transpiration for every unit of CO_2 fixed in photosynthesis, and so better WUE. But rising atmospheric CO_2 does not seem to offer unequivocal benefits for photosynthesis or WUE, including for C_3 plants.

Warming air temperatures are the most consistent effect of ACC, and since transpiration is important for cooling the plant, keeping stomata closed more may not be advantageous. Instead, in addition to mitigation efforts, heat-resistant species and varieties, new planting schedules, and more shading are good responses for gardeners.

5.5. Transpiration

Most CO_2 enters the plant through the stomata in the epidermis (Fig. 5.7). When the stomata are open, not only can CO_2 enter and diffuse into cells to reach the chloroplasts, but the water pulled up from the roots also evaporates into the air, the process of *transpiration*. The sum of transpiration plus the direct evaporation of water from soil, water and other surfaces together is *evapotranspiration*.

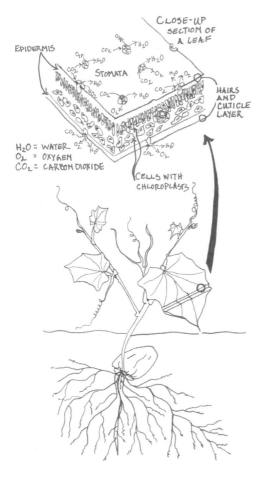

Figure 5.7. Leaf, stomata, and transpiration in chayote

About 95% of all water taken up by a plant is transpired, and only 1% is used directly in photosynthesis, with the remaining 4% absorbed osmotically by plant cells (Sperry, 2011). Most transpiration occurs through open stomata with a small proportion of transpired water lost directly through the walls of epidermal cells. In hot, dry environments, the tradeoff between water lost and CO_2 acquired in transpiration is critical and quantified by physiological WUE (Box 5.1), and reducing excess evapotranspiration is an important gardening strategy.

Even though transpiration removes a lot of water from the soil through the plant, it serves two important functions. First, it creates the pull that keeps water and nutrients moving up through the plant from the roots, providing the water for photosynthesis and other cellular processes. Second, transpiration cools the plant (as water evaporates through the stomata), the same way evaporation cools our skin when we sweat.

Transpiration rates vary depending on the growing environment and the plant, including evolutionary adaptations of their stomata. For example, one study compared the response to drought during the growing season of traditional varieties of fava bean originating in either higher (800 mm, >31 in year⁻¹) or lower rainfall (550 mm, <22 in year⁻¹) areas (Khazaei et al., 2013). Stomata of all varieties became smaller in response to drought, but compared to the varieties from wetter environments, varieties from low rainfall environments maintained functioning leaves with higher leaf water content, had greater physiological WUE, and had a significantly increased number of stomata per leaf area, likely indicating long-term selection for phenotypic plasticity for this trait in response to drought (Box 10.3). More and smaller stomata made the low-rainfall varieties more drought resistant, because the small stomata opened and closed more rapidly, minimizing the water lost per molecule of CO_2 taken in, making the plant more water efficient. As mentioned in Box 5.1, the fossil record shows that more and smaller stomata were selected under increasing drought (Franks and Beerling, 2009). Changes in number and size of stomata are ways in which plants adapt to dry conditions.

Where on leaves stomata occur can also reflect evolutionary adaptations to the growing environment. Grape, domesticated in the seasonally dry Mediterranean and West Asian region, have no stomata on the upper leaf surface, so transpiration is reduced because it occurs only from the shaded, cooler undersides of leaves. In contrast, the popular garden vegetable *verdolaga* only has stomata on the upper surface of its leaves, possibly because its low, creeping growth means hot soil surfaces would increase transpiration from stomata if they were on the underside of the leaves (Ren et al., 2011).

As already mentioned, photosynthesis increases with sunlight when water and CO_2 are not limiting, so under sunny conditions the stomata are open longer to supply the CO_2 needed, also increasing transpiration. But when it's hot, dry, or windy, the amount of water released in transpiration and thus the amount needed by the plant increases even more. So when it is warm and dry, plants face a major tradeoff. The stomata in some crops such as grape will close under extreme water stress, but reduced transpiration in response to water stress means less cooling of the plant, and is why water and heat stress often occur together in plants (Lopes et al., 2011). The stomata need to be open to acquire CO_2 for photosynthesis and for evaporative cooling, yet this also increases water loss. When stomata close to prevent water loss, photosynthesis may slow down and become less efficient due to lower CO_2 availability and heat, with high heat eventually killing cells.

Figuring out how to optimize evaporation and transpiration rates is a major challenge for gardeners, and one of the topics of Chapter 8 where we look at how to manage drought and water stress in the garden. But a first step is understanding how plants themselves respond to water and heat stress.

5.6. Plant Responses to Drought and Heat

Our garden plants need to be able to live and produce under hotter, drier, more stressful growing conditions due to ACC, but how plants experience and respond to these conditions differs. Understanding

these differences helps gardeners to make the most of those plant responses.

Drought is an unusually dry condition in the environment, and can result in insufficient water in the soil to keep up with the plant's transpiration needs, causing *water stress*—insufficient water inside the plant for it to maintain itself, grow, and produce a harvest (Section 8.4). Distinguishing between drought, a condition in the environment (especially the soil), and water stress, a condition in the plant, makes it easier to understand how plants respond to dry conditions. These different responses are adaptations that allow them to reproduce successfully under those conditions (Fig. 5.8).

Drought-adapted plants either escape drought or resist the water stress resulting from drought. *Drought-escaping* plants usually have short lifespans and are often called "short-cycle" varieties, allowing them to take advantage of brief periods when there is enough moisture so they are unlikely to experience drought and water stress. Early season, "famine" crops in dry regions of the world, like tepary beans and short-cycle (60–90 day) maize in the US Southwest and Mexico, and early millet in West Africa, follow this strategy, maturing before drought is likely.

Drought-resistant plants use one of two strategies when exposed to drought; they either avoid water stress or they tolerate it. *Water stress avoidance* means that the plant will not experience water stress and lower yields despite exposure to drought, because the plant adjusts to make more efficient use of the water that is available (Blum, 2009). For example, during periods of drought cowpea, also called black-eyed pea, avoids water stress because the orientation and movement of the leaves changes in relation to the sun, minimizing the amount of sunlight and heat it receives, and the amount of water lost from the leaves due to transpiration. Similarly, the leaf blades of maize plants roll up

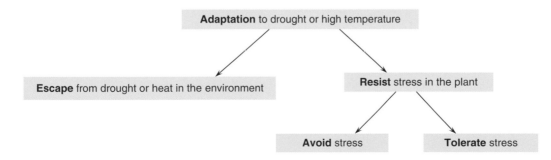

Figure 5.8. Drought and heat adaptation strategies in plants

when drought is present, minimizing water loss by exposing less leaf area to drying air and sunlight (Schmidt et al., 2011). In many plants, older leaves tend to dry out first, helping avoid stress by concentrating limited water in younger, more productive leaves (Blum, 2009). Gardeners can enhance this response by removing older leaves that are usually growing lower down on the plant. Drought deciduous plants lose their leaves when drought reaches a certain level, and regrow them when more water becomes available. These are all ways of decreasing the shoot:root ratio, or the ratio of plant parts losing water to plant parts taking up water from the soil.

Other physical characteristics acquired through evolutionary adaptation also reduce transpiration, making plants better able to avoid water stress. These include:

- small leaf surface area (e.g., garbanzo, jujube, pomegranate);
- stomata located mostly on the more protected (generally cooler), underside of leaves (e.g., grape);
- thick, waxy, or resinous layer or *cuticle* on the leaf surface reducing transpiration from epidermal cells (e.g., agave, nopal cactus, pineapple);
- light-colored, silvery leaves that reflect light (high albedo, Section 4.1.3), resulting in lower leaf temperatures and therefore less need for cooling by transpiration (e.g., olive, sage);
- hairs or other structures on leaves that reflect light, provide additional surface area for cooling the plant, and reduce air movement in the boundary layer (Section 4.1.4), thus decreasing transpiration (e.g., artichoke, cucurbits);
- self- or soil-shading canopy (e.g., grape, chayote, squashes).

Water stress tolerant plants may experience internal water stress due to drought but still produce a harvest. For example, some plants such as grape slow down photosynthesis and growth until the stress passes, others like wheat accumulate energy reserves in stems or roots as water stress increases, then use these reserves to produce grain, even under stress (Chaves et al., 2002).

For many plants, the ability to resist water stress depends on when in the plant's reproductive cycle and lifespan drought occurs. Many garden species are most sensitive to drought just after seeds have germinated and before seedlings have established root systems, and again when flowering, setting seed and forming fruit (Section 8.4).

Heat adaptation, like drought adaptation, can involve escaping or resisting heat (de la Peña et al., 2011). High temperatures can result in die-back, death of pollen or ovules, and *bolting*—premature flowering and plant death. Even without bolting, hotter temperatures can shorten plant reproductive cycles and reduce harvests. For example, a heatwave in central Europe during the summer of 2003 shortened crop reproductive cycles by 10–20 days; Italian maize yields dropped by 36% and French tree fruit production by 25%, all in response to high temperatures that are projected to be the new average in that region later this century due to ACC (Battisti and Naylor, 2009).

As long as daylength is not an issue, planting times or the species and varieties grown can be adjusted in response to ACC so that sensitive phases of the reproductive cycle are over before it gets too hot, in order to *escape* damaging heat. *Heat stress avoidance* occurs when it's hot but the plant doesn't experience heat stress. In plants that can *tolerate* heat stress, the temperature of the plant itself may rise without interfering with its growth, or at least not killing the plant, probably because the plant produces compounds that can repair tissue damaged by heat (Schwarz et al., 2010; de la Peña et al., 2011).

ACC is expanding the geographic range and growing season for some crops (Chapter 2) but drought and heat are already reducing plant growth in many places, and these will extend to more areas and for longer periods with ACC. Distinguishing between heat and drought adaptation, and when each is needed or not, is useful because these adaptations may divert plant resources away from growth, reproduction, and defenses against pests or pathogens. Drought and heat adaptation may not be needed in gardens making efficient use of a reliable water supply, and that are managed to avoid very high temperatures, for example by adjusting planting dates and providing shade. Ecological thinking in terms of benefit:cost can help. It is not a matter of selecting crops that will use the least water, but those that will produce the most food people want to eat for the amount of water available (highest agronomic WUE, Sections 5.4, 8.3.1).

Crop diversity can also help (Chapter 10). Figuring out what species and varieties fit your garden can make a big difference in how difficult, or how rewarding, gardening is. Different varieties of the same crop may be adapted to different conditions, showing a qualitative genotype x environment interaction (Box 10.3). That's what was found in evaluations of 10 cowpea varieties tested in Maryland in control and drought-stressed environments (Dadson et al., 2005). Several paired comparisons of the varieties showed one variety with higher yield under well-watered conditions, but a different variety with better yield under dry conditions (Fig. 10.4b).

Gardeners can take advantage of drought and heat adaptation by experimenting with growing plant species and varieties known to do well in places with the kind of water availability or temperatures present or projected in the near future in their location (Jarvis et al., 2014). This includes trying species or varieties that have traditionally been grown in drier, hotter areas (see the example from West Africa in Box 10.3).

5.7. Salt Tolerance

The accumulation of salts like sodium chloride (NaCl) in the soil can be a serious problem for gardens, especially in hot, dry environments. Salty soils reduce the water and nutrients available to plants because they have an osmotic effect on plant roots. That is, the movement of water from the soil to root cells is decreased, or even reversed (Zhu, 2001), and water is drawn out of the roots into the saltier soil water, resulting in poor yields, and even killing plants (Section 8.8.5). Sometimes plants take up dissolved NaCl through their roots, which can lead to browning of new leaf edges, a sign salt has accumulated in young plant tissues, where it interferes with photosynthesis and other cellular processes.

Some plants are less sensitive to salt than others, they are *salt tolerant*. The most salt-tolerant plants are *halophytic* (salt-loving). These species will produce more as salinity increases, within low levels. While no common garden plants are truly halophytic, some are salt tolerant, such as date palms, which can be grown using water that is 0.05% NaCl (Glenn et al., 1998); by comparison, sea water has about 3.5% salt. Other salt-tolerant crops include asparagus, barley, beet, fig, kale, New Zealand spinach, olive, spinach, Swiss chard, and some tomato varieties. *Halophobic* plants are particularly sensitive to salinity and should be avoided if your soil or water is known to be high in salts. These include some green bean varieties, citrus, onion, persimmon, radish, squash, and the stone fruits.

5.8. Seasonal Constraints to Plant Growth

No matter how well you manage your garden, there are times when certain plants just won't grow or thrive. One reason can be that they are not getting the daylengths or temperatures they need. Because of ACC some gardeners may need to adapt by using new species or varieties or new planting schedules as local weather changes. Also, when new migrants try to grow species or varieties never before grown in an area, they may have trouble if those plants are not adapted to local climates. Understanding a bit about seasonal constraints to plant growth improves the chances for healthy, vigorous garden plants.

5.8.1. Daylength requirements

The number of hours of sunlight and dark each day at a specific time and place (Section 4.1.2) won't change due to ACC, but understanding their impact on crop plants can help you anticipate how ACC will affect your garden. It is the combination of familiar daylengths with new temperatures, patterns of precipitation, and pests and pathogens that will make it more important than ever that gardeners observe, experiment, learn, and in some cases look for new varieties, species, and garden practices.

Some plants have a *daylength* requirement for a minimum number of continuous hours of darkness (also called *photoperiod requirement*) without which they cannot produce fruit and seeds. Short-day plants need nights equal to or longer than a particular threshold, depending on the species or variety; long-day plants need nights equal to or shorter than a particular threshold; day-neutral plants do not have photoperiod requirements. While many widely-used garden crop varieties are photoperiod-neutral, some important ones are not.

As you get closer to the equator, there is less difference between hours of darkness and hours of daylight, both daily and seasonally (Section 4.1). Moving north or south of the equator, latitudes increase and those locations have warm growing conditions during short-night summer months. It is not unusual to find crop varieties from the tropics and subtropics that need those warm temperatures, but also need more hours of darkness than are present during the warm season in higher latitudes.

Daylength problems can come up for gardeners new to an area who brought seeds from their former home at a very different latitude. Stockton, California (latitude 38°N) is farther north than Oaxaca, Mexico (17°N), and the longer nights that some common bean varieties from Oaxaca need to flower do not occur until late September in

Stockton, the beginning of the cool season. Because these beans are a warm-season crop (Table 5.3), they cannot be sown in the winter in Stockton to take advantage of the long nights of late spring. Therefore, the only time to plant those bean varieties is late in the warm season, in late July or early August, and this may not give some beans enough time to mature before the average first frost in late November, although a later date of first frost due to ACC is likely extending the season. Planting in garden locations that are protected from frost may help (Section 4.1.3). Overall, however, the way photoperiod sensitivity and warmer temperatures brought by ACC will interact isn't well understood, and can't be generalized. One study comparing a daylength sensitive and a daylength insensitive Mexican maize genotype found that projected temperature increases interfered with the rate of plant development and flowering of both (the temperatures exceeded the maximum threshold temperature TH_{high}, Section 5.8.2) (Craufurd and Wheeler, 2009). However, the daylength sensitive genotype was less negatively impacted, presumably because this sensitivity overcame some of the developmental delays caused by the high temperatures by triggering development even though temperatures were not optimal.

Another garden crop that can have daylength sensitivity is okra, thought to have originated in Southeast Asia or Africa (Kumar et al., 2010). While some common varieties are day neutral (e.g., Clemson spineless), others—probably from low-latitude areas—require long warm-season nights (≥13 hours) to begin reproductive growth, meaning they may not start producing a harvest until the fall in temperate regions.

5.8.2. Temperature requirements

Every food plant variety has optimal soil and air temperature ranges for germination and healthy growth, including successful flowering and fruiting. As with daylength, temperature requirements can vary both between and within species, a result of past evolution by selection in different growing environments. Temperature is especially important for seed germination and early seedling growth, and also affects the growth of pests and pathogens. When temperatures are too high or low, crop seeds may be slow, or unable, to germinate, lack vigor, or become more susceptible to the soil-borne pathogens that thrive at those temperatures (Section 9.8.2). A generalization is that the best air temperature for seed germination of many garden crops grown in the continental US is about 25–30°C (77–86°F) (Hartmann et al., 2002, 212), but Table 5.3 outlines three broad classes of temperature requirements and some garden crops in each. Germination speed tends to increase with warmer temperatures within each crop's range, up to its optimum, after which germination becomes slower as temperatures rise. In general, the temperature of the top 10 cm (4 in) of soil is highly correlated with air temperatures ($R^2 \geq 0.86$) (Zheng et al., 1993), so mean air temperature is used as an indicator of soil temperature, even though that relationship can vary.

Air temperature affects plant growth and the time it takes from planting the seeds until harvest—so for the same variety, days to maturity calculated

Table 5.3. Examples of air temperature requirements for seed germination in garden crops

Temperature class	Approximate temperature range		Examples of garden crops	Environments these species are often native to
	°F	°C		
Moderate season/cool temperature tolerant	40–86	4.5–30	Broccoli, cabbage, carrot, cauliflower	Temperate zones
Cool season/cool temperature requiring	≤77	≤25	Beet, celery, lettuce, onion	Mediterranean
Warm season/warm temperature requiring	>50	>10	Asparagus, some maize and tomato varieties, *verdolaga*	Subtropics or tropics
	>60	>15	Basil, chile and other peppers, cucurbits, eggplant, okra, most *Phaseolus vulgaris* and *Vigna* seed legumes	Subtropics or tropics

Based on Hartmann et al. (2002, 212)

in one place might not work somewhere else. That's why the "days to maturity" listed on most commercial seed packets may not be accurate for many locations. Imagine the temperature difference for a June day in central California and central Maine, close to where Johnny's Selected Seed breeds some of its varieties. An easy way to understand this is to think of air temperature requirements for plant development in terms of heat units, also called growing degree days. *Heat units* (HU) measure the accumulation of time at the range of temperatures that further the growth of an organism, in this case a garden plant, from one stage of development to the next, for example from vegetative to reproductive growth. For different species and varieties the temperature ranges and total HU requirements can both differ. Note that HU requirements and estimates may be in degrees Centigrade (HU-C) or Fahrenheit (HU-F), but converting between them is easy (HU-C × 1.8 = HU-F; HU-F × 0.5556 = HU-C). Some examples of HU required to a specific growth stage (all calculated using a slightly more complicated single sine formula, not the simple method we show) are: New Mexico-style chiles, ripe fruit, 1778 HU-C (3200 HU-F) (Brown, 2013); Black turtle bean, 50% flowering, 757.5 HU-C (1363.5 HU-F), sweet corn, fresh harvest, 855 HU-C (1539 HU-F), Thompson seedless grape, ripe fruit, 1600–1800 HU-C (2880–3240 HU-F) (UCANR, 2016); tomato, ripe fruit, 1214 HU-C (2185 HU-F) (Pathak and Stoddard, 2018). If you use published HU requirements, be sure those and any estimation of HU for your area use the same HU scale (C or F) and calculation method.

Heat units are used to assess accumulated time in a temperature range, and are calculated for a specific crop, by totaling HU for 24-hour periods for a given location. Of the several calculation methods, the simplest uses the maximum and minimum temperature over a 24-hour period at a location:

$$\text{Heat units} = [(T_{max} + T_{min})/2] - TH_{low}$$

T_{max} and T_{min} are the maximum and minimum temperatures observed in 24 hours, respectively, and TH_{low} is the lower temperature threshold for a particular crop species or variety, below which its growth suffers (Table 5.4). These thresholds are good estimates to start with, but remember they are usually approximations by crop species, and do not account for the differences that exist among varieties of the same species adapted to different environments.

Table 5.4. Low temperature thresholds (TH_{low}) for some common garden crops

Crop	Minimum threshold temperature	
	°F	°C
Asparagus	40	4.4
Bean, snap	50	10.0
Beet	40	4.4
Broccoli	40	4.4
Cantaloupe	50	10.0
Carrot	38	3.3
Collard	40	4.4
Cucumber	55	12.8
Eggplant	60	15.6
Lettuce	40	4.4
Onion	35	1.7
Okra	60	15.6
Pea	40	4.4
Pepper, sweet[a]	50	10.0
Potato	40	4.4
Squash, summer[a]	45	7.2
Strawberry	39	3.9
Sweet corn (maize)	48	8.9
Sweet potato	60	15.6
Tomato	51	10.6
Watermelon	55	12.8

[a]Assumed type, not given in original source.
Mullins (2010)

For some crops there are also high temperature thresholds (TH_{high}), above which plant growth is impaired, as discussed in the tomato example in Section 3.1.1. With average temperatures rising due to ACC, TH_{high} may become more important. When the maximum local temperature T_{max} is above the TH_{high}, then TH_{high} should be substituted for T_{max}, eliminating hot temperatures that harm crop growth. Table 5.5 illustrates how HU differ by location and date using the example of tomato. To estimate HU over a season you would need to do this for each day and sum them, or make a rough estimate for each month by multiplying the results of one day's HU calculations by 30.

With high and low temperature thresholds for tomato of 27.2 and 10.6°C, respectively, the daily heat units for June 1 in Bakersfield can be calculated as follows:

$$\text{Heat units, in °C (HU-C)} = [(T_{max} + T_{min})/2] - TH_{low}$$
$$\text{HU, using } TH_{low} \text{ value only} = [(30.9 + 15.9)/2] - 10.6$$
$$= 12.8 \text{ HU-C}$$

$$HU, using\ TH_{low}\ and\ TH_{high} = [(27.2+15.9)/2]-10.6$$
$$= 11.0\ HU\text{-}C$$

In this example, if only the temperatures supporting healthy growth for tomato are counted, eliminating too-high temperatures, then HU for an average June 1 in Bakersfield drop from 12.8 to 11.0; the reductions for July 1 and August 1 are even greater (Table 5.5). There is no change in Lewiston for any of the dates in Table 5.5 because T_{max} never exceeds TH_{high}. For example, if you are growing a variety that requires 1,214 HU-C to go from planting a seed to harvesting ripe tomatoes, you can start keeping track of the cumulative HU at your location—adding up the HU for each 24-hour period, and even calculate a rough estimate of future HU to have an idea of when your tomatoes will be ripe. Or you can use weather records to estimate days to harvest with different planting dates. Using the short-term climate prediction tool described in Section 8.7.2 you could calculate estimates for the next season based on current data that reflect the effect of ACC.

Looking at HU shows an important way that varieties can differ. For example, researchers developing Croptime at Oregon State University (OSU, 2018) compared six cucumber varieties and found HU-F to first harvest (calculated using the single sine method) to range from 525.6 to 672.8. Finally, it's good to remember that HU are useful for understanding plant development, but not necessarily yield, as in the examples from the summer of 2003 in Europe (Section 5.6).

Calculating HU isn't necessary in most cases, but it can give gardeners a rough idea of what to expect from new species or varieties they have not grown before, if the requirements and threshold temperatures are available. The online calculators listed in Section 5.9.2 can also be used for this.

In contrast to HU, some perennials require a minimum number of hours at cold temperatures for *vernalization*—literally a preparation for spring, and specifically for successful flower production. Vernalization helps synchronize plant development to seasonal environmental changes and resource availability like soil moisture and favorable growing temperatures. The need for this cold temperature period to produce viable flowers and therefore fruit is quantified as the *chilling requirement* (Baldocchi and Wong, 2008; Luedeling et al., 2009). Unlike HU, chilling requirement is not met solely through a single linear accumulation of time within a temperature range. Instead, it is made up of two dormancy stages (Section 6.1.2). First, internal, *physiological dormancy* is a species- and variety-specific accumulation of time below 7.2°C (45°F) required to complete the physiological processes that prepare the plant for producing flower and leaf buds. Once physiological dormancy is complete the plant is capable of new growth; when that occurs depends on the second stage, an external, *ecological dormancy*, which is based on exceeding a minimum air temperature, and other external factors, like moisture.

Many deciduous, temperate region perennials have a chilling requirement. For example, depending on the variety, pomegranate requires 100–200 chilling hours, fig, grape, and persimmon 100–500 hours, and almond 400–700 hours (Baldocchi and Wong, 2008). In regions with warm daytime temperatures in winter, chilling requirements are met during cold nights. However, warmer night-time temperatures due to ACC mean some species will no longer be able to produce fruit, or new varieties with shorter chilling requirements will be needed. For example, California's 2014 sweet cherry crop was 63% smaller than the 2013 harvest, which researchers attributed to both drought and warming that reduced winter chilling, and predictions are that by 2041 no locations in the primary production area (Central Valley) of that state will have adequate winter chilling for the crop due to ACC (Pathak et al., 2018). Winter varieties of cereals, such as wheat, barley, and rye typically planted in late summer or fall also require vernalization for flowering and grain development in the spring. For example, the cold deactivates a gene in winter wheat that prevents flowering, otherwise plants produce mostly vegetative growth (Yan et al., 2004).

Alternately, temperature increases will mean that some species and varieties formerly adapted to lower latitudes or altitudes will do well in your garden at a higher latitude or altitude, because average temperatures will be higher. However, for latitude changes this could be challenging if day-length requirements conflict with temperature requirements, for example, short-day species with warm temperature requirements (Section 5.8.1).

It is hard to predict the outcome of ACC on how plants live and grow in the garden. Overall, photosynthesis, respiration, evapotranspiration, and the effect of light and temperatures, together with

Table 5.5. Example of simple calculation of heat units on average summer days for tomato, 1949–1985

Average temperatures, 1949–1985	Bakersfield, CA 35°22'N, 119°1'W		Lewiston, ME 44°5'N, 70°12'W	
	°C	°F	°C	°F
June 1				
Minimum	15.9	60.6	10.8	51.4
Maximum	30.9	87.6	22.0	71.6
Heat units calculated, low threshold (10.6°C, 51°F) only	12.8	23.1	5.8	10.5
Heat units calculated, low and high (27.2°C, 81°F) thresholds	11.0	19.8	5.8	10.5
Change in HU when calculate using both low and high threshold	−1.8	−3.3	0.0	0.0
July 1				
Minimum	19.2	66.6	15.3	59.6
Maximum	35.8	96.4	26.0	78.9
Heat units calculated, low threshold (10.6°C, 51°F) only	16.9	30.5	10.1	18.2
Heat units calculated, low and high (27.2°C, 81°F) thresholds	12.7	22.8	10.1	18.2
Change in HU when calculate using both low and high threshold	−4.3	−7.7	0.0	0.0
August 1				
Minimum	21.3	70.4	15.9	60.7
Maximum	37.3	99.2	27.0	80.6
Heat units calculated, low threshold (10.6°C, 51°F) only	18.8	33.8	10.9	19.6
Heat units calculated, low and high (27.2°C, 81°F) thresholds	13.7	24.7	10.9	19.6
Change in HU when calculate using both low and high threshold	−5.0	−9.1	0.0	0.0

NCEI (2015). Low threshold (10.6°C, 51°F) from Maynard and Hochmuth (2007); high threshold (27.2°C, 81°F) from Peet et al. (1998); these will vary based on variety

plant genotypes, all interact in locally specific ways, producing different results in different regions and gardens, and these are shifting with the changing climate. It is likely that higher CO_2 concentration and temperatures and reduced water availability will have negative impacts (Table 2.4). A basic understanding of these processes helps us create strategies to respond effectively, reducing our exposure and sensitivity to these impacts (Chapter 3). This will include using crops and varieties that have some form of adaptation to the new conditions, developing simple, low-resource practices to create productive microenvironments, and as always, observing, experimenting, adjusting, and working together for best results.

5.9. Resources

All the websites listed below were verified on June 8, 2018.

5.9.1. Plants

For a western science approach to plant growth for gardeners, basic principles of botany can be found in many university textbooks. The perspective of a botanist or ecologist is often different from that of an agronomist. While the first two approach the subject with a biological and/or environmental outlook, agronomists emphasize production and economic value, and often focus on large-scale production and purchased inputs. This means different priorities and asking different kinds of questions. Frequently the information in botany or ecology books is more relevant to food gardens. This is especially true for low-resource, low-impact gardens, because agronomy books often assume that growing environments are optimal for crop production, or can be made optimal with inputs, instead of considering how to mitigate or work with negative changes and adapt to using limited resources. The good news is that this is changing, and there is a growing number of books looking at agriculture as an ecological system. Such books are useful to gardeners, and here are a few of the best:

> *Agroecology: The science of sustainable agriculture* (revised edn) (Altieri, 1995)
> *Agroecology: The ecology of sustainable food systems* (3rd edn) (Gliessman, 2015)
> *The ecology of agroecosystems* (Vandermeer, 2011)

For more information about food garden plants and pollinators, see the home landscaping book by Frey and LeBuhn (2016).

The USDA has a freely available report on agricultural pollinators that includes a table listing many crop species, whether they are attractive to honey, solitary or bumble-bees, and if there is evidence that honey bees obtain pollen or nectar rewards from them (USDA, 2015).

An amazing resource for gardeners curious about plant roots is the complete online edition of Weaver and Bruner (1927), with detailed chapters for many common garden crops, including drawings of young and mature root systems of most.
http://soilandhealth.org/wp-content/uploads/01aglibrary/010137veg.roots/010137toc.html

5.9.2. Plant temperature constraints

See the 4th US National Climate Assessment, listed in Section 2.7.2.

Information on HU and tools for calculating them using different methods, with links to HU/degree day calculators is available online.
http://www.ipm.ucdavis.edu/WEATHER/ddconcepts.html

Croptime (OSU, 2018) calculates HU/growing degree days for locations in the continental US using the simple method we describe, or a number of others including the common simple sine method. It also includes crop models for varieties of some garden crops, especially those grown in Oregon.
http://smallfarms.oregonstate.edu/croptime

The University of California Agriculture and Natural Resources has a web page linking to phe-nology models for agricultural pests, weeds, dis-eases, and also accumulated HU for development of a limited number of a crop varieties.
http://ipm.ucanr.edu/MODELS/index.html

Brown (2013) is a clear, illustrated explanation of HU and crop production by a University of Arizona Cooperative Extension biometerologist.

Knott's handbook for vegetable growers, 5th edition (Maynard and Hochmuth, 2007), is a widely used and cited reference (first edition 1980), and is avail-able free online from the University of Missouri Extension Service. Among many other things, this lists threshold (or "base") temperatures for some vegetables (Table 3.2, 106).
http://extension.missouri.edu/sare/documents/KnottsHandbook2012.pdf

This University of California Extension publication about fruit trees for home orchards includes low-chill varieties (Vossen and Silver, 2000), and is available free online.
http://homeorchard.ucdavis.edu/varieties.pdf

5.10. References

Altieri, M. A. (1995) *Agroecology: The science of sustainable agriculture*, revised edn. Westview Press and IT Publications, Boulder, CO and London, UK.

Atwell, B. J., Kriedemann, P. E. & Turnbull, C. G. N. (eds) (1999) *Plants in action: Adaptation in nature, performance in cultivation*, 1st edn. Macmillan Education Australia Pty Ltd, Melbourne, Australia.

Baldocchi, D. & Wong, S. (2008) Accumulated winter chill is decreasing in the fruit growing regions of California. *Climatic Change*, 87, 153–166, DOI: 10.1007/s10584-007-9367-8.

Battisti, D. S. & Naylor, R. L. (2009) Historical warnings of future food insecurity with unprecedented sea-sonal heat. *Science*, 323, 240–244, DOI: 10.1126/science,1164363.

Beerling, D. J. (2012) Atmospheric carbon dioxide: a driver of photosynthetic eukaryote evolution for over a billion years? *Philosophical Transactions of the Royal Society B: Biological Sciences*, 367, 477–482.

Bengough, A. G., Bransby, M. F., Hans, J., McKenna, S. J., Roberts, T. J. & Valentine, T. A. (2006) Root responses to soil physical conditions; growth dynam-ics from field to cell. *Journal of Experimental Botany*, 57, 437–447, DOI: 10.1093/jxb/erj003.

Bloom, A. J. & Lancaster, K. M. (2018) Manganese bind-ing to Rubisco could drive a photorespiratory pathway that increases the energy efficiency of photosynthesis. *Nature Plants*, 4, 414.

Blum, A. (2009) Effective use of water (EUW) and not water-use efficiency (WUE) is the target of crop yield improvement under drought stress. *Field Crops Research*, 112, 119–123.

Brenchley, R., Spannagl, M., Pfeifer, M., Barker, G. L. A., D'Amore, R., Allen, A. M., McKenzie, N., Kramer, M., Kerhornou, A., Bolser, D., et al. (2012) Analysis of the bread wheat genome using whole-genome shotgun sequencing. *Nature*, 491, 705–710, DOI: http://www.nature.com/nature/journal/v491/n7426/abs/nature11650.html, supplementary-information.

Brown, P. W. (2013) Heat units. University of Arizona, Cooperative Extension, Tucson, AZ, available at: https://extension.arizona.edu/sites/extension.arizona.edu/files/pubs/az1602.pdf (accessed Aug. 5, 2018).

Cassab, G. I., Eapen, D. & Campos, M. E. (2013) Root hydrotropism: An update. *American Journal of Botany*, 100, 14–24.

Chaves, M. M., Pereira, J. S., Maroco, J., Rodrigues, M. L., Ricardo, C. P. P., Osório, M. L., Carvalho, I., Faria, T. & Pinheiro, C. (2002) How plants cope with water stress in the field? Photosynthesis and growth. *Annals of Botany*, 89, 907–916, DOI: 10.1093/aob/mcf105.

Condon, A., Richards, R., Rebetzke, G. & Farquhar, G. (2004) Breeding for high water-use efficiency. *Journal of Experimental Botany*, 55, 2447–2460.

Craufurd, P. Q. & Wheeler, T. R. (2009) Climate change and the flowering time of annual crops. *Journal of Experimental Botany*, 60, 2529–2539, DOI: 10.1093/jxb/erp196.

Dadson, R., Hashem, F., Javaid, I., Joshi, J., Allen, A. & Devine, T. (2005) Effect of water stress on the yield of cowpea (*Vigna unguiculata* L. Walp.) genotypes in the Delmarva region of the United States. *Journal of Agronomy and Crop Science*, 191, 210–217.

Datta, S., Kim, C., Pernas, M., Pires, N., Proust, H., Tam, T., Vijayakumar, P. & Dolan, L. (2011) Root hairs: Development, growth and evolution at the plant-soil interface. *Plant and Soil*, 346, 1–14, DOI: 10.1007/s11104-011-0845-4.

de la Peña, R. C., Ebert, A. W., Gniffke, P. A., Hanson, P. & Symonds, R. C. (2011) Genetic adjustment to changing climates: Vegetables. *Crop Adaptation to Climate Change*, 396–410.

Dittmer, H. J. (1937) A quantitative study of the roots and root hairs of a winter rye plant (*Secale cereale*). *American Journal of Botany*, 417–420.

Edwards, E. J. & Ogburn, R. M. (2012) Angiosperm responses to a low-CO_2 world: CAM and C_4 photosynthesis as parallel evolutionary trajectories. *International Journal of Plant Sciences*, 173, 724–733, DOI: 10.1086/666098.

Esau, K. (1977) *Anatomy of seed plants*, 2nd edn. John Wiley and Sons, New York, NY.

Franks, P. J. & Beerling, D. J. (2009) Maximum leaf conductance driven by CO_2 effects on stomatal size and density over geologic time. *Proceedings of the National Academy of Sciences*, 106, 10343–10347.

Franks, P. J., Adams, M. A., Amthor, J. S., Barbour, M. M., Berry, J. A., Ellsworth, D. S., Farquhar, G. D., Ghannoum, O., Lloyd, J., McDowell, N., et al. (2013) Sensitivity of plants to changing atmospheric CO2 concentration: From the geological past to the next century. *New Phytologist*, 197, 1077–1094, DOI: 10.1111/nph.12104.

Frey, K. & LeBuhn, G. (2016) *The bee-friendly garden*. Ten Speed Press, Berkeley, CA.

Garibaldi, L. A., Steffan-Dewenter, I., Winfree, R., Aizen, M. A., Bommarco, R., Cunningham, S. A., Kremen, C., Carvalheiro, L. G., Harder, L. D. & Afik, O. (2013) Wild pollinators enhance fruit set of crops regardless of honey bee abundance. *Science*, 339, 1608–1611.

Glenn, E. P., Jed Brown, J. & O'Leary, J. W. (1998) Irrigating crops with seawater. *Scientific American* 279, August, 76–81.

Gliessman, S. R. (2015) *Agroecology: The ecology of sustainable food systems*, 3rd edn. CRC Press, Taylor & Francis Group, Boca Raton, FL.

Haberl, H., Erb, K. H., Krausmann, F., Gaube, V., Bondeau, A., Plutzar, C., Gingrich, S., Lucht, W. & Fischer-Kowalski, M. (2007) Quantifying and mapping the human appropriation of net primary production in earth's terrestrial ecosystems. *Proceedings of the National Academy of Sciences*, 104, 12942–12947.

Hartmann, H. T., Kester, D. E., Davies, F. T., Jr. & Geneve, R. L. (2002) *Plant propagation: Principles and practices*, 7th edn. Prentice Hall, Upper Saddle River, NJ.

Hohmann-Marriott, M. F. & Blankenship, R. E. (2011) Evolution of photosynthesis. In Merchant, S. S., Briggs, W. R. & Ort, D. (eds) *Annual review of plant biology*, 62, 515–548, DOI: 10.1146/annurev-arplant-042110-103811.

IPCC (Intergovernmental Panel on Climate Change) (2013) Climate change 2013: The physical science basis. Working group I contribution to the Fifth Assessment Report of the Intergovernmental Panel On Climate Change. IPCC, Geneva. Available at: http://www.climatechange2013.org/images/report/WG1AR5_SPM_FINAL.pdf (accessed Oct. 11, 2018).

Jarvis, A., Ramirez-Villegas, J., Nelson, V., Lamboll, R., Nathaniels, N., Radeny, M., Mungai, C., Bonilla-Findji, O., Arango, D. & Peterson, C. (2014) Farms of the future: An innovative approach for strengthening adaptive capacity. AISA Workshop on Agricultural Innovation Systems in Africa (AISA), 29–31 May 2013, Nairobi, Kenya, 119–124, available at: http://www.farmaf.org/images/documents/related_materials/AISA_workshop_proceedings_final__March_2014.pdf (accessed Oct. 11, 2018).

Khazaei, H., Street, K., Santanen, A., Bari, A. & Stoddard, F. (2013) Do faba bean (*Vicia faba* L.) accessions from environments with contrasting seasonal moisture availabilities differ in stomatal characteristics and related traits? *Genetic Resources and Crop Evolution*, 60, 2343–2357. DOI: 10.1007/s10722-013-0002-4.

Kumar, S., Dagnoko, S., Haougui, A., Ratnadass, A., Pasternak, D. & Kouame, C. (2010) Okra (*Abelmoschus* spp.) in West and Central Africa: potential and progress on its improvement. *African Journal of Agricultural Research*, 25, 3590–3598.

Lara, M. V., Disante, K. B., Podestá, F. E., Andreo, C. S. & Drincovich, M. F. (2003) Induction of a crassulacean acid like metabolism in the C_4 succulent plant, *Portulaca oleracea* L.: physiological and morphological changes are accompanied by specific modifications in phosphoenolpyruvate carboxylase. *Photosynthesis Research*, 77, 241–254, DOI: 10.1023/A:1025834120499.

Latshaw, W. & Miu, E. (1924) Elemental composition of the corn plant. *Journal of Agricultural Research*, 27.

Lopes, M. S., Araus, J. L., Van Heerden, P. D. & Foyer, C. H. (2011) Enhancing drought tolerance in C_4 crops. *Journal of Experimental Botany*, 62, 3135–3153.

Luedeling, E., Zhang, M. & Girvetz, E. H. (2009) Climatic changes lead to declining winter chill for fruit and nut trees in California during 1950–2099. *PLoS One*, 4, e6166.

Maynard, D. N. & Hochmuth, G. J. (2007) *Knott's handbook for vegetable growers*, 5th edn. Wiley & Sons, Hoboken, NJ.

McGregor, S. E. (1976) *Insect pollination of cultivated crop plants, agricultural handbook No. 496*. Agricultural Research Service, US Department of Agriculture, Washington, DC.

Melillo, J. M., Richmond , T. C. & Yohe, G. W. (eds) (2014) *Climate change impacts in the United States: The third national climate assessment*. US Global Change Research Program, Washington, DC.

Muller, M. & Schmidt, W. (2004) Environmentally induced plasticity of root hair development in Arabidopsis. *Plant Physiology*, 134, 409–419, DOI: 10.1104/pp.103.029066.

Mullins, D. (2010) Profitable vegetable production requires proper timing. *Panhandle Agriculture*, University of Florida, 2(2), 3–4.

NCEI (National Centers for Environmental Information) (2015) Climate of the US. NCEI, National Oceanic and Atmospheric Administration, available at: http://www.ncdc.noaa.gov/climate-information/climate-us (accessed June 15, 2018).

Okazaki, K., Kabeya, Y. & Miyagishima, S.-y. (2010) The evolution of the regulatory mechanism of chloroplast division. *Plant Signaling & Behavior*, 5, 164–167.

Olesen, J. E. & Bindi, M. (2002) Consequences of climate change for European agricultural productivity, land use and policy. *European Journal of Agronomy*, 16, 239–262.

Olesinski, A. A., Wolf, S., Rudich, J. & Marani, A. (1989) Effect of leaf age and shading on photosynthesis in potatoes (*Solanum tuberosum*). *Annals of Botany*, 64, 643–650.

OSU (2018) Croptime. Oregon State University, available at: http://smallfarms.oregonstate.edu/croptime (accessed June 8, 2018).

Pathak, T. B. & Stoddard, C. S. (2018) Climate change effects on the processing tomato growing season in California using growing degree day model. *Modeling Earth Systems and Environment*, 4, 765–775, DOI: 10.1007/s40808-018-0460-y.

Pathak, T. B., Maskey, M. L., Dahlberg, J. A., Kearns, F., Bali, K. M. & Zaccaria, D. (2018) Climate change trends and impacts on California agriculture: a detailed review. *Agronomy*, 8, 25.

Peet, M., Sato, S. & Gardner, R. (1998) Comparing heat stress effects on male, fertile and male, sterile tomatoes. *Plant, Cell & Environment*, 21, 225–231.

Reich, P. B., Hobbie, S. E., Lee, T. D. & Pastore, M. A. (2018) Unexpected reversal of C_3 versus C_4 grass response to elevated CO_2 during a 20-year field experiment. *Science*, 360, 317–320.

Ren, S., Weeda, S., Akande, O., Guo, Y., Rutto, L. & Mebrahtu, T. (2011) Drought tolerance and AFLP-based genetic diversity in purslane (*Portulaca oleracea* L.). *Journal of Biotech Research*, 3, 51–61.

Rennie, E. A. & Turgeon, R. (2009) A comprehensive picture of phloem loading strategies. *Proceedings of the National Academy of Sciences*, 106, 14162–14167, DOI: 10.1073/pnas.0902279106.

Sage, R. F. & Kubien, D. S. (2003) Quo vadis C_4? An ecophysiological perspective on global change and the future of C_4 plants. *Photosynthesis Research*, 77, 209–225, DOI: 10.1023/A:1025882003661.

Sage, R. F., Christin, P.-A. & Edwards, E. J. (2011) The C_4 plant lineages of planet Earth. *Journal of Experimental Botany*, 62, 3155–3169, DOI: 10.1093/jxb/err048.

Schmidt, J. J., Blankenship, E. E. & Lindquist, J. L. (2011) Corn and velvetleaf (*Abutilon theophrasti*) transpiration in response to drying soil. *Weed Science*, 59, 50–54, DOI: 10.1614/ws-d-10-00078.1.

Schwarz, D., Rouphael, Y., Colla, G. & Venema, J. H. (2010) Grafting as a tool to improve tolerance of vegetables to abiotic stresses: Thermal stress, water stress and organic pollutants. *Scientia horticulturae*, 127, 162–171.

Silvera, K., Neubig, K. M., Whitten, W. M., Williams, N. H., Winter, K. & Cushman, J. C. (2010) Evolution along the crassulacean acid metabolism continuum. *Functional Plant Biology*, 37, 995–1010, DOI: http://dx.doi.org/10.1071/FP10084.

Smith, W. K., Zhao, M. S. & Running, S. W. (2012) Global bioenergy capacity as constrained by observed biospheric productivity rates. *BioScience*, 62, 911–922, DOI: 10.1525/bio.2012.62.10.11.

Sperry, J. S. (2011) Hydraulics of vascular water transport. In Wojtaszek, P. (ed.) *Mechanical integration of plant cells and plants*, 303–327. Springer, Berlin, Heidelberg, DOI: 10.1007/978-3-642-19091-9_12.

Turgeon, R. (2010) The role of phloem loading reconsidered. *Plant Physiology*, 152, 1817–1823, DOI: 10.1104/pp.110.153023.

UCANR (2016) UCIPM models: crops. UC, available at: http://ipm.ucanr.edu/MODELS/index.html - CROPS (accessed July 17, 2016).

USDA (2015) *Attractiveness of agricultural crops to pollinating bees for the collection of nectar and/or pollen*. USDA, Washington, DC, available at: https://www.ars.usda.gov/ARSUserFiles/OPMP/Attractiveness%20of%20Agriculture%20Crops%20to%20Pollinating%20Bees%20Report-FINAL_Web%20Version_Jan%203_2018.pdf (accessed Dec. 6, 2018).

Vandermeer, J. (2011) *The ecology of agroecosystems*. Jones & Bartlett Publishers, Sudbury, MA.

Vossen, P. M. & Silver, D. (2000) *Growing temperate tree fruit and nut crops in the home garden*. University of California Cooperative Extension, Davis, CA, available at: http://homeorchard.ucdavis.edu/varieties.pdf (accessed Aug. 5, 2018).

Weaver, J. E. & Bruner, W. E. (1927) *Root development of vegetable crops*, 1st edn. McGraw-Hill, New York, NY.

Yamori, W., Hikosaka, K. & Way, D. A. (2014) Temperature response of photosynthesis in C_3, C_4, and CAM plants: Temperature acclimation and temperature adaptation. *Photosynthesis research*, 119, 101–117.

Yan, L., Loukoianov, A., Blechl, A., Tranquilli, G., Ramakrishna, W., SanMiguel, P., Bennetzen, J. L., Echenique, V. & Dubcovsky, J. (2004) The wheat VRN2 gene is a flowering repressor downregulated by vernalization. *Science*, 303, 1640–1644, DOI: 10.1126/science.1094305.

Zheng, D., Hunt Jr, E. R. & Running, S. W. (1993) A daily soil temperature model based on air temperature and precipitation for continental applications. *Climate Research*, 2, 183–191.

Zhu, J.-K. (2001) Plant salt tolerance. *Trends in Plant Science*, 6, 66–71.

6

Starting and Caring for Garden Plants

D. Soleri, D.A. Cleveland, S.E. Smith

Chapter 6 in a nutshell.

- For best results, plant healthy seeds and use simple seed treatments to eliminate disease if needed.
- Presoak and prime seeds before planting to prepare them for vigorous, synchronized germination and emergence, and to minimize time caring for them.
- Create microenvironments to encourage seed germination and emergence, and conserve water, nutrients, and time.
- Help young seedlings become established by avoiding temperature and water stress using light mulch and shade, or frost protection.
- Seedlings need special attention because their smaller root and shoot systems, and their tender tissues, make them more vulnerable to drought, pests, and diseases.
- Mixed garden polycultures increase plant density and diversity.
- Starting plants in nursery beds or containers can conserve resources and make it easier to give them special care.
- Transplanting requires preparing the site and the transplant to minimize stress and help the plant become established and vigorous.
- Vegetative propagation is the easiest or only method for propagating some garden crops; the progeny plant usually has the same genotype as the parent.

Buying seedlings and other plants for your garden is immediately gratifying—there they are, ready to transplant, with tantalizing hints of their potential. Put them in the soil and you have an instant garden! Still, there are lots of good reasons to start your own garden plants, whether from seeds or vegetative propagules: it can be a fun learning experience, costs much less than buying transplants, and uses less resources like containers and transportation. In addition, by starting your own garden plants it's possible to grow species, varieties, and local populations that are not available for purchase as plants, including those that are locally adapted, or have special meaning to you (Fig. 6.1).

Most annual garden crops, and some perennials, can be grown from seed. Vegetative propagation is the other way to start garden crops, especially longer-lived species like many fruit trees, but may also be useful for some annual plants. In this chapter we focus on starting garden plants from seeds both directly in the garden and in nursery beds and containers for transplanting into the garden. We also give a brief introduction to starting plants using vegetative propagation.

6.1. Seed Quality, Germination and Dormancy

Like many aspects of food gardening, starting garden plants from seeds immerses you in evolution. Garden crop genera have a long evolutionary history before and after domestication, which occurred thousands of years ago for most species (Box 10.1). (In *FGCW* we use "seed" in the popular sense, to include the grains of grass family cereals like maize, barley, and wheat that are actually dried fruits [*caryopses*], each containing one seed.) Seeds are small packages of genetic information that interact with the growing environment during plant growth to produce the plant from which your harvest comes. This genetic information is the *genotype* of each seed, and is complex and often unique to each plant. In Chapter 10 we look at genotypes, genetic diversity, saving seeds, and what it means for the gardener, her garden, and for all of us. In this chapter we discuss seed quality, planting, and early seedling care, especially under hot, dry conditions. We also suggest ways to diagnose a few common seed planting problems.

Figure 6.1. Certain varieties of food garden plants can have special meaning to us

A seed is *mature* when it has reached its maximum size, and is no longer dependent on the mother plant to survive—it contains a living plant embryo, and food reserves to fuel the seedling's growth until its root and shoot systems take over (Fig. 6.2). In most garden species, once the seeds have matured and dried, they are ready to be planted and grow into a new plant. We describe some exceptions below.

When water penetrates the dry, protective outer seed coat, the seed tissues swell, breaking the coat open. Breaking the seed coat is a physical process, not a biological one; it even happens in dead seed, so it is not a sign that the seed is alive. After this, when

a root and shoot start to grow, the seed has *germinated*. Seeds with *epigeal* germination—for example, seeds of many cucurbits, legumes, and onion—push their *cotyledon(s)* (seed leaves) above the soil surface where they start photosynthesizing until true leaves grow. In contrast, plants such as chard, maize, olive, pea, and the stone fruits have *hypogeal* germination, with the cotyledon(s) remaining underground and the shoot growing above that through the soil surface, with photosynthesis starting in the shoot and true leaves (Fig. 6.3). *Emergence* occurs when the cotyledons or shoot from a germinated seed first break the surface of the soil.

Planting seeds is easy, but it's only worth it if you have good quality seeds, which means they matured fully on the maternal plant and the embryos are alive. Good quality also means that the seeds have not been damaged, including by pests or disease. If there is evidence of pests but damage has not been substantial, freezing the dry seeds for a week or more kills most insect larvae and adults (Section 10.6.2). Saving seeds from healthy plants free of systemic pathogens (Section 9.8.1) is the best way to avoid diseased seed.

If you have reason to suspect seeds may be carrying a disease, a simple seed treatment before planting may help. There are seed-borne fungal, viral, and bacterial diseases that will infect the seedlings and plants grown from them (Section 9.2.4). Pathogenic bacteria and viruses inside the seed are there because they originally infested either ovules or pollen. Other pathogens like viruses, fungal spores, or cells of some systemic bacterial diseases are present on the seed coat. Seed-borne bacterial diseases can be controlled by treating seeds with hot water before planting (Miller and Lewis Ivey, 2005), an old method that is again being used by organic farmers and seed suppliers, and can be useful for gardeners too. Hot water seed treatment should not be used on large seeds like bean, cucurbits, pea, or maize kernels, and it must be done with care, for example, placing small seeds in cheesecloth bags. One study found that water treatment at 50°C (122°F) for 30 minutes eliminated or substantially reduced fungal (*Phoma* and *Alternaria* spp.) and bacterial (*Xanthomonas* spp.) pathogens on seeds of cabbage, carrot, and parsley (Nega et al., 2003). This treatment is only effective if the water is hot enough and treatment long enough to kill the pathogens, but not so hot that it kills the embryo. Treatment in water hotter than 52.7°C (127°F) damages seeds. To ensure an even distribution of heat it is best to do this in a

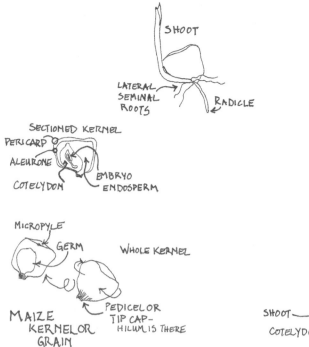
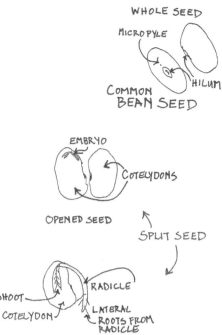

Figure 6.2. Parts of common bean seed and maize kernel

double boiler, or improvised version of one, and practice getting the temperature right before actually treating the seeds. More detailed information about hot water treatment is available (Ashworth, 1991; Miller and Lewis Ivey, 2005). One way to control viruses is soaking seeds in, or spraying young plants with, a 20% solution of powdered nonfat milk in water (Li et al., 2015), which can also be used to clean viruses off tools (Lewandowski et al., 2010).

6.1.1. Testing seed germination and vigor

If seeds are old, damaged, or of a variety you haven't grown before, you may want to do a germination test, especially if large quantities of the same seed are going to be planted. Testing seed germination can be a part of presoaking and priming to prepare seeds for planting (Section 6.2.1). The *germination percentage* is the proportion of a particular *seed lot*, a quantity of seed of one crop variety for planting, that can be expected to germinate, and is formally determined by testing. For example, in 1940 the US government passed a Federal Seed Act (FSA) to create quality standards, including for germination percentage, and greater accountability for interstate, commercial seed sales (Box 6.1). The

FSA, which you can read online (Section 6.9), includes a list of minimum germination percentage standards for different garden crops, for example: artichoke 60%, collard greens 80%, cowpea 75%, lettuce 80%, watermelon 70% (FSA §201.31). If the percentage for a seed lot is lower than the FSA minimum, it must be reported on the package; if at or above the threshold it does not need to be reported, but often is. All commercial seed companies in the US are required to conduct the "warm germination test" and maintain records of their results (ISU, 2016). The test involves four samples of 100 seeds each, all from the same original seed lot, kept moist at 25°C (77°F) for 7 to 10 days. The counts of germinated seeds for each of the four samples are summed and divided by four to give a mean germination percentage for that seed lot (ISU, 2016).

Seeds are said to lack vigor if they are very slow to germinate in the germination test, assuming the correct conditions are provided and dormancy isn't the issue (Section 6.1.2). Inadequate plant nutrition during seed production, poor postharvest handling, or aging can reduce seed vigor. When planted, seeds that lack vigor are more likely to die, or produce weak or deformed seedlings. One measure of seed vigor is *germination rate*, or the time it takes a seed

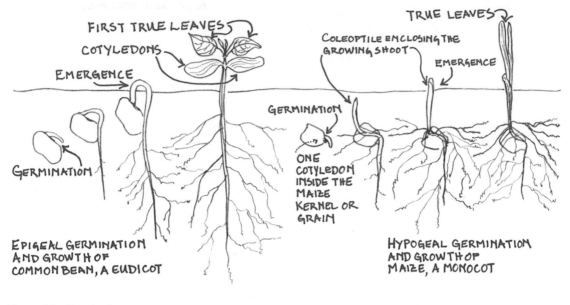

Figure 6.3. Germination types

lot to reach a particular germination percentage. For example, if it takes three days for 50% of the 400 basil seeds in a test to germinate, the 50% germination rate of that seed lot is three days, indicating it is more vigorous than another basil seed lot with a 50% germination rate of six days.

Seed size is often associated with seedling vigor. Generally, larger seeds will germinate faster and have more vigorous seedling growth. This is why small-scale farmers and traditional gardeners around the world save the largest seeds for planting (Box 10.4). Larger seed size is also favored in industrial agriculture systems where, for example, maize kernels are sorted by size and shape with the largest (>8.3 mm [0.3 in] diameter), flat grains being preferred (Popp and Brumm, 2003).

Healthy seeds of most annual food garden species have a germination percentage of 70% or greater, so if you don't know the percentage for the crop you are testing, you can assume that at least two-thirds of seeds in a test should germinate. However, germination in the garden is usually slightly lower than in germination tests, and this difference increases as the germination percentage drops. For example, if germination percentage in the test is 95%, in the garden it may be about 90%; if the test shows 70%, in the garden it may be only 50%.

In many cases a germination test is not necessary; you can just plant an area with the seed, water and watch it for a week or so and if few or no seedlings appear, replant with other seeds (Section 6.4). However, if you save seeds or exchange them with other gardeners, use seeds that were harvested a few years earlier, are planting a large area, or if the crop is really important (e.g., a major part of your diet, or you are committed to selling or trading it), then a germination test can be a good idea.

Most gardeners do not have 400 extra seeds to use in the standard germination test described above, but if a group of gardeners obtain seeds from the same seed lot they could each contribute some seeds for a cooperative test. Using a smaller sample size increases the possibility of test results being affected by error, so the results can't be compared with the standard 400-seed test. However, a small-scale test can help to see if there is a problem with the seed lot. In other words, is the germination percentage closer to 80% or 10%? If the germination percentage is closer to 10% there is a problem and it would be better to find other seeds. Directions for doing a germination test are widely available, including from some of the sources in Section 6.9, and *Food from dryland gardens* (Cleveland and Soleri, 1991).

6.1.2. Seed dormancy

Some garden seeds can be planted immediately after harvest if growing conditions are favorable. For example, the fruits of citrus, some cucurbits, and other crops produce chemicals that keep most of the seeds inside them from germinating, but when removed from the fruit, rinsed, and planted, they will germinate. Seeds of many other crops have a condition of *dormancy* that prevents germination until certain requirements are met (Finch-Savage and Leubner-Metzger, 2006).

The two different basic types of seed dormancy are external and internal, and a crop species can have seed with one or both types, or none (Fig. 6.4). *External*, including *physical dormancy* occurs when the embryo is capable of germinating, but does not because a thick or tough seed coat prevents moisture from reaching the living seed. Seed coats of some plants like carob are so tough that *scarification* (scratching or using hot water to help water penetrate the tough seed coat) is recommended. Scratching can be done easily by lightly rubbing the seed on a rough surface like a rock or sandpaper. To avoid injuring the embryo, don't scratch the *hilum*, where the seed was attached to the ovary, and the *micropyle*, the very small opening where the pollen tube entered to fertilize the ovule in most garden crops (Fig. 6.2). Once moisture can reach the embryo of a seed that only has physical dormancy it can germinate.

The seed coats of many small, hard seeds like okra or radish can be lightly scarified by putting them in a container with small, sharp-edged gravel and shaking vigorously. In this case, damaging the hilum is not usually a problem because the scratching is superficial and no pressure is applied. The hard, thick covering (*endocarp*) protecting seeds of some crops like almond, olive, pistachio, the stone fruits, and walnut can be carefully cracked just before soaking or direct planting to speed absorption of water and germination.

Some varieties have seeds with *internal dormancy* that require particular conditions to prepare the embryo for germination, which is an adaptation to environments where their growing requirements can be met. Some seeds are dormant for a period after fruit harvest, known as *after-ripening*, allowing the embryo to complete development, after which they germinate. Other internally dormant seeds require certain environmental stimuli before they are ready to germinate. The seed of some temperate fruit trees, such as peach, and biennial vegetables need exposure to a period of low temperatures when they are moist (*cold stratification*) to end embryo dormancy.

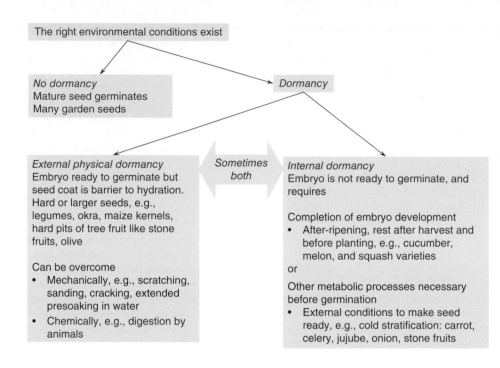

Figure 6.4. Basic types of seed dormancy

Zuni gardeners accomplish this by burying their peach pits next to their houses in summer and fall where the pits overwinter, and the seedlings emerge in spring and are transplanted out (Cleveland et al., 1994). Cold stratification should not be confused with the chilling requirement or *vernalization* of many deciduous fruit trees and some cereal grains, which is a period of low temperatures necessary for normal flower development (Section 5.8.2).

Most plants grown in food gardens have *orthodox* seeds, ones that can be dried to between 5–10% of weight as moisture, and stored for later use. But a few garden species have *recalcitrant* seeds, which must maintain ≥25% moisture to remain viable. This means that the seed should be planted soon after eating the fruit. Garden species with recalcitrant seeds tend to have originally evolved in moist subtropical or tropical regions, for example south and southeast Asia, (citrus, loquat, mango), or Mesoamerica and South America (avocado, mamey, passion fruit, sapotes).

6.2. Planting Seeds Under Hot, Dry Conditions

There are many resources with general information for planting seeds. Here we describe concepts with related examples of simple ways to focus essential resources on the needs of the seeds under the hot, dry conditions that are increasing in many areas with anthropogenic climate change (ACC).

6.2.1. Preparing the seeds

For centuries gardeners and farmers around the world have soaked their seeds in water before planting because it is an easy way to overcome physical dormancy and get a good start on germination. Soaking large or hard-coated seeds like bean, pea, squash, and maize kernels in water before planting to soften their seed coats is most common, but many smaller seeds like cilantro, okra, onion, and radish can be presoaked. The larger or older the seed, the longer it will take to be rehydrated. Where appropriate, seeds can be scarified before soaking (Section 6.1.2). Presoaking minimizes the time the seeds are in the ground before germination, and means the garden does not need to be watered as much to keep the seeds evenly moist before germination. Presoaking also saves time and resources because hollow, nonviable seeds will float to the surface when they are placed in water, and can be composted, instead of planting and tending them only to be disappointed.

A related process gardeners may want to experiment with is *priming*, which involves keeping orthodox seeds moist until the first signs of germination (Fig. 6.3), and then drying them, temporarily suspending germination (Hartmann et al., 2002, 213; Paparella et al., 2015). Similar to soaking, an advantage of priming is that all seeds are at the same stage of readiness for rapid, uniform emergence once planted. Priming also stimulates pre-germination metabolic processes that make primed seeds more vigorous and hardy when planted (Welbaum et al., 1998; Butler et al., 2009), and one study showed that priming is a way to start cool season lettuce seeds early in the fall, when they would otherwise remain dormant due to high soil temperatures (Cantliffe et al., 1984).

6.2.2. Preparing the planting site

The soil right where the seeds will be planted can be prepared to encourage early emergence and vigorous growth, for example by mixing in extra compost to add nutrients, making soil structure more open so that air and water can move through easily, encouraging root growth (Section 7.2.2), and discouraging

fungal diseases like damping off (Section 9.8.2). Compost also helps break up the soil surface, making it easier for new shoots to push through, and for water to infiltrate. In heavier, clayey soils, mixing in sand when planting root or bulb crop seeds such as beet, carrot, or onion not only helps the seedling get a good start, but also helps the bulb or root develop. But sometimes starting seeds in nursery beds or containers for transplanting later is better than planting directly into the garden (Section 6.5).

6.2.3. Planning planting density and diversity

Whether planting seeds directly, or transplanting seedlings or young trees, it's good to imagine how the garden will look in the future. Either from your own experience, talking with other gardeners, or reading, you can get a feeling for how plants will grow and change over time, to help decide what seeds to sow, where, and how densely. Conventional agricultural fields and gardens in the US are monocultures with only one crop, and plants surrounded by bare soil. In contrast, many traditional gardens are densely

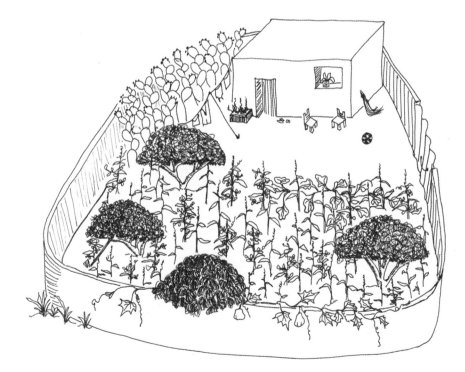

Figure 6.5. A mixed garden—a type of polyculture—in Durango, Mexico

---

Starting and Caring for Garden Plants

Box 6.2. Potential benefits of polyculture food gardens

A polyculture can provide *complementarity*, reducing competition for resources in space, for example for sunlight shared by maize and squash plants, or soil nutrients shared by arugola and the persimmon tree it is growing under, or by reducing competition in time by mixing plants of different ages and lifespans (Fig. 6.6).

One of the gardeners we visited in Durango, Mexico, pointed out that soon she would not be able to plant annual vegetables under her peach trees because it was getting too shady. For her, the peaches and the shade were very valuable, and so the trees were left to grow and she was finding other areas such as the edges of her garden where annuals could be grown, or planting more winter annuals like cilantro that could grow when the peach trees lost their leaves. She also encouraged chayote vines to climb up fences and over the rooftop where they got plenty of light and helped keep her house cooler.

Another way polyculture can be beneficial is *facilitation*, when one crop improves the environment of another crop. For example, plants with deep roots can access soil water well below the surface, including excess water applied to shallow-rooted annual plants. The deeper rooted plants can draw that water into and up their xylem tissues along a water potential gradient. This gradient is at least partially driven by the pull of transpiration from the plant's aboveground growth (Sections 5.3, 5.5). When the pull of transpiration is reduced, for example at night, or when the plants are shaded or pruned, then the water potential in the roots closer to the surface increases, and can become greater than the water potential of the dry soil around those roots in the upper soil layers. When this occurs, water from those shallower roots moves into the soil where it can be taken up by the roots of adjacent receiving plants, a process called *hydraulic lift*. Field studies of this process in intercropped pigeon pea (lifting) and maize (receiving) (Sekiya and Yano, 2004), and six different perennial forages (lifting) and a leaf *Brassica rapa* (receiving) (Sekiya et al., 2011), found hydraulic lift significantly improved the growth and yield of the maize and brassica crops, especially when the lifting and receiving roots were in close contact. For gardeners it may be worth experimenting with intercrops including lifting legumes like pigeon pea or kidney bean that have roots greater than 0.9 m (3.0 ft) deep (Weaver and Bruner, 1927). This may be especially useful in the weeks following the rainy season, when shallow layers of soil are drying out, but there is still ample water in deeper layers.

The yield of a polyculture and of monocultures of the same crops can be compared quantitatively using the *land equivalent ratio* (LER) (Vandermeer, 2011, 74–81; Cleveland, 2014, 174–177). Polyculture in gardens can make good use of space, resulting in high LERs, that is, greater total yields in poly- vs. monoculture. But polyculture can also have other objectives that are just as important, such as enabling gardeners to grow the diversity of species that are important to them.

Because the performance of a polyculture depends on so many environmental and crop-specific variables, it's best to observe what happens when you try a new mixture or intercrop, and then make adjustments as needed.

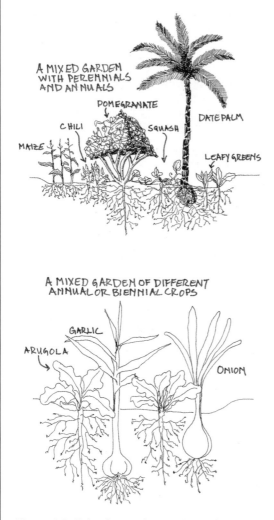

Figure 6.6. Polycultures of perennials and annuals, or annuals and biennials in mixed gardens

planted mixtures of many perennial and annual species and varieties, a type of *polyculture*. Figure 6.5 shows a garden we visited in Durango, Mexico, with fruit trees surrounded by maize, bean, and squash plants, and other crops, and containers planted with chiles, herbs, and flowers. Mixed gardens like this, or other forms of polyculture such as *intercropping*— planting alternating rows of different crops—can provide a number of benefits (Box 6.2).

Lots of garden books, online sources, and seed packets have planting density or spacing charts that you can consult. These provide a general idea, but are rarely based on mixed gardens, especially in hot, dry environments; they recommend densities between plants of the same species and varieties, not between different ones in polycultures; and they don't take into account variables such as the soil quality, water availability, temperatures, unique local varieties, and cultural and aesthetic values important to gardeners. To get started it can be useful to roughly "map out" leaf cover, growth habit (erect or vining; mature plant height) and lifespan as you plan, and then experiment with intercropping. If plants grow to be too close and compete for resources they can be thinned and transplanted, and leaf crop seedlings eaten.

6.2.4. Planting the seeds

Making depressions in the soil such as furrows or basins (we'll call all of them basins from now on) where seeds are planted concentrates and saves water and time (Fig. 6.7). You can better control seed planting density by sprinkling seeds into planting basins, not pouring them. Mixing presoaked or small seeds such as amaranth, chia, or basil with dry sand or compost prevents them from sticking together, making it easier to sow them evenly. If the seeds are primed and have started to germinate it's better not to mix them with anything, and handle them very gently.

Once sprinkled into planting basins, seeds need to be covered with enough soil to prevent them from drying out, but not so much that they have difficulty emerging. In heavy, clayey soils the covering should be thin (approximately two to three times the diameter of the seed) because these soils dry out more slowly, and emergence is harder because the soil is so dense. It is a good idea to make emergence easier in clayey soils by adding compost and even a little sand where seeds are planted to create a more open soil structure (Box 6.3). In sandy soils the soil cover should be thicker (about four to six times the diameter of the seed) because these soils dry out more

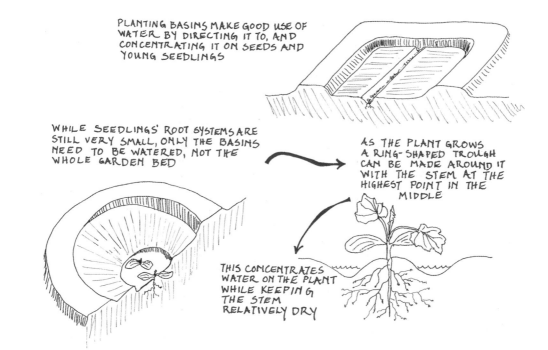

PLANTING BASINS MAKE GOOD USE OF WATER BY DIRECTING IT TO, AND CONCENTRATING IT ON SEEDS AND YOUNG SEEDLINGS

WHILE SEEDLINGS' ROOT SYSTEMS ARE STILL VERY SMALL, ONLY THE BASINS NEED TO BE WATERED, NOT THE WHOLE GARDEN BED

AS THE PLANT GROWS A RING-SHAPED TROUGH CAN BE MADE AROUND IT WITH THE STEM AT THE HIGHEST POINT IN THE MIDDLE

THIS CONCENTRATES WATER ON THE PLANT WHILE KEEPING THE STEM RELATIVELY DRY

Figure 6.7. Planting basins

In *FGCW* we outline concepts relevant to many aspects of food gardens, but we avoid recipes because the diversity of local conditions means that a good understanding of those conditions, along with basic concepts, produces the best strategies. This can be seen in how farmers and gardeners adapt their seed planting so that seeds have the moisture to germinate and grow, are protected, and can still emerge. For example, in their dry-farmed maize fields the Hopi Native Americans living in the high desert of southwestern North America plant maize kernels about 33 cm (13 in) deep and bean seeds at nearly that depth. This unusually deep planting method was developed by Hopi farmers in response to their hot, dry climate, and is made possible by varieties adapted to local soil and water conditions. In maize fields the top 30 cm (12 in) or so of soil is fine sand with a sandy clay loam layer beneath. In the spring, water from melted winter snow flows into the fields from surrounding slopes and is held in the sandy clay loam. The seeds are planted in this moist layer below the sand. The Hopi have selected maize

varieties over centuries for physical adaptations to this planting method such as a dominant, deep radicle—unusual in maize—and a shoot with an underground portion (mesocotyl) about twice as long as in other maize varieties (Collins, 1914; Soleri and Cleveland, 1993). These characteristics, and the sandy texture of the upper layer of soil, make it possible for Hopi maize to emerge from such a deep planting hole and grow successfully. Maize farmers in some parts of Mexico with similar soil profiles also plant this way, using a planting stick or *coa*.

Another Hopi example shows how experienced gardeners recognize the value of improving the environment immediately around the newly planted seeds. When ready to plant their terraced garden beds located just below the mesa tops, Hopi gardeners carry up buckets of fine yellow sand from the valley floor below (Soleri, 1989). They explain that they cover their garden seeds with this sand because the heavier, clayey soil in the terraces forms a crust when it dries out after irrigation. This crust can be so hard that it prevents seedlings from emerging.

quickly, and are easy for seedlings to grow through. For sandy soils, adding compost is a good way to improve the water-holding capacity. Pressing down the soil covering seeds helps retain moisture and protects them from being removed by wind, insects, and birds. A light mulch on the soil surface will reduce water loss to evaporation; placing thorny branches on top of newly planted seeds or young plants can give added protection and deter birds and small animals.

6.3. Caring for Seeds, Seedlings, and Transplants

When caring for newly planted seeds, young seedlings, and transplants, the goal is to avoid water and temperature stress (Section 5.6). Concentrating water as described earlier, and protecting plants from drying winds, strong direct sunlight, or cold night-time temperatures in spring, all help. We discuss diagnosing problems when they do occur in Section 6.4.

6.3.1. Watering

Seeds and young seedlings don't have extensive root systems for gathering water and so the soil around

them needs to be kept moist, but not saturated. If not watered enough, seedlings become water stressed and have slower shoot development and a larger root:shoot ratio than unstressed seedlings. This is because under water stress, development of roots to provide water takes priority over shoot development for photosynthesis, which increases water loss by transpiration. Overwatering, on the other hand, can lead to water-saturated soil, forcing the oxygen needed by roots out of the soil (Section 8.2.1), which can kill the seed or seedling, and the high moisture encourages damping-off fungi and other pathogens (Section 9.8.2). Once seedlings emerge and establish, avoiding frequent shallow watering in favor of less frequent, deeper watering encourages healthy root growth (Section 5.3).

Water use can be minimized by planting in basins that concentrate water on the roots (Fig. 6.7), and watering in the late afternoon when there will be less loss to evaporation. Once seedlings have emerged, the soil around them can be shaped into rings or furrows, leaving the seedlings on hills or ridges to concentrate water while protecting tender stems from disease.

All of these methods for watering seeds and seedlings assume gardeners are using some form of flood irrigation. But many gardeners use, and like, trickle or drip irrigation systems after seedlings are established. Drip irrigation (Section 8.8.4) can be convenient because those systems may be placed on timers, and have very high agronomic water use efficiency (WUE), although they may not be efficient in mixed gardens because of irregular spacing of plants with different water requirements.

6.3.2. Mulching and shading

Mulching and shading conserve moisture and reduce soil temperature around seeds and seedlings when the weather is hot and dry. Definitions and general information on mulching, shading, and windbreaks are given in Section 8.3.2. Things to keep in mind for mulching and shading seeds and seedlings include the following.

- Surface mulch, such as a sprinkling of fine compost after watering, reduces evaporation and soil temperature. Thicker mulches are more effective, but have a greater chance of harboring pests and encouraging disease, and can even smother and kill emerging seedlings. It's a good idea to periodically clear away mulch and check for insects or disease.
- Shading should allow enough sunlight to reach the young plants. When seedlings don't get enough sunlight they become pale and *etiolated*– having long stems and internodal spaces.

6.4. Diagnosing Seed Planting Problems

There are a number of reasons that seedlings fail to emerge or thrive. Figure 6.8 and Table 6.1 can help with basic troubleshooting, and germination tests can help identify seed problems (Section 6.1.1). See also Section 9.8.2.

6.5. Starting Garden Plants in Nursery Beds and Containers

Young seedlings and newly planted cuttings (Section 6.7) need protected growing conditions, extra care, and attention before their root systems are established, and *nursery beds*—garden beds devoted entirely to starting plants—are one way to do this. Water, improved soil, and gardeners' time can be used more efficiently, and favorable microclimates can be created for early- or late-season germination. Warm-weather plants can be started before the end of the cool season in sunny, protected areas receiving heat radiated from nearby objects like south-facing walls, or in a cold frame (Section 4.1.3). If you have a plot in a community or school garden that isn't visited daily, nursery beds where you live allow more frequent monitoring and care. Nursery beds can also be moved seasonally or yearly to take advantage of changing conditions or avoid problems with localized diseases

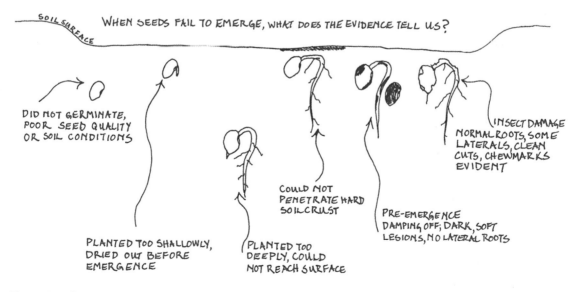

Figure 6.8. Seedling emergence problems

Table 6.1. Diagnosing and treating seed planting and seedling problems

Observations	Possible causes	Suggested actions
Failure to emerge		
Seeds did not germinate, or low germination percentage	Temperature too hot or cold; too little water; old, nonviable seed	Investigate environment; do germination test, look for other seed source
Seed missing	Insects, small mammals, birds	Plant again, try giving more protection, i.e., thorn branches, bird scaring, or plant in containers
Seed germinated but dried up before it emerged	Planted too shallow; did not get enough water, and/or too hot	Plant again, deeper; water more often, lightly mulch with fine compost
Seed germinated and grew but did not reach soil surface to emerge	Planted too deeply	Plant again, shallower
Seed germinated and grew curled up under soil surface	Soil crust too hard	Add organic matter and sand to improve soil structure around seeds
Pre-emergent seedling dead; normal roots with chew marks or clean cuts	Beetle larvae, other insects	Plant again using protective collars, or try planting in containers
Pre-emergent seedling dead; dark, soft lesions and dark root	Pre-emergence damping off	Add organic matter and sand to improve drainage around seeds; plant in containers with soil from other part of garden
Seedling problems		
Parts of seedling missing	Insects or wild or domestic birds	Look on/around plants, under mulch, especially at night; remove mulch; fence, bird scaring; plant in containers
Spindly, pale seedling, etiolated	Not enough sunlight; too much competition among seedlings	Increase exposure to sunlight, remove surface mulch; brush if plants not too weak (Section 6.6.2); thin, or replant less densely
Seedling deformed, abnormal	Damaged or infected seeds	Find new seed source

or pests like nematodes that can build up in continuously cultivated soil (Section 9.2.2).

A family we have worked with for many years in rural Oaxaca, Mexico, makes extensive use of nursery beds. In addition to being maize farmers, they are market gardeners growing long rows of flowers alternating with rows of herbs, leaf vegetables, and chiles. Rather than having to water and tend rows of vulnerable, hard-to-see seedlings, they start all plants in small (~0.9 x 1.2 m, ~3 x 4 ft) sunken nursery beds, often with overhead shading made by palm fronds and *carrizo* stalks (Fig. 6.9). These beds protect the tender, densely sown seedlings and make watering them fast and easy.

When selecting and planting a nursery bed:

1. Select a site that is convenient for daily care, easy to protect from wind, sun and pests, and has enough room for plants to grow to a good size for transplanting.
2. Prepare the soil and bed as you would any garden bed (Chapter 7, Section 8.9), except that the soil preparation doesn't need to be as deep because plants will not grow to maturity here. For most annuals 15–20 cm (6–8 in) is deep enough.
3. Make planting basins (Section 6.2.4) if the bed is larger than about 1 m² (10 ft²), otherwise the whole bed can be watered by flooding.
4. Sow seeds more thickly than in a permanent bed because the plants will be transplanted before maturity. However, leave enough room so that plants can be separated for transplanting without major root damage. Vegetative propagules like sweet potato stem cuttings can also be rooted in nursery beds.
5. Cover the sown seeds with soil and protect them from drying out with a light mulch, shades, or windbreaks. Protect from birds with thorn branches and bird-scaring devices.
6. When watering, make sure the soil about 2.5 cm (1 in) or so below the seeds or other propagules is wetted.

Planting in containers offers many of the benefits of nursery beds with the added advantage of not needing a location in the ground, and being easy to move around. Herbs, condiments, and other garden produce

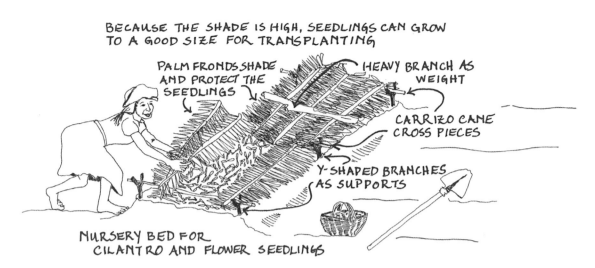

BECAUSE THE SHADE IS HIGH, SEEDLINGS CAN GROW
TO A GOOD SIZE FOR TRANSPLANTING

PALM FRONDS SHADE
AND PROTECT THE
SEEDLINGS

HEAVY BRANCH AS
WEIGHT

CARRIZO CANE
CROSS PIECES

Y-SHAPED BRANCHES
AS SUPPORTS

NURSERY BED FOR
CILANTRO AND FLOWER SEEDLINGS

Figure 6.9. A nursery bed in Oaxaca

can be grown permanently in containers on the ground, on rooftops, in windows, or on balconies. Some of the most ingenious examples of container planting we have seen were in a spontaneous settlement on the outskirts of Mexico City where people used everything, including plastic bottles and other discarded containers, for their garden plants (Fig. 6.10).

Many people use galvanized steel containers, but over time, especially if your soil or water is acidic (pH <7, Section 7.3.1), the galvanized coating will be corroded and zinc and cadmium from that coating can move into the soil and be taken up by plants, and can harm plants and people

(Section 4.2.1). Similarly, never plant in containers that have been used for paint, fuel, pesticides, or other toxic substances.

Baskets made of plant fibers can be used as containers for plants that will be transplanted, and don't need to be removed from tree seedlings if the fibers will decompose fairly quickly, which avoids disrupting the tree's root system. But the basket fibers should be cut through and pulled open in several places before planting the tree, especially if the basket is tightly woven or made of a tough, woody material. Gardeners or garden groups can also try making soil blocks to start transplants (Box 6.4).

Figure 6.10. Container planting in a Mexico City *colonia*

Starting and Caring for Garden Plants

Box 6.4. Soil blocks

Soil blocks avoid the waste and the possibility of spreading disease that go with containers, and eliminate root disturbance. An historical example comes from a system of canal-fed gardens in central Mexico where soil blocks were used to start grain amaranth during the dry season (Early, 1977; Wilken, 1987, 261). A shallow nursery bed measuring about 2 x 15 m (6.5 x 50 ft) and 4–5 cm (1.5–2 in) deep was filled with rich, silty-clay soil scooped up from the canal bottom. As the soil dried it was cut into 3 x 3 cm (1.25 x 1.25 in) blocks with a special slicing rake. Then, with a stick or their fingers, gardeners made a small hole less than 1 cm (0.5 in) deep in the middle of each block and dropped in amaranth seeds. Manure was sprinkled over the blocks, then swept off so only that falling into the planting hole remained, covering the seeds. Twenty to 30 days later the seedlings in their soil blocks were transplanted into garden beds.

Soil blocks have been used for some time by farmers and gardeners in Europe, and more recently in the US. There are many detailed resources about them and affordable tools for making them (Section 6.9), especially from the experienced farmer and educator Eliot Coleman (1995, Ch. 14).

Ecological and prosocial thinking is important when choosing the soil mixture to use for soil blocks, and any container planting. For example, many potting mixtures are made up of two-thirds or more *peat*, partially decomposed plants such as mosses, harvested from bogs or natural wetlands where it forms. These wetlands cover large areas, and are important C sinks, sequestering an estimated 510 petagrams of C (Strack, 2008, 14). Peat harvesting not only depletes this resource, but exposes that peat and the remaining peat at the site to oxygen, faster decomposition, and the resulting C release. Instead, we suggest using other fibrous plant material as part of the soil mixture for soil blocks, including finely shredded fiber from local palms, fiber from composted nopal cactus, ground pine bark (pine fines), and other organic material with good water-holding capacity. Coconut fiber (*coir*) is being used as a peat substitute in some soil mixes, but for gardeners outside the tropics it makes sense to explore local alternatives and avoid the environmental impact of transport.

Published recipes for soil block mixes containing peat always have added lime to raise the mixture's pH, counteracting the acidity of the peat. Other than pine-based material, most organic substitutes will not be so acidic and thus will not need the added lime.

Other common components of soil block and potting soil mixtures are vermiculite (a clay) and perlite (a volcanic glass), that both require mining and energy-intensive heating to transform them into the porous products used in soil mixes. Again, experimenting with less destructive alternatives including the substitutes listed above makes sense when thinking ecologically and prosocially.

Soil blocks need to be watered gently using a spray or watering can with very small holes so the blocks are not eroded. Keeping blocks close to each other helps them retain moisture.

A challenge with container planting is avoiding the water saturation zone that develops at the bottom of the container, in order to minimize damping-off diseases and suffocating the roots from lack of oxygen (Section 9.8.2). Having holes in the bottom of containers is a first step. However, water with lower potential (lower energy) will not move from the smaller pores of the potting soil into the much larger pores that are the holes in the pot bottom until the potting soil at the interface of the soil and the holes is saturated, raising the water potential until it is nearly positive (Sections 7.2, 8.2.2). No matter how open the soil structure, this change from smaller to larger pore size will occur at the container bottom, so after an adequate watering there will always be a zone of water saturation there. The best solution is to use containers tall enough to provide room for healthy root growth above the saturation zone, and not to overwater. The common practice of putting pot shards or gravel in the bottom of pots makes the situation worse, raising the zone of saturation to the soil above the large pores of the shard or gravel layer.

Containers also need to be big enough to avoid root binding. The roots of root-bound plants circle around each other in the limited space, and are not able to take up enough nutrients and water for vigorous growth, sometimes choking each other. Vertical ridges on the inner surfaces of containers help prevent roots from growing in this spiral pattern.

Because containers hold a relatively small amount of soil and are surrounded by air, they can dry out rapidly and experience more extreme temperatures compared with planting in the ground. The larger the container the more soil mass there is for holding water, and the less frequent the waterings need to be, and the temperature changes will be less extreme. Using containers with thick walls, or nesting containers in mulch or within other containers, are ways to provide better insulation and reduce temperature fluctuations. High albedo, light-colored containers reflect more sunlight, minimizing soil heat gain when it's hot, while low albedo, dark-colored containers have greater heat gain, helpful when it's cold (Section 4.1.3).

Whether in nursery beds or containers of some kind, starting garden plants ahead of time can be very helpful for making gardening easier for busy gardeners. Sometimes, in community and school gardens, or among a network of local gardeners, there is a person or group that takes responsibility for starting plants that everyone can use, or buy, something that Alemany Farm, a community garden in San Francisco, California does.

6.6. Transplanting

6.6.1. The site

A simple map of the garden layout and orientation (Chapter 4) is a good tool for thinking about transplant sites and anticipating the transplant's effect on space, shade, and sunlight, especially for perennials.

Seedlings of annuals such as chile or tomato are often transplanted into a spot in an existing garden bed, so no special soil preparation is needed, and the main thing to think about is nearby plants (Section 6.2.3). For transplants not placed in existing garden beds, or for trees and other large plants, the soil at the planting site will need preparation. If the soil quality is very poor, the hole should be made extra large so there is plenty of room for improved soil for the plant to become well established. As with planting containers, planting holes need good drainage so roots won't be suffocated in water-saturated soil. An impermeable soil layer like dense clay or caliche can prevent water from draining below the root zone, and salts will also accumulate there (Sections 7.1.1, 8.8.5). If such a layer exists you can try to dig through it to better-draining soil, or the hole must be dug deep enough to keep most of the roots out of the saturation zone.

Once the hole is dug you can refill it with a mix of compost and good soil just as you would when preparing a garden bed (Section 8.9). Use no more than 25–50% compost or other organic matter for refilling the hole, and the rest soil, or soil and sand. Make sure there is not an abrupt change in soil texture or organic matter content between the soil

in the transplant's root ball and the soil in the planting hole, because this change will slow the movement of water, discouraging the transplant's roots from growing out of the root ball, causing root-binding. A smooth sided planting hole will also impede water flow and root growth, so the walls of planting holes should be loose and rough textured, and mixed with the amended soil.

After filling, water the hole thoroughly, and allow a day or two for the soil to settle in large holes, in order to avoid problems later with sinking or shifting soil that can disturb roots or leave a transplanted tree growing in a hole.

6.6.2. The transplant

No matter how skillfully you do it, transplanting is stressful for plants. Minimizing that stress means preparing the young plant, and then transplanting it at the best time in its development, and the best time for the location where it will be planted.

Wait to transplant after the seedling has started to photosynthesize and feed itself. When seeds germinate and start growing the embryo is using the energy stored in the seed, either in the endosperm (e.g., maize, melon, onion, lettuce, and tomato and other solanaceous crops) or in the fleshy cotyledons (e.g., pea, bean, squash) (Chapter 5). This is the *heterotrophic* growth period, because it relies on organic compounds in those seed parts that were produced by the maternal plant, not by the embryo. After emergence and by the time the first true leaves (not the cotyledons) have developed, those seed reserves have been used up and the seedling must produce its own energy by photosynthesis, making it *autotrophic*, or self-feeding.

Seedlings should not be planted before they are autotrophic, and preferably only after they have about four or more true leaves, or more for perennials. At the other extreme, if you wait too long, a transplant may have extensive roots that can be more easily damaged, or if in a container, can become root-bound and stunted, resulting in plants that start reproductive growth early, and produce less harvest. It is also better that transplants do not have flowers or fruits because these use the energy needed to become established after transplanting, so remove these if present.

Published optimal transplant ages (Lorenz and Maynard, 1988; Vavrina, 1998) can be helpful for a general idea, but there will be variation depending on the crop variety and the environments where the seedling is started and where it is transplanted, so your own observations will be the best guide.

You can start transplants of warm-season crops early in a protected place and then plant them in the garden as soon as they are mature enough and the weather warm enough. For example, air temperatures below 12.5°C (54.5°F) generally limit germination and growth of warm-season garden vegetables (Snyder and de Melo-Abreu, 2005) (Table 5.3), so starting these indoors or in a cold frame produces seedlings that are ready when danger of frost has passed, and can extend the productive season. This also helps more vulnerable young plants to escape stressful temperatures or pests and diseases that can limit growth or kill plants, and gives them a head start on many seasonal weeds that emerge in the garden after the air temperatures warm.

When they have lost their leaves and are dormant, young deciduous perennials can be *bare root* transplanted, that is, without their root ball covered with soil. Evergreen perennials are best transplanted with a soil root ball at the beginning of a period of growth, for example, at the start of the rainy season.

If plants have been grown in a very protected place, preparing them for transplanting ideally begins about three to seven days ahead of time for annuals, and even longer for perennials. Plants are gradually exposed to light and temperatures similar to where they will be planted, and to reduced watering and slight water stress. This process of exposing plants to controlled stress is *hardening*, which decreases rates of transpiration and photosynthesis, slowing growth. The plant's tissues become denser and the concentration of solutes such as sugars increases as the available water decreases. Some plants may wilt slightly when the hardening process begins but usually recuperate at night. As long as the central stalk and growing tip remain green and firm they are not being harmed. After several days the plant will stop wilting, unless the hardening is too severe. Immediately before and after transplanting the plant should be well watered.

Reducing water to induce slight water stress before transplanting—the "drought" part of hardening—has been practiced by gardeners for a long time, but we are only now understanding that drought-hardened plants actually have a molecular "memory" of the experience of water stress so when they experience it again the genes that help them cope are activated faster than in plants that have not been hardened (Ding et al., 2012). Drought-hardened plants are more productive when subsequently exposed to drought.

When hardening includes cooler temperatures similar to the permanent planting site, it helps prepare warm-weather seedlings for planting out in early spring conditions. The increased solute concentrations in plant tissues that occurs with hardening can help prevent frost or freeze damage (Section 4.1.3). Whether done to prepare transplants for drier, sunnier, hotter or, colder conditions, hardening should never be too abrupt or severe as this could kill the plants, or stimulate them to bolt—mature rapidly and produce seed because of what to them feels like terminal stress.

Sometimes container-grown seedlings can become taller and weaker than is best for transplanting, a result of crowding or lack of light. In addition to being sure they have enough light, a way to avoid this is using *brushing* (Latimer, 1998) to prepare them for transplanting, which can be done at the same time as hardening. Once four or more true leaves have developed, the tops of the plants are literally brushed over with your hands, a wooden dowel or something similar 10–20 times consecutively once a day for 12–15 days before transplanting. Brushed seedlings are shorter and have thicker stems than unbrushed ones because the brushing slows growth in plants' growing tips. These shorter, stronger seedlings tend to be hardier, and have less problem with falling over, survive transplanting and resist some pests better (Garner and Björkman, 1999; Latimer, 1998). Brushing has been used successfully on seedlings of cabbage, cucumber, eggplant, squash, tomato, and watermelon.

Before planting, check the transplant to make sure it will not introduce pests or diseases into the garden. It should have a vigorous root system with no swellings or growths due to pests or pathogens (Sections 9.2, 9.8.2), or soft brown lesions, and no harmful insects or their eggs on the stems or leaves.

Except for bare-root transplanting of deciduous perennials, and untangling a root-bound transplant, a goal in transplanting is to disturb the roots as little as possible. Plants with damaged roots have a harder time obtaining water and nutrients, and are more susceptible to disease and environmental stress. So handle seedlings as little as possible, and when you need to, do it by gently holding the soil-covered root ball. Avoid holding the plant by the stem because young stems bruise easily, creating entry points for disease. Keep the roots pointing downward; if pushed upward during transplanting the plant will take longer to recover and grow more slowly because the roots will be closer to the soil surface where temperatures are more extreme and there is less moisture.

No matter how carefully the transplanting is done, some root hairs will be damaged. Though microscopic, these root hairs make up the majority of the plant's total root surface area and are vital for the intake of water and nutrients (Section 5.3). If protected and well watered, the transplant will soon grow new root hairs.

Some garden plants will produce an earlier and larger harvest if transplanted deeper than the seedling was originally growing. Studies of cabbage, sweet pepper (Vavrina et al., 1994), and tomato (Vavrina et al., 1996) all found benefits from deeper planting, up to the cotyledon leaves or to the first true leaves. In one season, peppers transplanted to their first leaves had 58% greater yield than ones planted at root-ball depth (Vavrina et al., 1994). One reason for the advantage is the production of adventitious roots (Sections 5.3, 6.7) from nodes on the buried stem, especially in tomato. Other than these garden crops, we are not aware of others that can be transplanted more deeply than originally growing.

When transplanting a grafted plant (see Table 6.3), keep the graft union well above the soil surface or it may stay wet when watered, encouraging microorganisms that can destroy the graft, killing the plant.

Under hot, dry conditions perennial and mature annual transplants can be pruned to increase the root:shoot ratio and decrease the total leaf surface area and therefore the amount of water lost through transpiration. When transplanting perennials, about one-third of their aboveground growth can be pruned back to compensate for root damage. By reducing total plant size with pruning, the plant can focus its resources on survival. Drought-deciduous plants self-prune under extreme drought for the same reason.

6.6.3. Water

A good watering is important to end the stress of hardening off, and bring fine soil particles into close contact with the transplant's roots. To hold water, make a basin with a diameter about two times larger than the root ball with the plant in the center on a slightly raised area to keep the stem or trunk from sitting in water (Fig. 6.11).

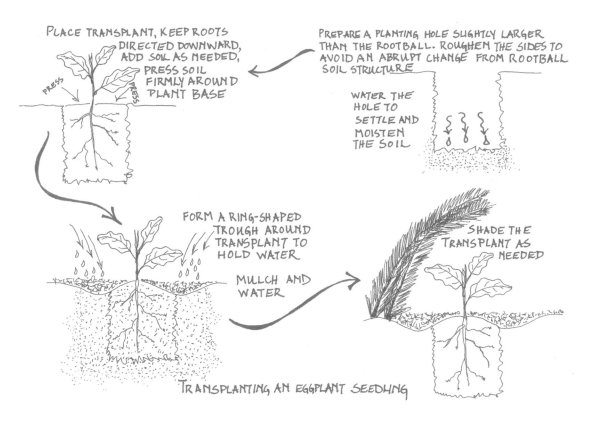

Figure 6.11. Steps in transplanting an annual garden plant

6.7. Vegetative Propagation

Propagating plants from parts other than seeds is *vegetative propagation*. Vegetative propagules, including tubers, cuttings, and layered branches are usually *clones*, or genetic replicas of the parent plant that the propagule is taken from, products of *somatic* (nonreproductive) cell division. Grafting is a form of vegetative propagation that combines in one plant the root qualities of one genotype and the shoot qualities of another genotype.

Vegetative propagation of fruit trees and vines was known and practiced at least 2,500 years ago in Western Asia and the Mediterranean Basin, and is mentioned by a number of writers including Aristotle (Ramón-Laca and Mabberley, 2004, 434). Because of a long-standing bias in western archaeology and history that favors western European people and their crops (Allaby et al., 2008), and because plant remains are not well preserved under humid conditions, less is known about vegetative propagation in other areas, especially in the tropics. However, as is true of crop domestication in general (Section 10.1, Fig 10.2), important perennials that are often vegetatively propagated originated all over the world, for example cashew and cacao (South America), coffee (northeastern Africa), avocado and nopal cactus (Mesoamerica), persimmon and kiwi (Asia), banana (Oceania), and many more (Miller and Gross, 2011).

Knowing the characteristics of the new plant is one of the main reasons for using vegetative propagation, especially for trees that do not produce fruit for several years. For example, the dioecious date palm can be propagated both from seed or vegetatively, but if using seed, the new plant will have a unique combination of the characteristics of its two parents, and it will be 8 to 10 years after planting before the "sex" (whether seed- or pollen-producing) and other qualities will be known. However, if propagated vegetatively, in this case as an offshoot, the new plant will be a clone of the parent plant, with the same sex, growing characteristics, and fruit qualities, although there are some minor exceptions.

Vegetatively propagated plants generally have the advantage of predictability but, due to genotype by environment interaction (Box 10.3), sometimes the same genotype may produce different phenotypes in different environments. In grafted plants there can be interactions between the two parts that affect the plant phenotype, which in some cases are desirable such as when using dwarfing rootstock with grafted

Table 6.2. Examples of hot water treatments for two common soil nematodes in garden crop propagules

Crop	Water temperature		Minutes in water
	°F	°C	
Root knot nematode, *Meloidogyne* spp.			
Cherry and peach, rootstock	122.0–124.0	50.0–51.1	5
Grape, rootstock	125.6	52.0	5
Potato, tuber	114.8–117.5	46.0–47.5	120
Strawberry, root	127.0	52.8	5
Sweet potato, root	122.0	50.0	3–5
Sweet potato, root	116.2	46.8	65
Yam, tuber	123.8	51.0	30
Stem nematode, *Ditylenchus dipsaci*			
Garlic, cloves	120.0	48.9	20
Onion and shallots, setts	110.3	43.5	120

Bridge (1975, Table 1)

fruit trees. Finally, mutations during somatic cell division are rare, but can result in genetic and phenotypic changes compared to the parent plant, and so occasionally a vegetative propagule can have a new genotype (Section 10.3.1).

In addition to predictability, there are other reasons for vegetative propagation. Some garden plants such as banana, garlic, Jerusalem artichoke, nopal cactus, potato, sweet potato, and sugarcane are also typically propagated vegetatively because it is fast and easy, or because they produce few viable seeds.

As the climate changes, grafted perennials can be used to tailor varieties to changing conditions. For example, putting a preferred variety *scion* (shoot part) on *stock* (root part) better adapted to the new climate or pests in its customary growing area, or onto stock adapted to the soils of a new area where that species must now be grown due to ACC. Grafting in Solanaceae (eggplant, pepper, tomato) and Cucurbitaceae (cucumber, melon, squash) garden annuals can also be a response to current and anticipated changes. For example, a study found watermelon grafted onto a squash rootstock produced 60% greater yield under drought compared to ungrafted watermelon, and grafting one eggplant variety onto another heat-adapted variety increased yields by 10% under high temperatures (Albacete et al., 2015). This is pretty interesting, and may be fun to try, but changing to a new variety is much easier.

Hygiene is important for preventing the spread of disease in vegetative propagation. For best success with all of the propagation methods outlined below, use a well-sharpened cutting tool (knife, pruning clippers) to minimize tissue damage around the cut, and disinfect the blade between use on different plants. For viral diseases you can soak the cutting blade in a 20% powdered nonfat dry milk and water solution for at least one minute; your hands can also be cleaned with this (Section 6.1). Soaking or wiping the blade with a 10% bleach solution reduces the chances of spreading all types of plant disease.

In some cases, a hot water treatment similar to that described for seeds (Section 6.1) can be used. For example, if propagules are, or might be, infected by some common nematodes (Section 9.2.2), and no other materials are available, propagules can be immersed in water hot enough to kill nematodes, but not hot enough to kill the plant (Table 6.2). See the description and resources (Sections 6.1, 6.9) for hot water treatment of seeds for methods to use to ensure stable water temperature, and for a cool water plunge after the treatment period is over.

In most vegetative propagation methods, separating the propagule from its parent plant cuts and wounds both. This is why newly started propagules, and sometimes their parent plants, need protection from sunlight, wind, and heat until they are established.

In Table 6.3 we outline some common types of vegetative propagation using shoot growth (tuberous stems, cuttings, bulbs, layering, offshoots), roots (tuberous roots, root suckers), and by grafting.

Vegetative propagation with shoots is based on the growth of adventitious roots (Section 5.3). Jerusalem artichoke, potato, and yam are garden plants that can be propagated from their edible *tuberous stems*, enlarged stems that grow underground. Whole potatoes and yams can be planted,

Table 6.3. Common types of vegetative propagation in food garden plants

Propagule	Plant part	Process	Food garden examples
Establish new plant by growth from nodes on parent stem			
Tuberous stem	Enlarged underground stem	Use whole or cut into setts each including several leaf nodes, or eyes. Let cut surface suberize, then plant	Jerusalem artichoke, potato
	Rhizomes: enlarged underground stems that grow horizontally	Same as above	Bamboo, ginger, tumeric; some perennial herbs, including *hoja santa*, mints, oregano
Stem cutting (*slip*)	Piece of young stem with leaf nodes. In case of nopal, leaf "pad" including base	Plant cutting directly in well-draining soil with multiple nodes underground, or started in water and planted when roots visible. Suberize nopal pad before planting with base and part of pad underground	Nopal cactus, passion fruit, sweet potato; perennial herbs including marjoram, mints, oregano, rosemary, sage
Shoot for layering	Length of a stem with leaf nodes still attached to parent plant	Before major period of growth, i.e., before deciduous perennials leaf out. A section of stem with leaf nodes is bent and buried in soil or wrapped in light soil and cloth. Once roots are growing, gradually cut rooted stem off of parent and plant in soil	Carob, fig, grape, many perennial herbs
Bulb	Enlarged underground/semi-underground leaves, roots grow from short stem or crown beneath leaves	Carefully separate individual cloves, parts, including base where roots will grow. Plant root-end down	Bunching onion (*Allium fistulosum*), garlic, lemongrass

Continued

Table 6.3. Continued.

Propagule	Plant part	Process	Food garden examples
Offshoot (*offset*)	Lateral shoot growing from base of stem in monocots	Separate shoot with part of parent plant and with roots, plant this	Banana, date palm, pineapple, sugarcane
Establish new plant by growth from root of parent			
Tuberous root	Enlarged fleshy root	Cut piece of root including several nodes. Allow to suberize, plant with same orientation as original root	Horseradish, sweet potato; (tuberous roots grown from seed and NOT propagated this way: beet, carrot, radish, turnip)
Sucker	Shoot growing from underground root	Dig up and cut from main root, keeping roots of sucker intact, plant. Preferably use suckers farthest from main plant, minimizing impact on parent when cut	Jujube, olive, stone fruits
Grafting: combine shoot and root of two different plants through shared vascular system			
Stock and scion	Stock: plant with preferred root characteristics Scion: bud, juvenile shoot from plant with preferred fruit or other harvest characteristics, and compatible with stock	Align cambial tissue in stock and scion, secure alignment, give time for growth of callus tissue bridging vascular system between two parts. Many different forms	Only eudicots, no monocots; many fruit trees: almond, avocado, grape, olive, persimmon, pistachio, stone fruit, walnut; also Solanaceae and Cucurbitaceae annual crops

but they are usually grown from *setts*, or pieces of tubers, which is much more economical. You can make setts by cutting up the tuber so that each piece has at least one or more *eyes* or nodes, where the roots and shoots will grow. The setts must be large enough to provide the energy for shoot and root establishment, weighing at least about 30 g (1 oz). After cutting a tuber into setts, leave them in a dry place out of the sunlight for a day or two to suberize before planting. *Suberization* is the formation of a thin layer of dry cells over the cut surface that helps prevent disease that can develop when a fleshy, wet cutting is planted.

6.8. Pruning

Pruning is the selective removal of plant parts to promote different patterns of growth and development. Pruning is also done as part of transplanting or grafting to reduce water loss through transpiration. What you hope to harvest from the plant and the way the plant grows determine which parts are pruned. For example, flowering is undesirable in annual leaf crops such as basil or leaf amaranth because it diverts energy and resources away from leaf production. In these crops the growing tips are harvested before they flower, or flower heads are pruned off to encourage more lateral shoots that will produce more leaves. In fruit-bearing plants such as tomato and eggplant, however, flower production is essential for producing the fruit, so some vegetative growth may be pruned to encourage flowering. Pruning is also a way to shape perennial plants to allow better ventilation and light penetration into fruit-bearing trees, to better fit the place where they are growing (vines creating arbors, espaliered fruit trees), or to keep plants from growing where you don't want them, as in the case of root pruning aggressively growing perennials at their drip line. Table 6.4 lists some reasons to prune food garden plants.

Table 6.4. Some reasons to prune food garden plants

Reason	Example
Delay reproductive (flower and seed) growth to encourage leaf quality and production	Remove flowers, leaf amaranth, basil, cilantro, mint, and many other herbs
Reduce vegetative growth to produce an earlier, more concentrated fruit harvest	Remove lateral shoots, tomato (Davis and Estes, 1993; Navarrete and Jeannequin, 2000)
Improve the quality and size of fruits	Remove some fruits when immature, or prune some reproductive buds in dormant season, stone fruit. Pruning two-thirds of the flower buds on each pad of nopal cactus has no effect on total weight of yield, but increases fruit size significantly (Zegbe and Mena-Covarrubias, 2009)
Even out alternate bearing when an extremely heavy fruit crop one year will be followed by a very small one the next, and in some cases no crop at all; thinning in heavy crop year stimulates reproductive bud growth the following year, evening out fruit production year to year. In garden fruit trees bearing fruit on previous year's growth, the fruit present in one year send biochemical messages to young shoots affecting whether their buds the following year will be vegetative or reproductive	Thin fruit in years of heavy production. Thin early in fruit development for greatest impact (Dag et al., 2010; Martínez-Fuentes et al., 2010; Park, 2011). Citrus, loquat, olive, persimmon, stone fruit
Create a strong structure and reduce the risk of wind damage, allow sunlight into center of tree	Selectively thin branches of perennials; fig, persimmon, pistachio, olive, stone fruit
Improve air circulation to reduce conditions that encourage fungal disease	Thin dense foliage; basil, chile, loquat, olive, tomato
Shape plants for specific purposes, e.g., to provide space for other crops	Chayote, grape, kiwi, pomegranate, stone fruit
Reduce stress, and direct the plant's energy into becoming established and vigorous	Remove flowers or fruit on very young plants, most flowering plants
Decrease leaf surface area to increase root:shoot ratio and reduce water loss, concentrating resources during times of stress such as grafting or transplanting	Many perennials and some annuals
Harvesting of nonedible plant parts	Date fronds for shading or building, banana leaves for shading and wrapping foods
Control tree height to make it easier to harvest fruits	Prune central stem to remove apical dominance and encourage lateral growth, jujube, olive, stone fruit
Root pruning to stimulate deeper, downward growth and discourage root growth into the area of other plants, or under buildings	Dig trench where the roots are to be cut, at least 0.9 m (3 ft) away from the tree's trunk, or 50 cm (20 in) beyond the drip line, if this is farther. If root invasion is a recurring problem it may help to leave the trench open without refilling it, or refill with rocks or sand so the roots are discouraged and any that do grow are easy to prune. Berries, garden trees with invasive root growth like jujube, or whose roots produce allelopathic substances, like walnut. Non-garden perennials whose roots may intrude into the garden like bamboo, or produce allelopathic compounds like eucalyptus (Section 9.7.2)
Encourage fresh, full growth and greater leaf production.	Prune back to the ground or to a few major branches, perennial herbs that have become woody, *hoja santa*, lemon verbena, marjoram, mint, oregano

6.9. Resources

All the websites listed below were verified on June 11, 2018.

A number of the resources listed in Chapter 10 have information about seed physiology, health, and planting.

Botany and biology textbooks are good resources for explaining the process of sexual reproduction in plants, for example Baskin and Baskin (2001), Hartmann et al. (2002).

More information about hot water treatment of seeds can be found on websites of a number of state agricultural extension services, including the following.
http://vegetablemdonline.ppath.cornell.edu/NewsArticles/HotWaterSeedTreatment.html
http://www.extension.org/pages/18952/organic-seed-treatments-and-coatings#.VbJwKUXb0sd
https://ag.umass.edu/news/hot-water-treatment-of-seeds

For a description of a home seed germination test see Chapter 6 of *Food from dryland gardens* (Cleveland and Soleri, 1991). There are also lots of resources online (search for "ragdoll germination test") such as the following from University of California Davis.
https://epakag.ucdavis.edu/vocational_training/factsheets/fs-seed-germination-test-ragdoll.pdf

Colorado State University Extension has many accessible, informative "garden notes."
http://cmg.colostate.edu/GardenNotesUpdate.shtml

Similarly, the University of California Cooperative Extension Service has a number of web pages and publications for gardeners, including the California Backyard Orchard site with information about perennial fruit tree propagation, pruning and general management.
http://homeorchard.ucanr.edu/

Among the many resources about soil blocks are these from regional seed companies:
http://www.johnnyseeds.com/growers-library/tools-supplies/soil-block-makers-eliot-coleman.html;
http://cdn.territorialseed.com/downloads/soilblocker.pdf

California Rare Fruit Growers (CRFG) is a nonprofit organization of rare fruit enthusiasts, many of whom are very knowledgeable about species and varieties suitable for the western US, and areas of similar growing conditions worldwide. The website has useful information about plants and methods, including some on vegetative propagation. There are CRFG chapters throughout California, and in Texas and Arizona.
http://www.crfg.org

The publicly available website for Cornell University Professor Ken Mudge's grafting class has information about the history, physiology and practice of grafting.
http://www.hort.cornell.edu/grafting/index.html

6.10. References

Albacete, A., Martínez-Andújar, C., Martínez-Pérez, A., Thompson, A. J., Dodd, I. C. & Pérez-Alfocea, F. (2015) Unravelling rootstock × scion interactions to improve food security. *Journal of Experimental Botany*, 66, 2211–2226.

Allaby, R. G., Fuller, D. Q. & Brown, T. A. (2008) The genetic expectations of a protracted model for the origins of domesticated crops. *Proceedings of the National Academy of Sciences of the United States of America*, 105, 13982–13986, DOI: 10.1073/pnas.0803780105.

Ashworth, S. (1991) *Seed to seed*. Seed Saver Publications, Decorah, IA.

Baskin, C. C. & Baskin, J. M. (2001) *Seeds: Ecology, biogeography, and evolution of dormancy and germination*. Academic Press, San Diego, CA.

Bridge, J. (1975) Hot water treatment to control plant parasitic nematodes of tropical crops. *Mededelingen – Faculteit van de Landbouwwetenschappen, Rijksuniversiteit, Ghent*, 40, 249–259.

Butler, L. H., Hay, F. R., Ellis, R. H., Smith, R. D. & Murray, T. B. (2009) Priming and re-drying improve the survival of mature seeds of *Digitalis purpurea* during storage. *Annals of Botany*, 103, 1261–1270, DOI: 10.1093/aob/mcp059.

Cantliffe, D. J., Fischer, J. M. & Nell, T. A. (1984) Mechanism of seed priming in circumventing thermo-dormancy in lettuce. *Plant Physiology*, 75, 290–294.

Cleveland, D. A. (2014) *Balancing on a planet: The future of food and agriculture*. University of California Press, Berkeley CA.

Cleveland, D. A. & Soleri, D. (1991) *Food from dryland gardens: An ecological, nutritional, and social approach to small-scale household food production*. Center for People, Food and Environment (with UNICEF), Tucson, AZ. https://tinyurl.com/FFDG-1991

Cleveland, D. A., Eriacho, D., Soleri, D., Laate, L. & Keys, R. (1994) Zuni peach orchards, part III. *Zuni Farming*, 3, 25–28.

Coleman, E. (1995) *The new organic grower*. Chelsea Green, White River Junction, VT.

Collins, G. N. (1914) A drought-resisting adaptation in seedlings of Hopi maize. *Journal of Agricultural Research*, 1, 293–392.

Dag, A., Bustan, A., Avni, A., Tzipori, I., Lavee, S. & Riov, J. (2010) Timing of fruit removal affects concurrent vegetative growth and subsequent return bloom and yield in olive (*Olea europaea* L.). *Scientia Horticulturae*, 123, 469–472.

Davis, J. M. & Estes, E. A. (1993) Spacing and pruning affect growth, yield, and economic returns of staked fresh-market tomatoes. *Journal of the American Society for Horticultural Science*, 118, 719–725.

Ding, Y., Fromm, M. & Avramova, Z. (2012) Multiple exposures to drought "train" transcriptional responses in *Arabidopsis*. *Nature Communications*, 3, available at: http://dx.doi.org/10.1038/ncomms1732.

Early, D. (1977) Cultivation and uses of amaranth in contemporary Mexico. 1st Amaranth Seminar, Emmaus, PA, Rodale.

Finch-Savage, W. E. & Leubner-Metzger, G. (2006) Seed dormancy and the control of germination. *New Phytologist*, 171, 501–523.

Garner, L. C. & Björkman, T. (1999) Mechanical conditioning of tomato seedlings improves transplant quality without deleterious effects on field performance. *HortScience*, 34, 848–851.

Hartmann, H. T., Kester, D. E., Davies, F. T., Jr. & Geneve, R. L. (2002) *Plant propagation: Principles and practices*, 7th edn. Prentice Hall, Upper Saddle River, NJ.

ISU (2016) Iowa State University Seed Laboratory. ISU, Ames, IA, available at: http://www.seedlab.iastate.edu/testing-methods (accessed Dec. 17, 2015).

Latimer, J. G. (1998) Mechanical conditioning to control height. *HortTechnology*, 8, 529–534.

Lewandowski, D. J., Hayes, A. J. & Adkins, S. (2010) Surprising results from a search for effective disinfectants for tobacco mosaic virus-contaminated tools. *Plant Disease*, 94, 542–550.

Li, R., Baysal-Gurel, F., Abdo, Z., Miller, S. A. & Ling, K.-S. (2015) Evaluation of disinfectants to prevent mechanical transmission of viruses and a viroid in greenhouse tomato production. *Virology Journal*, 12, 5.

Lorenz, O. & Maynard, D. (1988) *Knott's handbook for vegetable growers*, 3rd edn. John Wiley & Sons, New York, NY.

Martínez-Fuentes, A., Mesejo, C., Reig, C. & Agustí, M. (2010) Timing of the inhibitory effect of fruit on return bloom of 'Valencia' sweet orange (*Citrus sinensis* (L.) Osbeck). *Journal of the Science of Food and Agriculture*, 90, 1936–1943, DOI: 10.1002/jsfa.4038.

Miller, A. J. & Gross, B. L. (2011) From forest to field: Perennial fruit crop domestication. *American Journal of Botany*, 98, 1389–1414, DOI: 10.3732/ajb.1000522.

Miller, S. A. & Lewis Ivey, M. L. (2005) *Hot water treatment of vegetable seeds to eradicate bacterial plant pathogens in organic production systems*. Ohio State University, Columbus, OH, available at: http://www.oardc.ohio-state.edu/sallymiller/Extension Outreach/informationtransfer/Factsheets/Vegetable/organicseedtrt.pdf (accessed Oct. 12, 2018).

Navarrete, M. & Jeannequin, B. (2000) Effect of frequency of axillary bud pruning on vegetative growth and fruit yield in greenhouse tomato crops. *Scientia Horticulturae*, 86, 197–210.

Nega, E., Ulrich, R., Werner, S. & Jahn, M. (2003) Hot water treatment of vegetable seed: An alternative seed treatment method to control seed borne pathogens in organic farming. *Journal of Plant Diseases and Protection*, 110, 220–234.

Paparella, S., Araújo, S. S., Rossi, G., Wijayasinghe, M., Carbonera, D. & Balestrazzi, A. (2015) Seed priming: State of the art and new perspectives. *Plant Cell Reports*, 34, 1281–1293, DOI: 10.1007/s00299-015-1784-y.

Park, S. J. (2011) Dry weight and carbohydrate distribution in different tree parts as affected by various fruit-loads of young persimmon and their effect on new growth in the next season. *Scientia Horticulturae*, 130, 732–736.

Popp, J. C. & Brumm, T. J. (2003) The effect of corn seed sizing methods on seed quality. Paper 46. *Agricultural and Biosystems Engineering Conference Proceedings and Presentations*, available at: http://lib.dr.iastate.edu/abe_eng_conf/46 (accessed Oct. 11, 2018).

Ramón-Laca, L. & Mabberley, D. J. (2004) The ecological status of the carob-tree (*Ceratonia siliqua*, Leguminosae) in the Mediterranean. *Botanical Journal of the Linnean Society*, 144, 431436, DOI: 10.1111/j.1095-8339.2003.00254.x.

Sekiya, N. & Yano, K. (2004) Do pigeon pea and sesbania supply groundwater to intercropped maize through hydraulic lift? Hydrogen stable isotope investigation of xylem waters. *Field Crops Research*, 86, 167–173, DOI: http://dx.doi.org/10.1016/j.fcr.2003.08.007.

Sekiya, N., Araki, H. & Yano, K. (2011) Applying hydraulic lift in an agroecosystem: forage plants with shoots removed supply water to neighboring vegetable crops. *Plant and Soil*, 341, 39–50.

SELC (2014) Setting the record straight on the legality of seed libraries. Available at: https://www.shareable.net/blog/setting-the-record-straight-on-the-legality-of-seed-libraries (accessed Dec. 6, 2018).

Snyder, R. L. & de Melo-Abreu, J. P. (2005) Frost protection: Fundamentals, practice, and economics, Vol. 1. In Natural Resources Management and Environment Department (ed.) *Environment and natural resources series 10*. FAO, Rome.

Soleri, D. (1989) Hopi gardens *Arid Lands Newsletter*, 29, 11–14.

Soleri, D. & Cleveland, D. A. (1993) Hopi crop diversity and change. *Journal of Ethnobiology,* 13, 203–231.

Strack, M. (ed.) (2008) *Peatlands and climate change*. International Peat Society, Saarijärvi, Finland.

Vandermeer, J. (2011) *The ecology of agroecosystems*. Jones & Bartlett Publishers, Sudbury, MA.

Vavrina, C. S. (1998) Transplant age in vegetable crops. *HortTechnology*, 8, 550–555.

Vavrina, C. S., Shuler, K. D. & Gilreath, P. R. (1994) Evaluating the impact of transplanting depth on bell pepper growth and yield. *HortScience*, 29, 1133–1135.

Vavrina, C. S., Olson, S. M., Gilreath, P. R. & Lamberts, M. L. (1996) Transplant depth influences tomato yield and maturity. *HortScience*, 31, 190–192.

Weaver, J. E. & Bruner, W. E. (1927) *Root development of vegetable crops*, 1st edn. McGraw-Hill, New York, NY.

Welbaum, G. E., Shen, Z., Oluoch, M. O. & Jett, L. W. (1998) The evolution and effects of priming vegetable seeds. *Seed Technology*, 209–235.

Wilken, G. C. (1987) *Good farmers: Traditional agricultural resource management in Mexico and Central America*. University of California Press, Berkeley, CA.

Zegbe, J. A. & Mena-Covarrubias, J. (2009) Flower bud thinning in 'Rojo Liso' cactus pear. *Journal of Horticultural Science & Biotechnology*, 84, 595–598.

PART III GARDEN MANAGEMENT

7 Soil, Nutrients, and Organic Matter

D.A. CLEVELAND, D. SOLERI, S.E. SMITH

Chapter 7 in a nutshell.

- The soil is where plant roots obtain water, nutrients, and oxygen, and interact with a large number of living organisms.
- To provide the biggest harvest and other benefits for the time and resources invested (benefit:cost) in a manner consistent with your values and the key ideas described in the Introduction, manage garden soils by:
  - choosing plants that are best adapted to the conditions in your garden, including the easiest- and lowest-impact management practices;
  - feeding the microorganisms in the garden soil with organic matter, which they decompose and recycle, improving soil structure and making nutrients available to plants;
  - managing soil pores so that plants can obtain adequate water, air, and nutrients.
- The characteristics of the soil in your garden, and everywhere, are the result of parent material (original rocks), climate, organic matter and soil organisms, topography, and time.
- Soil can be divided into six main layers or horizons with specific characteristics that affect crop production.
- Soil texture is determined by the relative proportions of sand, silt, and clay, and is a key soil property; in combination with organic matter it determines soil structure, including the size of soil aggregates.
- Soil structure determines the amount of pore space in the soil for air needed by plant roots, and the water and nutrients that roots take up and transport to the rest of the plant.
- The soil contains the mineral nutrients plants need to grow, which are taken up by the roots as electrically charged ions.
- These nutrients cycle between plants and other organisms and the soil, and need to be replaced as food is harvested and plants are removed from the garden.

- Garden soil can be tested for nutrient levels, biological activity, and possible contaminants.
- Soil organic matter is a major source of plant nutrients, improves soil structure and water-holding capacity, and can sequester carbon, helping to mitigate anthropogenic climate change (ACC).
- Gardeners can also mitigate ACC by turning lawns into gardens, managing organic waste well, and replacing store-bought fruits and vegetables with garden produce.
- Identify and use the best methods, location, and management for making compost from organic waste, and use compost and natural minerals in the garden as the main source of plant nutrients.

Soil covers much of the Earth and is the foundation for plant growth, including in our gardens. Soil is a combination of minerals from decomposed rock, organic matter, and a wide range of living things, from bacteria to plants to gophers, and plays a key part in global cycles of carbon, nitrogen, phosphorus, water, and other resources that are critical for all life on Earth. Soil is the primary source of nutrients and water for plants, including air for the roots, and where many beneficial and some harmful organisms live. The overall goal for managing garden soil is to create an environment that optimizes these factors for the crops you are growing in order to produce the most harvested food and other benefits for the amount of time and resources you invest (benefit:cost), and in a manner consistent with your values and reflecting the key ideas described in the Introduction.

Maintaining adequate soil moisture, organic matter, and plant nutrients for food production can be challenging, and gardeners and farmers have invented many methods over the centuries that are still useful today. For example, in the southwestern US, native people have been using rainfall runoff, rock mulches, rock borders, and terraces for millennia in ways that enhance soil physical and biological properties for plant growth (Homburg and Sandor, 2011). But environmental and social changes, the increasing demand for food, and the decreasing availability of land, water, and other resources, mean that gardeners, farmers and scientists need to work together to adapt and improve many of these traditional methods through experimentation, often combined with new methods. This is why current scientific research on soils and crops can be so useful for gardeners.

7.1. Where do Soils Come From, and Why does it Matter?

The journey soil takes through space and time to wind up at the end of your garden spade determines most of its characteristics. The five main variables affecting soil during this journey are the parent material (original rocks), climate, organic matter and soil organisms, topography, and time (Brady and Weil, 2010, 32).

There is a great deal of local variation in soils as a result of the many different combinations of these five variables. For example, as a result of climate, topography, and time, soil deposited in valleys by seasonal river flooding has a much higher clay and nutrient content, and higher water-holding capacity than adjacent higher elevation soils. This is the case for the soils in the valleys of the Santa Cruz River and its tributaries in Arizona, which make them much better for gardening than nearby upland desert soils even though they both experience the same climate.

Urban soils are a special case because they are the product not only of these five variables, but are often heavily affected by human activities. People often settled first in areas with good soils for obvious reasons. In many parts of the world, people conserved good agricultural soils for food production, and sited houses and urban development on higher, less fertile land, as did the Pueblo peoples of North America. However, as populations grew and food production was moved to distant locations and imported to urban centers, settlement often spread over the adjacent farmland, as in the western US, where there has been accelerating loss of farmland to urbanization, for example in California (Thompson, 2007). The result is that urban soils can be very good, or degraded and contaminated as a result of construction, grading, excavation, and dumping (Section 4.2).

There is much information on classification of soils, including their potential uses. However, probably the best place to begin understanding your local soil in terms of food gardening is by talking with local gardeners, Cooperative Extension agents, and researchers who have experience with them. The value of understanding local soil properties and their distribution from the perspective of those using them was made dramatically clear to us when we were working with the Zuni tribe's Zuni Conservation Project in New Mexico. The Bureau of Indian Affairs (BIA) soil scientists were classifying soils on the reservation in terms of irrigation potential, for a water rights claim the tribe was working on. DAC arranged a meeting between the chief BIA scientist and several Zuni farmers, where the scientist displayed maps and described his findings. In one part of the irrigation district he had classified the soils as "not irrigable" because of lack of drainage. When he explained this to the farmers they were surprised, telling him that that area had been irrigated year after year throughout the growing season for many, many generations. After some discussion, the scientist began to understand how the Zuni farmers were using lateral drainage to keep these soils productive, and therefore changed the classification to recognize their management approach.

7.1.1. The soil profile

The unique combinations of the five variables that create soil result in a sequence of soil layers from bedrock to the surface in each location (Brady and Weil, 2010, 52–55; Soil Science Division Staff, 2017, 91–114). A *soil profile* is a vertical section or slice through the soil displaying its layers. Soil scientists divide soil profiles into six main layers or *horizons* using specific characteristics (color, texture, structure, and density), beginning with O at the surface, and A, E, B, C, continuing downward to the bedrock layer, or R (Figure 7.1). Often only a subset of these horizons are present in a given soil profile. Several of them, particularly the B horizon, are subdivided—indicated with small letters, for example By or Bk (see below)—based on specific properties that can be important for crop production.

The O is a layer of organic material such as leaf litter or mulch, often found under vegetation, on top of the soil proper, and usually doesn't contain mineral soil.

The A horizon, sometimes referred to as the *topsoil*, contains most of the organic matter in the soil profile below the O horizon, often giving it a dark color. However, in arid areas there may be very little organic matter, and the topsoil can be a light color. The depth of the topsoil is important because this is where most of the roots of annual

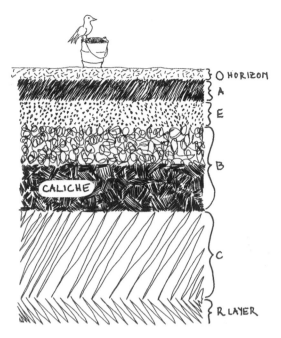

Figure 7.1. The soil profile

plants grow, so for gardens, shallow topsoils need to be deepened by replacing or improving the underlying soil, or by making mounds or raised beds of improved soil.

The E horizon is formed below the A horizon as the result of *eluviation*, the leaching by water of finer material like clays and iron oxides into the B horizon below. This tends to make the E horizon sandy, and due also to a lack of organic matter, lighter in color than either the overlying A or underlying B horizon.

The B horizon, called *subsoil*, accumulates some material washed out of the horizons above, such as clay, carbonates and other salts, and iron. In alkaline soils of arid and semiarid areas *caliche*, a white, nonporous layer composed mainly of calcium or magnesium carbonates, can form a Bk horizon. If it is too close to the surface it can cause problems for root growth and drainage, as in many areas of Arizona and New Mexico. In somewhat drier regions the layer of salt accumulation can be dominated by calcium sulfate (gypsum) forming a By horizon, that can also limit root growth. Poor drainage resulting in salt build-up, waterlogging, or nutrient deficiencies in the B horizon, can be a problem that needs to be solved so that your crops can flourish, especially deeply rooted, large perennials (Sections 7.2.2, 8.2.2, 8.9).

The C horizon is formed by decomposition of the parent material or *bedrock* over which it lies, so it contains little organic matter, and usually relatively few roots. Like the B horizon, the C horizon is likely to accumulate concentrations of carbonates and calcium sulfate in dry areas. Under the C horizon is physically decomposed rock that has been transported by geological or hydrological processes, or the parent rock itself, which is referred to as the R layer.

In Section 7.7.2 we describe resources for obtaining maps of soil classification, including soil profiles, for specific locations.

7.2. Soil Physical Properties

7.2.1. Soil texture

Texture describes the proportion of different-sized mineral particles (sand, silt, and clay) in the soil (Brady and Weil, 2010, 97–104). Texture is important in the garden because it has a strong influence on nutrient availability (Section 7.3), and on soil structure (Section 7.2.2) which determines the quantity and movement of air and water, and the ease of bed preparation and cultivation, as discussed below. Knowing the soil texture helps in deciding how the soil can be improved, and the best watering strategies.

Clay comprises the smallest soil particles, less than 0.002 mm (0.00008 in) in diameter, and plays a big role in soil fertility and structure. Clay is the most important mineral portion of the soil for plant nutrition because its very small charged particles provide a large area for binding plant nutrients.

The other categories of soil mineral particles are *silt* (0.05–0.002 mm; 0.002–0.00008 in), and *sand* (2.0–0.05 mm; 0.08–0.002 in). *Gravel* and *cobbles* are larger than 2.0 mm (0.08 in), are called *fragments*, and are not considered in determining texture. *Loam* is soil containing sand, silt, and clay (0–40% clay by weight, and 10–80% each sand and silt). Most soils are mixtures, and are referred to as sandy clay, silt loam, sandy clay loam, and so on. In general, sandy soils are those with 70–80% or more sand, and clayey soils have more than 40% clay.

Soils with a loam texture are best for gardens because they combine the properties of sand, silt, and clay. Sandy soils are very permeable, giving them good drainage, but they are low in nutrient and water-holding capacities. Clayey soils have high water- and nutrient-holding capacity, but have poor drainage, and are hard when dry, and sticky when wet, making them difficult to work, and difficult for young plants to grow in. Silty soils tend to be intermediate in these properties. Soils of all textures can be improved by adding organic material (Section 7.5). While it would be too much work to change the texture of the soil in most fields because of their size, it can be practical and worthwhile for a garden, especially if suitable materials are nearby, such as uncontaminated sand or silt from a streambed. We give an example in Box 6.3. When combining sand with clayey soil, experiment first with a small quantity. In the wrong proportions and with the wrong type of sand the mixture can become compacted and hard like adobe (Brady and Weil, 2010, 102).

For most purposes, an adequate idea of soil texture can be obtained by simply looking at the soil and rubbing a sample between your thumb and finger. The hard particles of sand can be distinctly felt, and if the sample is dry the crunching sound it makes can be heard when you hold it near your ear. Silt cannot be felt as individual particles but gives a smooth or soapy feeling. Clay is hard when dry, and slippery and sticky when wet. If a more precise idea is needed, two simple ways to determine the proportion of sand, silt and clay are the settling out test, in which a soil sample is mixed with water and left to settle out, and the ribbon test, in which a moistened soil sample is squeezed out into a ribbon (Cleveland and Soleri, 1991, 155–157). The soil texture triangle can be used to translate proportions into a texture class. All of this is clearly described in Whiting et al. (2011), with more detail in Soil Science Division Staff (2017, 120–126). There is even an online calculator where you only need to enter the percentage of sand and clay (Section 7.7.2).

7.2.2. Soil structure

Soil structure is the way in which soil is grouped into separate units to form *aggregates* or clusters of mineral and organic particles with spaces between them, and determines soil porosity and permeability. *Porosity* is the amount of spaces or pores in the soil that can be filled with either air or water; *permeability* is a measure of the ease with which water and air move through the soil pores. Soil structure is mainly determined by the content of clay, and of humus (Section 7.5.1) from the breakdown of organic material, cementing soil particles together to create aggregates.

A good soil structure for gardens is one with many aggregates, with high permeability due to the many larger spaces in the soil for the movement of water and air, and small pores that provide water-holding capacity. Two experienced soil scientists have suggested that it's more useful to focus on the pores instead of the particles, because pores are so important for healthy root growth (Shaxson and Barber, 2003). They make the analogy with a home—the walls and roof are important but it is what occurs under and between these that is of real interest.

Clayey soils can hold more water than sandy soils because they have higher porosity due to many small pores that add up to much more total pore space than the fewer but larger pores of sandy soils. Clayey soils provide more soil surface per volume of soil, so more water is held to soil particles by adhesion, but because of the smaller pores a higher proportion of water is held tightly and is unavailable to plants (Brady and Weil, 2010, 157–158). Small pore size also means that clayey soils have low permeability. In soils with small pore size (less than 0.03 mm; 0.0012 in) characteristic of clay soils, drainage of excess water is impeded, making waterlogging more likely (Section 8.2.1). The large pore size of sandy soils gives them high porosity and permeability, and poor water-holding capacity.

The structure of both sandy and clayey soils can be improved by adding organic material. For soil structure with a good balance between small pores that retain a lot of water, and large pores that allow easy movement of water out of the root zone, medium-textured loam soils are best, especially when they contain adequate amounts of organic matter (Section 7.5). Soil porosity is also improved by earthworms, termites, and other small animals that make tunnels in the soil. Larger animals like gophers mix and aerate the soil as well, but can also do a lot of damage to plants.

Cultivating the soil by hoeing, tilling, or plowing increases exposure to air, speeding the microbial breakdown of organic matter and the loss of the carbon in it to the atmosphere, and breaks up worm tunnels, as well as mycorrhizal and other fungal hyphae. This reduces soil structure, porosity, and fertility, and encourages wind and water erosion. However, cultivation to loosen compacted soil can also increase porosity, and is needed to mix organic material directly into the root zone, especially in heavier, clay soils, and soils with low organic matter content—it's a tradeoff. The more the garden soil is cultivated, the more organic material needs to be added, and the more attention needs to be given to controlling erosion. So whether or not, and when, to cultivate depends on balancing pros and cons for your local conditions; identifying "optimal strategies that exploit both the benefits of cultivation, while maintaining earthworm and fungal communities" (Spurgeon et al., 2013, 9). Gardeners can experiment with different methods that strike a balance, for example, plugging transplants into a garden bed, and mixing compost into soil in the planting hole instead of cultivating the entire bed. The tradeoff between till, and no till or minimal till, is another illustration of why there is often no unequivocal "right" answer to the question "What is the best way to manage the garden?" Tradeoffs are common, but by understanding basic concepts, observing, experimenting, talking, and working with others you can figure out strategies that strike the right balance for you, your community and the Earth.

7.3. Soils and Plant Nutrition

As a result of our common evolutionary history, all living cells on Earth, whether in animals, plants, or microbes, contain mostly the same elements in approximately the same proportions, reflecting the availability of these elements as life evolved (Fedonkin, 2009). Eighteen elements are essential nutrients for plants. The six *macronutrients*, required in large amounts, are carbon (C), hydrogen (H), and oxygen (O) that plants obtain from air and water, and nitrogen (N), phosphorus (P), and potassium (K) that plants absorb from the soil. The *secondary nutrients* calcium (Ca), magnesium (Mg), and sulfur (S) are required in smaller amounts. The *micronutrients* are copper (Cu), zinc (Zn), boron (B), iron (Fe), manganese (Mn), molybdenum (Mo), cobalt (Co) chlorine (Cl), and nickel (Ni), and are required in very small amounts (Epstein and Bloom, 2004). Plants get N, P, K, secondary nutrients, and micronutrients primarily through their roots. Humans need to obtain these same mineral nutrients from food, and some that plants do not require: sodium (Na), chromium (Cr), selenium (Se), and iodine (I), as well as a number of vitamins (Quintaes and Diez-Garcia, 2015) (see Appendix 1A, Table 1A.1). In this section we discuss some of the important plant nutrient concepts for food gardens. Most state Cooperative Extension websites have more detailed information on managing specific nutrients (Section 7.7.3).

Nutrient deficiencies often show up as changes in leaf color including *chlorosis*, a change from dark green to lighter green and yellow due to a lack of nutrients that makes the plant unable to synthesize new chlorophyll to replace chlorophyll lost through natural degradation. These symptoms need to be distinguished from those caused by pests and diseases (Section 9.8).

Plant nutrients in the soil are mostly available to plants in the form of electrically charged atoms or molecules called *ions*, although N and P can also be absorbed as soluble organic compounds (Brady and Weil, 2010, 375). *Salts* are compounds composed of electrically charged *ions*, which separate from each other, for example in water: *cations* with a positive charge, and *anions* with a negative charge. Ions are held in the soil electrostatically on the surface of soil *colloids*, very small clay and humus particles (Section 7.5.1) whose surfaces are covered with positive and negative charges. The nutrient ions become available for uptake by plants when they enter the soil water solution. Absorbing nutrients from the soil is by active transport, requiring energy obtained from the carbohydrates formed in photosynthesis. Once nutrients and water have been absorbed into the xylem tissues of the roots, they move throughout the entire plant, carried in the flow of water or sap in the vascular system (Section 5.1).

Positively charged cations, for example ammonium (NH_4^+), calcium (Ca^{++}), potassium (K^+), ferrous and ferric iron (Fe^{++} and Fe^{+++}), and zinc (Zn^{++}) are attracted to negatively charged sites on colloids. Negatively charged anions, for example nitrate (NO_3^-), phosphates ($H_2PO_4^-$ and HPO_4^{--}), and sulfate (SO_4^{--}), are attracted to positively charged sites on soil colloids. The *cation exchange capacity* (CEC) is a quantitative measure of the amount of cations, measured by their electrical charge, that can be absorbed by soil colloids in a specified amount of soil. The *anion exchange capacity* (AEC) is the equivalent measure for anions. The CEC and AEC of soils are determined by the amount of electrical charges on the colloids, a function of how much humus and different types of clay they contain. Montmorillonite and vermiculite are the clays with the highest CEC, illite has less, and kaolinite and sesquioxides have a very low CEC. Basically, if all else is equal, the greater the CEC and AEC the better the soil is for garden plants because more nutrients can be available. The lower the CEC and AEC the more nutrients will be lost to either leaching or volatilization.

Plants absorb most of the water and minerals they need through their microscopic root hairs, either directly or via mycorrhizal fungi (Box 7.1, Section 5.3). The mucigel coating on root hairs is a combination of root-exuded mucilages, microbial cells, and clay particles that helps make contact with soil particles, creating a bridge between root cells and soil, increasing their ability for absorption (Brady and Weil, 2010, 338–340). As water increases to near field capacity (Section 8.2.1), nutrient uptake generally increases because transport of nutrients through the soil by water flow and *diffusion* (the unimpeded movement of solutes from areas of high to low concentrations) increases, as does root growth.

7.3.1. Soil pH and plant nutrition

Soil acidity and alkalinity influences many soil properties, especially the availability of plant roots to take up nutrients, as well as toxins. The degree of acidity and alkalinity is measured in terms of *pH*—the negative of the base 10 logarithm of hydrogen ion (H^+) concentration in *moles* (molecular weights) per liter (where each unit change is by the power of 10). Therefore, as H^+ concentration increases, pH decreases, and visa versa. In pure water a *very* small number of molecules come apart ($HOH \rightarrow H^+ + OH^-$), with both cations ($H^+$) and anions ($OH^-$) at a concentration of 10^{-7} moles liter^{-1}, for a pH of 7. A pH of 7 is therefore *neutral* because the ions are at the same concentration. A soil with a pH of 8 has one-tenth as much H^+ and is 10 times as alkaline (basic) as a neutral soil; a soil with a pH of 6 has 10 times as much H^+ and is 10 times more acidic than a neutral soil, a soil with a pH of 5 has 100 times as much H^+, and so on.

Soils in dry areas tend to be neutral to alkaline (pH 7 and higher), especially in the topsoil, because of the relative lack of precipitation, which is acidic due to reaction with CO_2 (forming carbonic acid, H_2CO_3) and other molecules in the atmosphere. The lack of precipitation also means that non-acidic cations like Ca^{2+}, Mg^{2+}, K^+, and Na^+ from the weathering of minerals accumulate, because they are not washed out of the soil, and therefore dominate colloid exchange sites (Section 7.3) and the soil solution. The drier the region, the higher the pH tends to be. In areas with less than 600 mm (24 in) of annual rainfall, like many of the drier areas of the western and southwestern US, pH can be up to 8 or 8.5. Topography also affects pH; the middle of slopes tend to be better drained and therefore more leached of cations and more acidic than adjacent uplands or valleys.

Alkaline soils often contain relatively large amounts of exchangeable (available to plants) calcium and magnesium, but the availability of many micronutrients is limited. In general, a pH between 6 and 7 is best for phosphorus availability, and overall promotes availability of all plant nutrients (Brady and Weil, 2010, 289). It is easy and inexpensive to test your garden soil pH yourself (Section 7.3.3).

Box 7.1. Mycorrhizae

Mycorrhizae are relationships between members of a group of soil fungi and the roots of flowering plants, including most garden crops, with a few exceptions such as the brassicas (cabbage family) and chenopods (beet family) (Brady and Weil, 2010, 343–345, 421–422, 448). These symbiotic relationships are *mutualistic* because they benefit both organisms. Arbuscular mycorrhizae are the most important for food crops—fungal hyphae grow inside plant roots and make some nutrients, especially P, available to plants, help plants resist some soil-borne diseases, protect plants from uptake of toxic heavy metals like cadmium and toxic levels of essential mineral elements like Zn (Ferrol et al., 2016), and may even make water more available to plants. In return, the fungi receive sugars and fats from the plant that they cannot synthesize themselves (Jiang et al., 2017; Luginbuehl et al., 2017). Without mycorrhizal fungi most plants grow poorly.

From what is known now, garden practices that can help to establish and maintain successful mycorrhizae include introducing soil from the root zone of nearby healthy, successful, established gardens; not using P amendments, because root colonization drops with rising soil P concentrations; minimizing cultivation to avoid disturbing growing mycorrhizae; maintaining optimal soil moisture (Deepika and Kothamasi, 2015); and using surface mulch to slow evaporation of that moisture (Berruti et al., 2015).

Commercial mycorrhizal fungi soil inoculants are widely available, but some may not contain live spores, and studies have shown that even those that do may contain species that aren't compatible with the local soil, or can increase competition among fungal species (Verbruggen et al., 2013). This is why inoculating with soil from a local garden is a better strategy as long as you know it won't also introduce pests or pathogens into your garden.

Gardeners can reduce the pH of alkaline soils by adding organic material, especially more acidic organic matter like composted pine needles (Brady and Weil, 2010, 297). Elemental sulfur can also be added, which lowers pH by first being oxidized to sulfate by bacteria, and eventually becoming sulfuric acid. To increase bacterial action, the surface area of the sulfur should be maximized by applying it as a powder. Although sulfur is nontoxic to humans, other animals and beneficial insects, and allowed under the USDA organic standards, it is still an irritant; don't inhale it or let it contact your eyes or skin.

Acidic soils have a pH of less than 7 and are relatively rare in dry areas, for example where the parent material is an acidic rock like granite, or when a formerly wet climate leached the soils. While a pH of 6–7 is optimal, when it falls below 6, some nutrients become less available, but manganese and aluminum become more available and can be toxic to plants. The pH of acidic soils can be increased through the addition of alkaline substances like lime (e.g. $CaCO_3$), ashes, or potash (alkaline potassium compounds). Ashes are a valuable soil amendment in many parts of the world where acid soils dominate, including western Australia.

7.3.2. Nutrient cycles

Nutrients are constantly cycling through the living organisms in the soil, including bacteria, fungi, plant roots, and animals like earthworms and millipedes (Fig. 7.2). We can think about nutrient cycles in terms of what chemical compounds they are part of, and where these are located in relation to plants. When fruits, stems, roots, or leaves are harvested, the nutrients they contain are removed from the nutrient cycle in the garden, and will need to be replaced at some point. The need to replace harvested nutrients has been recognized in the practices of many traditional agriculture systems. For example, the Zuni Native Americans of New Mexico have maintained soil fertility and yields in their maize fields on alluvial fans for about 3,000 years, by practicing runoff agriculture, with seasonal water flows carrying debris from forests higher in the watershed, and depositing nitrogen, phosphorus, and organic carbon in their fields (Norton et al., 2007; Sandor et al., 2007).

Polyculture and crop rotation can help to balance availability of nutrients with the diverse demands for nutrients among different crops. Nutrients are also cycled between plants within the garden, for example, deep-rooted plants such as many fruit trees can absorb nutrients at deep soil levels, making them available later to shallow-rooted plants when their leaves drop to the soil surface and are decomposed (Box 6.2). Gardeners also recycle plant nutrients by returning garden food scraps and garden waste to the garden after composting.

However, the export of organic garden waste and human waste to municipal treatment facilities removes nutrients in the garden plants we eat from the garden nutrient cycle, so some nutrients will need to be replaced. This can be done by importing them from outside that cycle, for example, in manure, yard waste, municipal compost, and waste from purchased food that is composted and applied to the garden, or commercial organic fertilizers. If you diagnose a nutrient deficiency in your garden plants (Section 9.8), you can add nutrients, but may also want to follow up with a soil test (Section 7.3.3), and contact local resource persons for advice. Table 7.1 shows the approximate N, P, and K content of some organic and natural mineral fertilizers, but actual content can vary a lot by source and processing method, especially for organic material, and most contain variable amounts of many other plant nutrients or other growth supporting factors. If specific nutrient deficiencies are diagnosed it is good to investigate the composition of available fertilizers (Cogger, 2014; UGE, 2014).

Nitrogen is a key nutrient that, unlike other mineral nutrients, is originally from the air in the form of nitrogen gas (N_2) which is inert and can't be used by plants (Fig. 7.3). Nitrogen in the air can be made available to plants as ammonia (NH_3) by N-fixing bacteria living freely in the soil or in the root nodules in a mutualistic relationship, typically with leguminous plants (Box 7.2). Nitrogen supplies can be increased by growing N-fixing leguminous cover crops and composting them, and growing N-fixing food legumes mixed with other crops, or in rotation, and leaving their roots in the soil to decompose. For plants without N-fixing root nodules, N becomes available as plants and microorganisms die, and their tissues are decomposed, producing soluble organic N compounds, NH_4^+ (ammonium) and NO_3^- (nitrate), which can be taken up by plant roots.

The N cycle is influenced by oxygen (O) supply. When excess water fills air spaces in garden soil or compost piles, the garden loses N because anaerobic bacteria proliferate and break down nitrate to form nitrous oxide (N_2O) and nitric oxide (NO), both powerful greenhouse gases, as well as N_2. When O

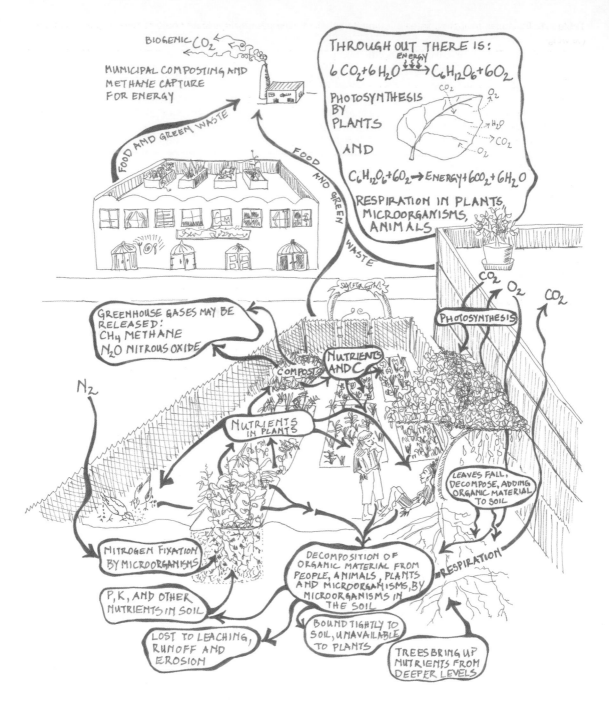

Figure 7.2. Nutrient cycles in the garden

Table 7.1. Examples of approximate N, P, and K content of materials for organic and natural mineral garden fertilizers (% weight)

Material	N	P_2O_5 (phosphate)	K_2O (potash)	Relative availability[c]	Source
Alfalfa meal	3.0	1.0	2.0	Medium-slow	[a]
Brewers' grain (wet)	0.9	0.5	0.1	Slow	[a]
Cocoa shell meal	2.5	1.0	2.5	Slow	[a]
Coffee grounds (dry)	2.0	0.4	0.7	Slow	[a]
Compost (not fortified)[d]	1.5	1.0	1.5	Slow	[a]
Compost from kitchen waste					
Vermicompost	0.9	0.3	0.0	Not available (NA)	[b]
Regular compost	0.7	0.2	0.0	NA	[b]
Compost from green waste and pruning residues					
Vermicompost	2.1	1.0	1.5	NA	[b]
Regular compost	2.2	0.9	1.0	NA	[b]
Cottonseed meal (dry)	6.0	2.5	1.7	Slow-medium	[a]
Eggshells	1.2	0.4	0.1	Slow	[a]
Grape pomace	3.0	0.0	0.0	Slow	[a]
Granite dust	0.0	0.0	6.0	Very slow	[a]
Greensand	0.0	1.0–2.0	5.0	Slow	[a]
Guano (bat)	5.7	8.6	2.0	Medium	[a]
Guano (Peru)	12.5	11.2	2.4	Medium	[a]
Kelp[e]	0.9	0.5	1.0	Slow	[a]
Manure[f] (fresh or as is)					
Broiler litter	3.1	3.1	2.8	Medium-rapid	[a]
Cattle	0.5	0.2	0.4	Medium	[a]
Horse	0.6	0.3	0.6	Medium	[a]
Sheep/goat	0.6	0.3	0.8	Medium	[a]
Swine	0.6	0.2	0.4	Medium	[a]
Manure[f] (dry)					
Cricket	3.0	2.0	1.0	Medium-rapid	[a]
Dairy	0.5	0.2	0.5	Medium	[a]
Rabbit	2.0	1.3	1.2	Medium	[a]
Marl (a calcium carbonate mixture, often with clay)	0.0	2.0	4.5	Very slow	[a]
Soybean meal	6.7	1.6	2.3	Medium-slow	[a]
Wood ashes[g]	0.0	1.0–2.0	3.0–7.0	Rapid	[a]

[a]UGE (2014).
[b]Lim et al. (2015).
[c]Rapid = <1 month; medium = 1 to 4 months; slow = 4 months to 1 year; very slow = >1 year.
[d]Nutrient content varies considerably with feedstock used for compost.
[e]Primarily a micronutrient source.
[f]Plant nutrients available during year of application vary with amount of straw/bedding and storage method.
[g]Potash content depends on the tree species burned. Wood ashes are alkaline, containing approximately 32% CaO, and should not be used in alkaline soil.

is present but limited, the aerobic process of nitrification can also produce these gases. When undecomposed organic material with high C:N is added to the garden, resulting in soil with C:N higher than 8:1, soil microorganisms can use up available N, resulting in N deficiencies in crop plants (Section 7.5.1). N is also lost when NH_4^+ in the soil water solution is volatilized as NH_3 (ammonia gas). Ammonia volatilization increases with high pH and low levels of soil colloids that can bind NH_4^+ (as in sandy soils), and also occurs when organic material is left on the surface under dry, hot conditions.

Soil nutrient management is a balancing act with the goals of a) maintaining a supply of nutrients for the soil microorganisms that support high soil quality, b) ensuring a nutrient supply that provides good yields of nutritious vegetables and fruits, and c) minimizing pollution of the local environment from

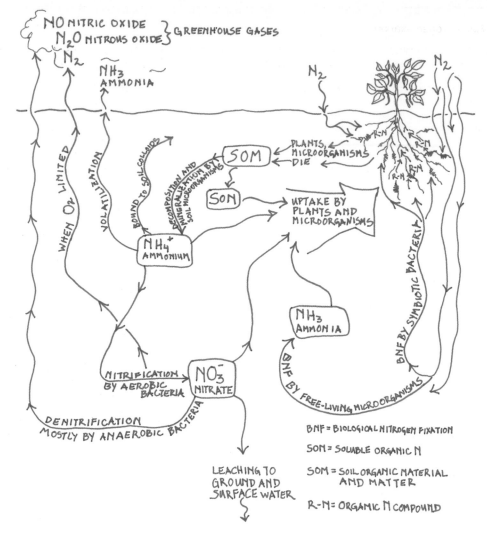

Figure 7.3. Simplified nitrogen cycle in food gardens

excess nutrients, for example, N and P pollution of water (cf. Brady and Weil, 2010, Ch. 13). Diagnosing any soil nutrient deficiencies by the plant symptoms they create (Section 9.8) and by soil testing, is important for maintaining good garden harvests.

7.3.3. Soil testing

In addition to talking to local experts about local soils, it's a good idea to test the soil in the sites being considered before establishing your garden (Section 4.2.1). Soil nutrient testing is also worth doing if you have persistent problems with plant growth, vigor, or other symptoms (Section 9.8).

Government agencies, like state Cooperative Extension programs, have information about how to do this, and what the results mean (e.g., Cogger, 2014; Whiting et al., 2014). Soil test results can tell you if there are any major imbalances in available plant nutrients. Very inexpensive kits that test for pH, nitrogen, phosphorus, and potassium are available and easy to use. A review of do-it-yourself soil test kits found that two gave results that were over 90% accurate compared to analytical lab test results, including the RapiTest kit that was recommended for being easy to use (Faber et al., 2007). In addition, some Cooperative Extension Services, universities, and companies offer soil tests at reasonable prices.

Box 7.2. Nitrogen fixation

Bacteria that convert nitrogen (N) in the air into forms that can be used by plants are a critical link in our food system. Some bacteria establish mutualistic relations with legumes, and a type of bacteria called actinomycetes establish mutualistic associations with non-legumes, both of them in root nodules. Some free-living bacteria and many species of cyanobacteria (formerly known as blue-green algae) in the soil also fix N.

Many garden crops in the legume family, such as acacia, carob, cowpea, fava bean, garbanzo, lima bean, mesquite, snap bean, soybean, and tepary bean can form relationships with *rhizobia*, various bacteria of the genus *Rhizobium* that fix N in root nodules. A smaller number of legumes, including cowpea and soybean, form symbiotic relationships with bacteria in the *Bradyrhizobium* genus. In all of these the bacteria enter the root hairs when the plants are young, and subsequently cause the legume root to form tumor-like growths or *nodules.* These nodules are different than the tumors or knots formed by root knot nematodes (Section 9.2.2), because the bacterial nodules can be rubbed off without breaking the root apart. The legume protects the bacteria inside the nodule, and supplies it with carbohydrates for energy, the bacteria fix atmospheric N into forms that the legume can use.

Specific *Rhizobium* species induce nodules only in certain species or even varieties of legumes. The seeds of many commercially available legumes may need to be inoculated with the appropriate variety of rhizobia that colonize them if those varieties have not been grown in a location in five years or more. Legume seeds may also need inoculation if there were signs of N deficiency the last time they were grown, since the appropriate rhizobia may not be present in local soils, or not in large enough numbers (Durst and Bosworth, 2008). Mutualistic relationships like these evolve over time in an area, and are one reason to plant garden varieties traditional to your area. The effect of ACC and resulting changes in crop species and varietal distributions on *Rhizobium* and other soil microorganisms is not well understood, but will probably vary across different locations and landscapes (Classen et al., 2015).

When N-fixing crops die, the N they fixed can become available to future crops that do not have mutualistic N-fixing bacteria, reducing the need to add N in the garden. Most N is probably contributed by legumes in the season after they die, when their leaves and roots decompose in the soil. It is a good idea to leave the roots of annual legumes along with their rhizobia in the soil where they can provide N as they decompose, as well as serve as a source of bacteria for later plantings. So instead of pulling annual legumes like beans out of the garden at the end of the year, cut them off at ground level and compost just the aboveground growth. Legume seeds are rich in N (Rogers et al., 2009), and you can also harvest N for your gardens by gathering materials from beyond its borders, like seeds pods from N-fixing leguminous trees like acacia, catclaw, mesquite, or palo verde. In addition, legumes can be interplanted with other crops as in the familiar example from Mesoamerica of maize, which has a high N requirement, and *Phaseolus* beans, which fix N.

The three common types of soil tests are for: nutrient availability, biological activity, and contamination. Testing for soil-borne contaminants is essential if you live in an area known or suspected to be contaminated (Section 4.2.1). Otherwise, the most useful test, and the easiest and cheapest, is for nutrient availability. Because the balance of nutrients in your soil is key to the health of the microbes and plants, this test can help you to understand what your soil needs to be more productive.

7.4. Nutrient Cycles and Anthropogenic Climate Change

Ecological thinking is the key to understanding the relationship between the food system and atmospheric chemistry, and subsequent effects on the climate system. In this section we follow some of the paths that C and N take among these systems related to ACC. Because the benefits of nutrient cycling are mostly to the individual gardens and gardeners, and the costs, in terms of climate change, are generalized to everyone else on Earth, the possibilities for reducing costs depend on creating the social structures that encourage effective nutrient cycling that reflects empathy, equity, and prosocial goals.

Humans increased CO_2 in the atmosphere well before the industrial period as a result of land use change for food acquisition and production (burning and clearing vegetation and cultivating soil, grazing animals), beginning up to 7,000 years ago (Kaplan et al., 2011; Ruddiman, 2013). This released CO_2 through the oxidation of organic C compounds in plants and soil. The Industrial Revolution and the increased burning of fossil fuels beginning in the late 1700s led to an increase in emissions, and the food system is currently a major consumer of fossil energy for manufacturing inputs, crop production, food processing and storage, transportation and preparation. In the US in 2007, for example, the food system produced 13.6% of all CO_2 emissions from fossil fuels (Canning et al., 2017). In Sections 7.5 and 7.6 we'll look in more detail at the impact on ACC of gardening and organic waste management, including composting.

Agriculture, and organic agriculture in particular, has been promoted as having important potential to sequester C in the soil, but evidence shows that many factors are important in C sequestration, especially the amount of cultivation. A recent analysis of global data suggests that most existing models have overestimated the C sequestration potential of soils by up to 40% (He et al., 2016). Reducing or eliminating soil tillage can increase C sequestration, but the amount of C that can be sequestered in soil is limited—especially under dry conditions, the variables determining the rate of sequestration at different depths are not well understood, and the process can easily be reversed, releasing CO_2 back into the atmosphere (Powlson et al., 2014). However, compared with using synthetic fertilizers that contribute to ACC, both no/low till methods and organic methods can help to mitigate ACC, not only through C sequestration, but lower nitrous oxide emissions (see below) (Robertson et al., 2014). To help reduce greenhouse gas emissions (GHGE) from your garden, use organic fertilizers like compost and green manure, and minimize cultivation.

CO_2 emitted from the decomposition of plants that were recently alive, whether decomposition is on the ground in the garden, in compost piles, or in landfills, is referred to as *biogenic* CO_2 (Section 5.4), and is usually not included in estimates of GHGE (EPA, 2018). This is because the C in it has been recently removed from the atmosphere as CO_2 and fixed by photosynthesis into carbohydrates. When the carbohydrates are metabolized for energy during respiration, the C is released back into the atmosphere as CO_2 again; its sequestration in the plants is very brief, and its effect on ACC is negligible.

However, under *anaerobic* (lacking oxygen) conditions the CO_2 from bacterial respiration of carbohydrates can be converted into CH_4 (methane) by methanogenic *archaea* (the single-celled organisms formerly called archaebacteria). CH_4 has a *global warming potential* (GWP, the capacity to trap heat in the atmosphere compared to that of CO_2 which has a GWP = 1) of 86 in the first 20 years when it does the most damage, after which most of it is eliminated from the atmosphere. However, most models and projections use the 100-year GWP for CH_4 which is currently 34 (IPCC, 2013, 714, Table 8.7), or even lower earlier estimates of 21 or 25. Due to this, it is likely that some estimates of methane's contribution to ACC may be too low. GWP is expressed in the common unit of CO_2 *equivalents* (CO_2e), the amount of warming caused by a greenhouse gas like CH_4 in terms of the concentration of CO_2 that would cause the same warming.

About 2% of the CO_2 fixed by photosynthesis each year ends up as methane, and about 20% of this methane ends up in the atmosphere as a greenhouse gas (Thauer et al., 2008). The majority of CH_4 emissions are directly or indirectly from the food system. For example, based on EPA data, of total US anthropogenic CH_4 emissions in 2014, 31% was from fermentation by ruminant animals and manure management (dairy and red meat production), 2% from irrigated rice and burning of field wastes, all part of the food system, and 23% from landfills, composting, and wastewater treatment, much of which is part of the food system, for a total of 55%, with fossil fuels accounting for the remaining 45% (over half of which was from natural gas) (EPA, 2016). That 55% of CH_4 emissions is equal to almost 6% of total US GHGE.

The other major greenhouse gas produced by the food system is nitrous oxide (N_2O). There was a dramatic increase in human intervention in the nitrogen cycle with the invention of an artificial means of fixing inert N_2 from the atmosphere into reactive N that plants can use, the industrial Haber-Bosch process (Leigh, 2004; Smil, 2004). Today the amount of reactive N used in food production each year (153 Tg, 1 teragram = 10^{12} g) is about 1.8 times the amount of terrestrial N from natural biological N fixation annually occurring outside of food production. Of the 153 Tg, 72% is from synthetic and mined fertilizers, and 28% from N fixation by crops, dominated by soybean (calculations based on Battye et al., 2017). Because more N fertilizer, in the form of ammonium or nitrate, is commonly applied to crops than can be taken up by microorganisms and plants, a large proportion of it is quickly converted to other forms, including N_2O, or lost through volatilization as ammonia or leaching as nitrate which pollutes ground and surface waters

(Fig. 7.3). Most N_2O is emitted from agricultural fields (79% of the US total in 2014) as a result of the breakdown of synthetic N fertilizers, but also from decaying organic material and organic matter (EPA, 2016). Over a period of 100 years N_2O has a GWP almost 300 times greater than CO_2 (IPCC, 2013, 714, Table 8.7), and, including manure management, agricultural production accounted for 83% of US N_2O emissions in 2014, equal to 5% of total US GHGE (EPA, 2016).

Food gardens can make important contributions to fighting ACC through good soil and organic material management, by reducing purchases of conventionally produced and distributed fruits and vegetables, and by encouraging more fruits and vegetables in the diet to replace some animal foods (Section 1.1.2), especially red meat and dairy, the main dietary source of GHGE in the global north.

7.5. Managing Soil Organic Matter

Soil organic matter, the decomposed remains of living organisms, contains carbon, the key element of life on Earth.

7.5.1. Soil organic matter and plant nutrition

Soil is full of living organisms. A handful of most garden soils contain billions of microorganisms including nematodes, bacteria, and fungi. Larger organisms living in the soil include insects, millipedes, earthworms, lizards, snakes, and mammals like moles, mice, gophers, and rabbits. Roots of growing plants are another living part of soils. When these organisms die, microorganisms decompose them to obtain nutrients and energy, making the nutrients in the dead organism available for further plant growth. Nutrients are constantly recycling between living organisms, *organic material* (dead organisms), *organic matter* (decomposed organic material), and non-organic mineral forms attached to soil particles, and in soil solution in the pore spaces where they can be taken up by plants (Fig. 7.4). Organic material from many different local sources is a high-quality, low-cost resource for maintaining garden soil fertility.

Organic matter content of as little as 1.5% of the soil weight improves soil quality and plant growth, but there is really no ideal amount (Brady and Weil, 2010, 386). When you are first making garden beds, 25–50% of the soil volume can be compost, or higher in holes for transplants, although care must be taken to avoid abrupt changes in organic matter content (Section 6.6.1).

Soil scientists classify soil organic matter in natural (not managed) soils into three pools (Brady and Weil, 2010, 378–379). The *active pool* makes up about 10–20% of soil organic matter, and consists of easily decomposed particulate matter, mostly carbohydrates, and lasts from days to less than a decade. Most of the soil organic matter in the slow and passive pools is *humus*—organic matter that decomposes more slowly—the major source of storage sites for plant nutrients in the soil. The majority (60–90%) of soil organic matter is in the *passive pool* and is very stable, lasting hundreds to thousands of years, because it is protected from decomposition in clay-humus complexes, and so can make an important contribution to combating ACC by increasing the amount of carbon sequestered in the soil. Inorganic carbon, like biochar, remains in the soil for a very long time, and can also improve plant growth (Box 7.3). The passive pool provides most cation exchange and water-holding capacity in soil (Brady and Weil, 2010, 372–374). Organic matter in the *slow pool* is intermediate between active and passive pools in terms of time to decompose, and is an important source of nitrogen and other plant nutrients as it is decomposed by microorganisms. In your garden, the proportion of organic matter in the active pool can be very large after compost or organic material like green manure is added, and while most of this decomposes fairly rapidly, it also contributes to the amount converted to slow and passive pools.

Soil microorganisms use the carbohydrates in organic material and matter as food, metabolizing it to obtain energy and nutrients, which further decomposes it into simpler forms that plants can take up. In organic gardens this process supplies most of the nutrients needed by plants. We can think about plant nutrition in terms of feeding the soil not the plants. Feeding the soil means feeding the beneficial soil microorganisms that decompose organic substances, making nutrients available to plants (Kirkby et al., 2014). Ideal conditions for microorganisms, and therefore fast decomposition of organic material and matter, include dense plantings, warm temperatures, adequate water, good soil structure with plenty of air circulation, a pH between 6 and 8, and plenty of other nutrients. The growth of soil microorganisms is also closely related to the *carbon to nitrogen ratio* (C:N) in

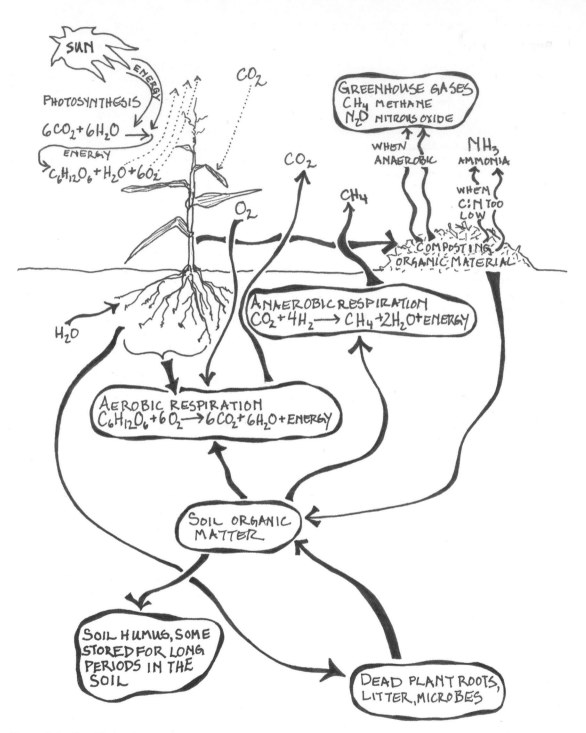

Figure 7.4. Simplified carbon cycle in food gardens

Box 7.3. Biochar

Biochar is inorganic carbon created by incomplete burning of organic material at high temperatures, under low oxygen conditions, known as *pyrolysis*. Interest in biochar as a soil amendment to boost crop yield has increased dramatically in recent years, due to its role in producing the benefits of *terra preta* (dark earth) soils (Steinbeiss et al., 2009). These soils were created by indigenous Amazonian peoples beginning up to 3,000 years ago, and are found scattered throughout the Amazon (Glaser and Birk, 2012). They are dark-colored and have high nutrient content due to high levels of biochar and soil organic matter, and contrast with adjacent soils that are lighter with low organic matter content and nutrient-holding capacity.

Biochar can increase water storage and cation exchange capacity (Section 7.3) of the soil because of the large surface area of its molecules (Steinbeiss et al., 2009). Some research has shown that biochar in concentrations of at least 30% by volume in the root zone of tomato seedlings can increase their resistance to drought in sandy soils (Mulcahy et al., 2013). However, another study using pine biochar in pots growing spinach and amaranth found that while biochar increased water-holding capacity it reduced crop biomass, although if enriched by soaking in mineral fertilizer solution, plant growth increased (Wang et al., 2016).

Research so far on the effects of biochar, especially in gardens, is scarce. One study in Europe evaluated reports of 144 home gardeners who compared two 1-m^2 plots, one with 1 kg (2.2 lb) of biochar combined with 1 kg of either compost or manure, and the other with the equivalent amount of compost or manure only (Schmidt and Niggli, 2012). Although they found a 7.5% increase in average crop yield with biochar, some crops did much better than others, while some grew more poorly with biochar. Overall, there was also a lot of variation in results between gardeners, which the researchers thought could be due to the wide range in quality of compost and manure used in combination with the biochar, and differences in soils among gardens. This variation in results shows how important it is for gardeners to not simply follow recipes, but to observe and ask questions about the variables that affect their own situation.

Because biochar is a form of inorganic carbon that can persist for thousands of years in the soil, it has also been proposed as a way of combatting ACC by sequestering carbon. One experiment found that biochar produced from the straw of cereal crops reduced GHGE by 80% over 8.5 years compared with just allowing the straw to decompose in the field (Thakkar et al., 2016). Biochar also appears to reduce the production of N_2O from denitrification in favor of N_2 (Fig. 7.3) (Cayuela et al., 2013).

Several companies now produce biochar from local organic materials for sale to gardeners and farmers.

---

their environment. Soil microorganisms have an average C:N of 8:1, and incorporate into their bodies about a third of the C in the C compounds they eat, while the rest is emitted as CO_2 during respiration to obtain energy. The C:N in organic material added to the soil should be about 25:1 (Brady and Weil, 2010, 370). If the C:N is higher, nitrogen becomes limiting, so it will be tied up in the bodies of microorganisms, making it unavailable to your garden crops.

This is why adding good quality organic material to the garden soil is the most important part of soil management. This is especially true if the soil is cultivated regularly, because this increases the level of oxygen in the soil, increasing metabolism of organic material and matter by soil microorganisms, making nutrients rapidly available for further growth of microorganisms and plants. However, these nutrients will be lost if they are available in amounts greater than plant requirements. There are tradeoffs between degree of cultivation and nutrient availability from the decomposition of soil organic matter, similar to the tradeoff between cultivation and soil structure (Section 7.2.2).

7.5.2. Gardens and organic matter in an urbanized world

Ecological thinking about organic matter and urbanization reveals two related problems very relevant to food gardens—how to supply fresh, healthy food to the large and growing number of people living at high densities in cities (Section 2.5), and what to do with the organic waste they generate, including yard trimmings, food waste, and their human manure. In many ways, these problems are like the well-documented problems of the *CAFOs* (concentrated animal feeding operations), and we can think of them as *CHuFOs*, concentrated human feeding operations. CHuFOs result in the spatial and structural disconnect in resource cycles between food production and food consumption, and between the nutrients in harvested crops that were obtained from the soil and end up in food and human waste, and the growing crops that need those nutrients.

There were 254 million tons of municipal solid waste ("garbage") in the US in 2013; 14.6% of this was food, 13.5% was yard trimmings, 37 and 34 million tons, respectively, but of those, less than 2 million tons of food and 21 million tons of yard waste were composted (EPA, 2015). Obviously there's lots of room to improve on that—for example, the EPA estimated that in 2008, 38% of waste going to landfills annually could have been recycled using some form of composting. Cities are beginning to address the CHuFO problems, like San Francisco, which has a leading program for treating organic waste, including food waste.

When crops are harvested, the plant nutrients they contain are removed from the local cycle. With continued harvesting and without adding nutrients back, levels of nutrients in the soil will fall to levels below those needed for healthy plant growth and good harvests. At the same time, in the cities where the food is eaten, food waste from stores, restaurants, schools and homes, and human waste, along with plant waste from parks, yards, and gardens is generated in large quantities. This organic waste is far from the fields where the soil microbes and crops need it. The mainstream system has solved this problem by supplying nutrients to crops through synthetic manufactured fertilizers, while food and plant waste is dealt with in landfills, and human manure in septic tanks and sewage treatment operations. But "solutions" like this create new problems.

Manufacturing, transporting, and applying synthetic fertilizers require energy and other resources, and also result in water and air pollution, including large dead zones in fresh water and oceans, and synthetic nitrogen fertilizer is the largest source of the greenhouse gas nitrous oxide (N_2O) (Section 7.3.2). The management of urban organic waste also requires energy and resources for its transportation and treatment, and can pollute water and air, and emit CH_4 and N_2O.

The large and growing quantity of urban organic waste has led to better ways of managing it, including capturing methane from landfills and anaerobic composting operations, and burning it to generate electricity, creating organic soil amendments like compost from organic waste for use in gardens and fields, processing green waste for mulch, and campaigns to reduce food waste. Recycling organic waste for use in food production can also sequester a small amount of C in the soil, preventing it from being quickly re-emitted as CO_2, and thus helping to mitigate ACC (Section 7.4). Although landfills may be able to sequester a higher proportion of C over the long run (Kong et al., 2012; Morris et al., 2013), they also sequester the nutrients and energy-containing C compounds crops need to grow.

As described in Section 7.3.2, if gardeners are recycling only garden waste from the garden, they will have to supplement their soil and plants with other sources of nutrients because nutrients in their food and human waste are being removed from the system. Urban food gardens can be part of a low-impact solution that deals not only with the urban organic waste management problem, but also with the food supply problem—producing food using organic material from urban organic waste as the main soil amendment. This can be done, for example, by growing N-fixing leguminous crops, and adding compost made using organic material from outside the garden (e.g., yard waste, organic waste from purchased food, animal manure), or adding concentrated amendments such as pelleted organic fertilizers or natural mineral fertilizers, like crushed granite, or marl (Section 7.3.2, Table 7.1).

Ecological thinking about our kitchen and garden waste, and about growing food in our gardens, makes clear that they embody stages in the life cycle of water, C, nutrients, and energy. In combination with a prosocial response to ACC, this helps us to frame questions like: "What are the properties of the organic material, and how are these affecting the microorganisms that turn it into compost? Should I compost in piles, containers, trenches, or worm bins? Should I send organic waste to a landfill that captures methane produced by decomposition for energy production, which could offset the GHGE from burning fossil fuels? What compounds are the C and N part of, and what effect do these have on people and the environment? Why does all this matter in terms of food production, my community, and ACC?" We'll address these questions in the next section, using a study we did as an example. Sometimes the answers can lead us to learn new, or different ways of doing things, including how we handle organic waste in food gardens, and how we organize our activities.

7.5.3. Turning lawns into food gardens to fight climate change

Could food gardens, including using compost made on site from household organic waste, help to mitigate ACC? To answer this question, DAC and his students modeled the potential for organic household vegetable gardens in Santa Barbara County (SBC), California, to reduce GHGE (Fig. 7.5) by converting an area of lawn to a vegetable garden

(#3); and then replacing some purchased vegetables with ones grown in the garden (#1); diverting some greywater from export to treatment plants to water the garden (#2), and diverting some organic waste from export to landfills and large-scale composting to making compost on site for use in the garden (#4) (Cleveland et al., 2017).

We found that in the scenario described above, switching from a lawn to a garden could reduce the

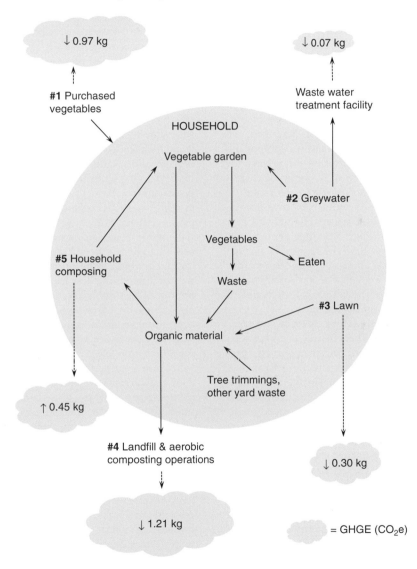

Figure 7.5. Results of a scenario modeling household vegetable gardens for lower GHGE. Increase or decrease is in kg CO_2e kg^{-1} vegetables (= lb CO_2e lb^{-1} vegetables)
┄► Change in GHGE based on changes described in research scenario
Based on Cleveland et al. (2017)

household's GHGE, with the biggest reduction (48%) from diverting some organic waste, followed by replacing purchased vegetables (38%), replacing lawn area (12%), and diverting some greywater (3%). However, composting organic waste (diverted from export) at the household level for use in the garden also created some GHGE (CH_4 and N_2O) (#5). Those emissions reduced the total emissions savings of this scenario by 18%. Still, the net reduction in GHGE was 2 kg CO_2e kg^{-1} of vegetable, so that if 50% of single family households grew 50% of their vegetables in these gardens, they would contribute 3% to the official GHGE mitigation targets for SBC, and if the same results were applied to all of California, it would contribute 8% of the state's mitigation target.

We also looked at some alternative scenarios to see how different options might affect the results. For example, in the scenario just described, we assumed that SBC landfills and aerobic composting operations had the same GHGE as the average for California found in one study (Kong et al., 2012). In an alternative scenario, households sent all their organic waste to landfills and aerobic composting operations that had the lowest GHGE found in a meta-analysis of 82 studies in a range of countries, which was much lower than the California average (Morris et al., 2013). The result was that the net reduction in GHGE, 4.3 kg CO_2e kg^{-1} of vegetable, would be more than twice that for the original scenario with household composting.

These results illustrate why gardeners and communities need to balance benefits and costs between individual composting to improve garden food production, with the benefits and costs to society and the environment, that is, balancing individualism with prosocial communalism. For example, the easiest thing to do with your organic waste might be to simply accumulate a pile of weeds, dead garden plants, and food waste in the corner of your garden and just leave it until it decomposes. But this could mean that it generates a lot of CH_4 and N_2O due to lack of aeration or a less than optimal C:N (Section 7.5.1). A more prosocial option, in other words, one with social as well as personal benefits, would be to take the time to aerate the pile regularly, for example by turning it, or using an aerating tool, and to make sure to have the right amount of moisture and C:N. Another prosocial solution that could involve less time, and provide even greater benefits, would be working with fellow gardeners to find out what the options are for efficient off-site organic waste composting, and to use them, or if they are not already available, to organize and advocate as a community for these.

7.6. Composting

Compost is a soil amendment made from decomposed organic material (Gajalakshmi and Abbasi, 2008; Brady and Weil, 2010, 391–393;). Composting has become increasingly popular in the US (Platt et al., 2014a), with many publications now available describing how to do it (Platt et al., 2014b). Compost, and other organic soil amendments, can be better than mineral fertilizers for plant growth (e.g., Ahmad et al., 2014) even though nutrient concentration is lower, because they release nutrients more slowly, contain a large number of different nutrients, other beneficial compounds, and living microbes, can contribute to long-term carbon sequestration, and improve the soil structure. Interest in composting in recent years has risen because of the need to greatly reduce waste going to landfills, the increasing scarcity and cost of the plant nutrients like phosphorus that organic material contains, the many benefits of compost for soil health, and the potential contribution of composting and compost application for mitigating ACC. If we don't make the best use of these resources it is not only a loss of benefits for our gardens, but also for the world. We also have to think of our own composting in the larger context in other ways, for example how it can be affected by large-scale antisocial behavior (Box 7.4).

Because organic material management can produce a wide range of results for your garden, your community, and beyond, it is important to understand the principles of composting, whether you are already doing it, or contemplating it for the future, in order to develop and improve techniques best suited to your needs and values. There are many excellent guides to composting in print and online that you can refer to for details about planning or improving your composting of food, garden and yard waste, and even your own manure in composting toilets (Section 7.7.4), and we won't repeat that information here. In the following sections we briefly discuss vermicomposting, and landfills and large-scale composting operations to help you consider if these could be a good option for your situation.

If you look beyond your garden for organic material for your composting, be sure you know a bit about its history—think ecologically about the lifecycle of what you compost. Recently, a class of widely used persistent herbicides (pyridine carboxylic acids) for controlling broad leaf (mostly eudicot) weeds in grass, hay, straw, and some vegetable production, were found in the manure of animals after they ate the treated plant material, and even in compost made with the treated material (Davis et al., 2015). These herbicides are so persistent that it has even been suggested that clippings of plants that have been sprayed could be used as a weed control mulch (Lewis et al., 2014). The herbicides are picloram, clopyralid, and aminopyralid, manufactured by Dow AgroSciences®, and found in a range of brand name products, such as Curtail®, Forefront®HL, and GrazonNext®HL, all manufactured by Dow, and many of which are combined with other herbicides such as 2, 4-D, a known carcinogen (Davis et al., 2015) (Chapter 9).

Food gardeners and farmers in several states have had problems with deformed or dying plants due to compost contaminated with these herbicides, which appear to be able to persist for years, including in soil and water supplies. Clopyralid can harm garden crops in the Fabaceae (bean), Solonaceae (tomato, chile, potato), and Asteraceae (sunflower) families (Michel and Doohan, 2003). It can stunt plants when it is present in compost in concentrations of only 10 parts per billion, and since the concentration of clopyralid when applied to grass to control broad leaf weeds is 10,000 to 50,000 parts per billion, it is not surprising that even a small amount of contaminated grass added to compost can cause problems.

Commercial composting companies have had to develop and implement tests for these herbicides at their own expense in order to protect their customers and their businesses, since it is so difficult to trace the possible contamination routes of compost inputs (Coker, 2015). The US EPA started a review of picloram, clopyralid, and aminopyralid in 2013, but they can continue to be sold during the review period, and were still easily available in 2018, for example on Amazon.com. This is a good example of antisocial corporate behavior driven by greed rather than empathy, resulting in the public paying the costs for the negative consequences.

7.6.1. Vermicomposting

Vermicomposting, which is composting with worms, can be an alternative or additional method for composting your food scraps to create nutrient-rich worm castings and worm tea (Munroe, 2011). *Epigeic* (surface dwelling) earthworms eat organic material and microbes, and excrete casts, which make nutrients more available to plants and promote beneficial microbe activity. A study comparing two composting systems, one with worms and one without worms, for five weeks found evidence that the addition of worms increased the rate of decomposition of fruit and vegetable wastes, and increased numbers and diversity of microbes, especially bacteria (Huang ct al., 2014). There is also evidence that when vermicompost is applied to soil it increases the number and diversity of bacteria and fungi populations and microbial activity compared to compost made without worms, and contains plant growth-enhancing materials not present in regular compost (Lim et al., 2015). One study found that microbes in vermicompost castings decreased the rate of bacterial infections in planted seeds of cucumber and other crops, resulting in healthier plants (Jack, 2012).

It is relatively easy to create your own vermicompost in your yard, garage, or even under your kitchen sink or desk, and when properly managed, it creates no unpleasant odors. There are many detailed guides to vermicomposting (Section 7.7.4), including ones for teachers, that can be used to create activities for children helping in the garden (e.g. Pagan and Steen, 2004).

7.6.2. Landfills and large-scale composting operations

As we already mentioned, detailed information about composting is widely available. However, many sources don't consider possible broader environmental and social impacts, especially in relation to other options because they assume that gardeners have the space, time, knowledge, and motivation for optimal composting (Section 7.5.3). Composting in a way that minimizes GHGE and optimizes quality for the garden requires attention, and not all gardeners or garden projects have the time or space to do this.

The alternative to composting at the garden site is to send the organic waste from the garden, yard, and kitchen to centralized organic waste management operations. Landfills and large-scale composting operations vary a lot in the way they process organic material, and how efficient they are, leading to a wide range of environmental impacts, including increasing or decreasing GHGE, and water and air pollution, compared with typical home composting (Morris et al., 2013). To make a well-informed decision requires comparing the estimated benefits and costs of the alternatives in terms of the value to your garden, community, and the larger world. For example, composting in the garden includes the benefits of the educational experience of making compost and the convenience of proximity, and the costs of the resources, space, and time to compost with minimal GHGE.

Exporting organic waste typically requires fossil fuel for trucks to transport it to the landfill or composting facility and to return the finished compost or mulch for use by gardeners, and to manage the landfill or composting facility, although this is a small part of their total environmental impact. On the other hand, a possible climate benefit of landfills is their potential for long-term C sequestration, although unlike sequestering in soil, landfill sequestration has no other environmental benefit. If landfills have systems to efficiently capture methane from the buried waste and burn it for energy generation, this can offset the much greater GHGE per unit energy from energy generation using fossil fuels, because the C in the CO_2 from burning organic waste is biogenic, i.e., it was only recently removed from the atmosphere. This means that landfills could have a net positive effect on emissions (Kong et al., 2012), assuming the energy generated actually replaces fossil fuel burning, rather than just adding to energy consumption. However, many factors that could reduce or eliminate these benefits need to be considered: the efficiency of methane capture in landfills, which can be far from 100%; limited room for landfills; environmental injustices including air, water, and noise pollution due to these often being located near minority and low-income communities.

If the organic waste is composted anaerobically to produce methane for energy generation, and the finished compost is also used by gardeners, this can offset emissions from peat or other fertilizers that it replaces, and be more environmentally friendly than landfills. This increases the positive effects of moving the processing of organic waste from households and communities to larger, more centralized operations (Levis and Barlaz, 2011). Community involvement can play an important role in getting municipalities to organize organic waste management at a larger scale, and for helping decide on waste pick-up and compost delivery systems that would make participation easy and worthwhile, and community-level composting is a growing movement (ILSR, 2018).

7.7. Resources

All the websites listed below were verified on August 10, 2018.

7.7.1. General

Teaching organic farming and gardening is a manual from the University of California, Santa Cruz, that has a lot of good, practical information on managing soils (Miles and Brown, 2005).

An excellent, well-written textbook on soils that we use as a reference is *Elements of the nature and properties of soil* by Brady and Weil (2010) which is a condensed version of their standard textbook.

A simple, clearly written article giving an overview of soil science for watershed management is available for free online (Schoonover and Crim, 2015).

A resource for the home gardener in the US, including for soil questions, is the local Master Gardener program of each state's Cooperative Extension Service, typically housed in state universities. Master Gardeners are regionally focused, and are available to answer most gardening questions by phone, or in person in their county offices. They also hold local programs to inform and involve the public and local gardening community. Master Gardener programs can be fairly conventional about the inputs they recommend, including seeds, pesticides, and fertilizers, so you will want to compare their suggestions with those from other sources.
http://articles.extension.org/mastergardener

The USDA soils homepage has links to educational materials, online textbook-style information about soil quality and biology, current versions of soil taxonomy, graphics, and more.
http://soils.usda.gov/

7.7.2. Soil types and soil maps

The USDA provides a free online soil texture calculator.
http://www.nrcs.usda.gov/wps/portal/nrcs/detail/soils/survey/?cid=nrcs142p2_054167

The University of California, Davis has interactive soil maps for the US that can be viewed in Google Maps (SoilWeb) or 3-D in Google Earth (SoilWeb Earth) (http://casoilresource.lawr.ucdavis.edu/soilweb-apps/). Clicking on the mapped units for your location gives detailed information, including soil profiles. It is also possible to download 1:24,000 scale soil survey maps for use with Google Earth, including smartphone apps.
https://casoilresource.lawr.ucdavis.edu/

The USDA's Natural Resources Conservation Service has maps of soils available online (http://www.nrcs.usda.gov/wps/portal/nrcs/detail/soils/survey/class/?cid=nrcs142p2_053589), including detailed maps at the county and subcounty level (http://www.nrcs.usda.gov/wps/portal/nrcs/soilsurvey/soils/survey/state/), as well as detailed descriptions and interactive maps (http://websoilsurvey.nrcs.usda.gov/app/).

7.7.3. Nutrients and nutrient management

The 2005 FAO report, *The importance of soil organic matter: Key to drought-resistant soil and sustained food production* (Bot and Benites, 2005), is a good resource for information on soil organic matter and its role in production.

The University of California has a list of its downloadable extension publications on nutrient management for organic crop production.
http://ucanr.edu/sites/nm/Organic_production/

A condensed guide to diagnosing crop nutrient deficiencies from the University of Arizona extension is available here:
https://extension.arizona.edu/sites/extension.arizona.edu/files/pubs/az1106.pdf

The University of Hawai'i at Manoa has an excellent nutrient management website, with detailed information on most nutrients, though some is Hawai'i specific.
http://www.ctahr.hawaii.edu/MauiSoil/manage.aspx

Extension.org is part of the USDA's Cooperative Extension Service and publishes accessible, short bulletins about many different subjects including organic agriculture, all written by public university scientists, and some other experienced practitioners.
http://articles.extension.org/organic_production

Useful articles about soil tests and more can be found on the website of the Soil Testing Lab at Colorado State University.
http://www.soiltestinglab.colostate.edu/index.html

Helpful guides to soil nutrient testing and fertilizer application are published by Washington State University (Cogger, 2014) and the University of Georgia (UGE, 2014).

The USDA's National Organic Program (NOP) site includes links to NOP's regulations and the national list of allowable and prohibited substances including soil amendments, for USDA organic certification.
http://www.ams.usda.gov/AMSv1.0/nop

A short primer on biochar for gardeners was published by Washington State University.
http://cru.cahe.wsu.edu/CEPublications/FS147E/FS147E.pdf

7.7.4. Composting, vermicomposting

Cornell Waste Management Institute has useful information about composting including at the household scale with clear extension leaflets and scientifically-based information.
http://cwmi.css.cornell.edu/index.html

Vermicomposting information is available from Cornell University Cooperative Extension.
http://cwmi.css.cornell.edu/vermicompost.htm

A thorough online book from Canada's Organic Agriculture Centre about vermicomposting is:
https://www.eawag.ch/fileadmin/Domain1/Abteilungen/sandec/E-Learning/Moocs/Solid_Waste/W4/Manual_On_Farm_Vermicomposting_Vermiculture.pdf

The Institute for Local Self-Reliance provides resources for organizing small-scale community composting (ILSR, 2018), and has a number of useful publications (Platt et al., 2014a, 2014b).
https://ilsr.org/composting/

7.8. References

Ahmad, A., Hue, N. & Radovich, T. (2014) Nitrogen release patterns of some locally made composts and their effects on the growth of Chinese cabbage (*Brassica rapa*, Chinensis Group) when used as soil amendments. *Compost Science & Utilization*, 22, 199–206, DOI: 10.1080/1065657X.2014.920282.

Battye, W., Aneja, V. P. & Schlesinger, W. H. (2017) Is nitrogen the next carbon? *Earth's Future*, 5, 894–904, DOI: doi:10.1002/2017EF000592.

Berruti, A., Lumini, E., Balestrini, R. & Bianciotto, V. (2015) Arbuscular mycorrhizal fungi as natural biofertilizers: Let's benefit from past successes. *Frontiers in Microbiology*, 6, available at: http://www.ncbi.nlm.nih.gov/pmc/articles/PMC4717633/ (accessed July 1, 2018), DOI:10.3389/fmicb.2015.01559.

Bot, A. & Benites, J. (2005) *The importance of soil organic matter: Key to drought-resistant soil and sustained food production, FAO Soils Bulletin 80*. FAO Land and Plant Nutrition Management Service, Food and Agriculture Organization of the United Nations, Rome.

Brady, N. C. & Weil, R. R. (2010) *Elements of the nature and properties of soil*, 3rd edn. Prentice Hall, Upper Saddle River, NJ.

Canning, P., Rehkamp, S., Waters, A. & Etemadnia, H. (2017) *The role of fossil fuels in the U.S. food system and the American diet, ERR-224*. US Department of Agriculture, Economic Research Service, available at: https://www.ers.usda.gov/publications/pub-details/?pubid=82193 (accessed Oct. 12, 2018).

Cayuela, M. L., Sanchez-Monedero, M. A., Roig, A., Hanley, K., Enders, A. & Lehmann, J. (2013) Biochar and denitrification in soils: When, how much and why does biochar reduce N_2O emissions? *Scientific Reports*, 3, DOI: 10.1038/srep01732.

Classen, A. T., Sundqvist, M. K., Henning, J. A., Newman, G. S., Moore, J. A. M., Cregger, M. A., Moorhead, L. C. & Patterson, C. M. (2015) Direct and indirect effects of climate change on soil microbial and soil microbial-plant interactions: What lies ahead? *Ecosphere*, 6, art130, DOI: 10.1890/ES15-00217.1.

Cleveland, D. A. & Soleri, D. (1991) *Food from dryland gardens: An ecological, nutritional, and social approach to small-scale household food production*. Center for People, Food and Environment (with UNICEF), Tucson, AZ. https://tinyurl.com/FFDG-1991

Cleveland, D. A., Phares, N., Nightingale, K. D., Weatherby, R. L., Radis, W., Ballard, J., Campagna, M., Kurtz, D., Livingston, K., Riechers, G., et al. (2017) The potential for urban household vegetable gardens to reduce greenhouse gas emissions. *Landscape and Urban Planning*, 157, 365–374, DOI: http://dx.doi.org/10.1016/j.landurbplan.2016.07.008.

Cogger, C. (2014) *A home gardener's guide to soils and fertilizers*. Washington State University, available at: https://extension.arizona.edu/sites/extension.arizona.edu/files/pubs/az1106.pdf (accessed July 21, 2018).

Coker, C. (2015) Coping with persistent herbicides in composting feedstocks. *BioCycle*, 56, 44, available at: https://www.biocycle.net/2015/01/14/coping-with-persistent-herbicides-in-composting-feedstocks/ (accessed Dec. 6, 2018).

Davis, J., Johnson, S. E. & Jennings, K. (2015) Herbicide carryover in hay, manure, compost, and grass clippings: Caution to hay producers, livestock owners, farmers, and home gardeners. North Carolina Cooperative Extension, available at: http://www.nccgp.org/images/uploads/resource_files/Herbicide_Carryover_NCSU.pdf (accessed July 1, 2018).

Deepika, S. & Kothamasi, D. (2015) Soil moisture-a regulator of arbuscular mycorrhizal fungal community assembly and symbiotic phosphorus uptake. *Mycorrhiza*, 25, 67–75, DOI: 10.1007/s00572-014-0596-1.

Durst, P. & Bosworth, S. (2008) Inoculation of forage and grain legumes. *Agronomy Facts*, 11. Penn State College of Agricultural Sciences, available at: https://extension.psu.edu/inoculation-of-forage-and-grain-legumes (accessed Dec. 6, 2018).

EPA (US Environmental Protection Agency) (2016) *Inventory of U.S. greenhouse gas emissions and sinks: 1990–2014*. EPA, available at: https://www3.epa.gov/climatechange/ghgemissions/usinventoryreport.html (accessed Oct. 11, 2018).

EPA (US Environmental Protection Agency, Office of Resource Conservation and Recovery) (2018) Composting. In: *Documentation for greenhouse gas emission and energy factors used in the waste reduction model (WARM), management practices chapters*. EPA, Washington, DC, available at: https://www.epa.gov/sites/production/files/2016-03/documents/warm_v14_management_practices.pdf (accessed July 19, 2018).

EPA, U. (2015) *Advancing sustainable materials management: 2013 fact sheet*. US EPA, Washington DC, available at: https://19january2017snapshot.epa.gov/sites/production/files/2015-09/documents/2013_advncng_smm_fs.pdf (accessed Oct. 11, 2018).

Epstein, E. & Bloom, A. J. (2004) *Mineral nutrition of plants: Principles and perspectives*, 2nd edn, Sinauer Associates, Sunderland, MA.

Faber, B. A., Downer, A. J., Holstege, D. & Mochizuki, M. J. (2007) Accuracy varies for commercially available soil test kits analyzing nitrate–nitrogen, phosphorus, potassium, and pH. *HortTechnology*, 17, 358–362.

Fedonkin, M. A. (2009) Eukaryotization of the early biosphere: A biogeochemical aspect. *Geochemistry International*, 47, 1265–1333, DOI: 10.1134/s0016702909130011.

Ferrol, N., Tamayo, E. & Vargas, P. (2016) The heavy metal paradox in arbuscular mycorrhizas: From mechanisms to biotechnological applications. *Journal of Experimental Botany*, 67, 6253–6265, DOI: 10.1093/jxb/erw403.

Gajalakshmi, S. & Abbasi, S. A. (2008) Solid waste management by composting: State of the art. *Critical Reviews in Environmental Science and Technology*, 38, 311–400, DOI: 10.1080/10643380701413633.

Glaser, B. & Birk, J. J. (2012) State of the scientific knowledge on properties and genesis of anthropogenic dark earths in Central Amazonia (*terra preta de Índio*). *Geochimica et Cosmochimica Acta*, 82, 39–51, DOI: 10.1016/j.gca.2010.11.029.

He, Y., Trumbore, S. E., Torn, M. S., Harden, J. W., Vaughn, L. J. S., Allison, S. D. & Randerson, J. T. (2016) Radiocarbon constraints imply reduced carbon uptake by soils during the 21st century. *Science*, 353, 1419–1424, DOI: 10.1126/science.aad4273.

Homburg, J. A. & Sandor, J. A. (2011) Anthropogenic effects on soil quality of ancient agricultural systems of the American Southwest. *CATENA*, 85, 144–154, DOI: http://dx.doi.org/10.1016/j.catena.2010.08.005.

Huang, K., Li, F., Wei, Y., Fu, X. & Chen, X. (2014) Effects of earthworms on physicochemical properties and microbial profiles during vermicomposting of fresh fruit and vegetable wastes. *Bioresource Technology*, 170, 45–52, DOI: http://dx.doi.org/10.1016/j.biortech.2014.07.058.

ILSR (Institute for Local Self-Reliance) (2018) Composting for community, available at: https://ilsr.org/composting/ (accessed July 21, 2018).

IPCC (Intergovernmental Panel on Climate Change) (2013) *Climate change 2013: The physical science basis. Working Group I contribution to the Fifth Assessment Report of the Intergovernmental Panel On Climate Change*. IPCC, Geneva, available at: http://www.climatechange2013.org/images/report/WG1AR5_SPM_FINAL.pdf (accessed Oct. 11, 2018).

Jack, A. L. H. (2012) *Vermicompost suppression of Pythium aphanidermatum seedling disease. Practical applications and an exploration of the mechanisms of disease suppression*. PhD dissertation, plant pathology and plant microbe biology. Cornell University, Ithaca, NY.

Jiang, Y., Wang, W., Xie, Q., Liu, N., Liu, L., Wang, D., Zhang, X., Yang, C., Chen, X., Tang, D., et al. (2017) Plants transfer lipids to sustain colonization by mutualistic mycorrhizal and parasitic fungi. *Science*, 356, 1172–1175, DOI: 10.1126/science.aam9970.

Kaplan, J. O., Krumhardt, K. M., Ellis, E. C., Ruddiman, W. F., Lemmen, C. & Goldewijk, K. K. (2011) Holocene carbon emissions as a result of anthropogenic land cover change. *Holocene*, 21, 775–791, DOI: 10.1177/0959683610386983.

Kirkby, C. A., Richardson, A. E., Wade, L. J., Passioura, J. B., Batten, G. D., Blanchard, C. & Kirkegaard, J. A. (2014) Nutrient availability limits carbon sequestration in arable soils. *Soil Biology and Biochemistry*, 68, 402–409.

Kong, D., Shan, J., Iacoboni, M. & Maguin, S. R. (2012) Evaluating greenhouse gas impacts of organic waste management options using life cycle assessment. *Waste Management & Research*, 30, 800–812, DOI: 10.1177/0734242x12440479.

Leigh, G. J. (2004) *The world's greatest fix: A history of nitrogen and agriculture*. Oxford University Press, New York, NY.

Levis, J. W. & Barlaz, M. A. (2011) What is the most environmentally beneficial way to treat commercial food waste? *Environmental Science & Technology*, 45, 7438–7444, DOI: 10.1021/es103556m.

Lewis, D. F., Jeffries, M. D., Gannon, T. W., Richardson, R. J. & Yelverton, F. H. (2014) Persistence and bioavailability of aminocyclopyrachlor and clopyralid in turfgrass clippings: Recycling clippings for additional weed control. *Weed Science*, 62, 493–500.

Lim, S. L., Wu, T. Y., Lim, P. N. & Shak, K. P. Y. (2015) The use of vermicompost in organic farming: Overview, effects on soil and economics. *Journal of the Science of Food and Agriculture*, 95, 1143–1156, DOI: 10.1002/jsfa.6849.

Luginbuehl, L. H., Menard, G. N., Kurup, S., Van Erp, H., Radhakrishnan, G. V., Breakspear, A., Oldroyd, G. E. D. & Eastmond, P. J. (2017) Fatty acids in arbuscular mycorrhizal fungi are synthesized by the host plant. *Science*, 356, 1175–1178, DOI: 10.1126/science.aan0081.

Michel, F. C. & Doohan, D. (2003) *Clopyralid and other pesticides in composts. Extension FactSheet AEX-714-03*. The Ohio State University Extension. https://www.global2000.at/sites/global/files/Clopyralid_Factsheet.pdf (accessed Jan. 16, 2019).

Miles, A. & Brown, M. (eds) (2005) *Teaching organic farming and gardening*, 2nd edn. UCSC Farm & Garden Apprenticeship, Center for Agroecology & Sustainable Food Systems, University of California, Santa Cruz, CA.

Morris, J., Scott Matthews, H. & Morawski, C. (2013) Review and meta-analysis of 82 studies on end-of-life management methods for source separated organics. *Waste Management*, 33, 545–551, DOI: http://dx.doi.org/10.1016/j.wasman.2012.08.004.

Mulcahy, D. N., Mulcahy, D. L. & Dietz, D. (2013) Biochar soil amendment increases tomato seedling resistance to drought in sandy soils. *Journal of Arid Environments*, 88, 222–225, DOI: http://dx.doi.org/10.1016/j.jaridenv.2012.07.012.

Munroe, G. (2011) *Manual of on-farm vermicomposting and vermiculture*. Organic Agriculture Centre of Canada, Dalhousie University, Truro, NS.

Norton, J. B., Sandor, J. A. & White, C. S. (2007) Runoff and sediments from hillslope soils within a Native American agroecosystem. *Soil Science Society of America Journal*, 71, 476–483, DOI: 10.2136/sssaj2006.0019.

Pagan, T. & Steen, R. (2004) *The worm guide: A vermicomposting guide for teachers*. California

---

Integrated Waste Management Board (CIWMB), Office of Education and the Environment, available at: https://www2.calrecycle.ca.gov/Publications/Details/912 (accessed 22 Oct. 2018).

Platt, B., Goldstein, N., Coker, C. & Brown, S. (2014a) *State of composting in the US: What, why, where & how*. The Institute for Local Self-Reliance (ILSR), available at: http://www.ilsr.org/state-of-composting/ (accessed Oct. 11, 2018).

Platt, B., McSweeney, J. & Davis, J. (2014b) *Growing local fertility: A guide to community composting*. Highfields Center for Composting and The Institute for Local Self-Reliance (ILSR), available at: http://www.ilsr.org/growing-local-fertility/ (accessed Oct. 11, 2018).

Powlson, D. S., Stirling, C. M., Jat, M. L., Gerard, B. G., Palm, C. A., Sanchez, P. A. & Cassman, K. G. (2014) Limited potential of no-till agriculture for climate change mitigation. *Nature Climate Change*, 4, 678–683, DOI: 10.1038/nclimate2292.

Quintaes, K. D. & Diez-Garcia, R. W. (2015) The importance of minerals in the human diet. In de la Guardia, M. & Garrigues, S. (eds) *Handbook of mineral elements in food*. John Wiley & Sons, DOI:10.1002/9781118654316.ch1.

Robertson, G. P., Gross, K. L., Hamilton, S. K., Landis, D. A., Schmidt, T. M., Snapp, S. S. & Swinton, S. M. (2014) Farming for ecosystem services: An ecological approach to production agriculture. *BioScience*, 404–415, DOI: 10.1093/biosci/biu037.

Rogers, A., Ainsworth, E. A. & Leakey, A. D. B. (2009) Will elevated carbon dioxide concentration amplify the benefits of nitrogen fixation in legumes? *Plant Physiology*, 151, 1009–1016, DOI: 10.1104/pp.109.144113.

Ruddiman, W. F. (2013) The Anthropocene. In Jeanloz, R. (ed.) *Annual review of earth and planetary sciences*, 41, 45–68. Palo Alto, CA, DOI: 10.1146/annurev-earth-050212-123944.

Sandor, J. A., Norton, J. B., Homburg, J. A., Muenchrath, D. A., White, C. S., Williams, S. E., Havener, C. I. & Stahl, P. D. (2007) Biogeochemical studies of a Native American runoff agroecosystem. *Geoarchaeology: An International Journal*, 22, 359–386, DOI: 10.1002/gea.20157.

Schmidt, H.-P. & Niggli, C. (2012) Biochar gardening: Results 2011. *Ithaka Journal*, 265–269, available at: www.ithaka-journal.net/druckversionen/e042012-bc-gardening.pdf (accessed July 5, 2018).

Schoonover, J. E. & Crim, J. F. (2015) An introduction to soil concepts and the role of soils in watershed management. *Journal of Contemporary Water Research & Education*, 154, 21–47, DOI: 10.1111/j.1936-704X.2015.03186.x.

Shaxson, F. & Barber, R. (2003) Optimizing soil moisture for plant production: The significance of soil porosity. *FAO Soils Bulletin*, 79, Food and Agriculture Organization of the United Nations, Rome.

Smil, V. (2004) *Enriching the earth: Fritz Haber, Carl Bosch, and the transformation of world food production*. MIT Press, Cambridge, MA.

Soil Science Division Staff (2017) *Soil survey manual, USDA Handbook 18* (Ditzler, C., Scheffe, K. & Monger, H. C. eds), Government Printing Office, Washington, DC.

Spurgeon, D. J., Keith, A. M., Schmidt, O., Lammertsma, D. R. & Faber, J. H. (2013) Land-use and land-management change: relationships with earthworm and fungi communities and soil structural properties. *BMC Ecology*, 13, DOI: 10.1186/1472-6785-13-46.

Steinbeiss, S., Gleixner, G. & Antonietti, M. (2009) Effect of biochar amendment on soil carbon balance and soil microbial activity. *Soil Biology and Biochemistry*, 41, 1301–1310, DOI: 10.1016/j.soilbio.2009.03.016.

Thakkar, J., Kumar, A., Ghatora, S. & Canter, C. (2016) Energy balance and greenhouse gas emissions from the production and sequestration of charcoal from agricultural residues. *Renewable Energy*, 94, 558–567, DOI: http://dx.doi.org/10.1016/j.renene.2016.03.087.

Thauer, R. K., Kaster, A.-K., Seedorf, H., Buckel, W. & Hedderich, R. (2008) Methanogenic archaea: Ecologically relevant differences in energy conservation. *Nature Reviews Microbiology*, 6, 579–591.

Thompson, E., Jr. (2007) Paving paradise: A new perspective on California farmland conversion. American Farmland Trust, available at: https://www.farmlandinfo.org/paving-paradise-new-perspective-california-farmland-conversion (accessed Oct. 22, 2018).

UGE (University of Georgia Extension) (2014) How to convert an inorganic fertilizer recommendation to an organic one, Available at: http://extension.uga.edu/publications/detail.html?number=C853&title=How%20to%20Convert%20an%20Inorganic%20Fertilizer%20Recommendation%20to%20an%20Organic%20One (accessed July 21, 2018).

Verbruggen, E., Heijden, M. G., Rillig, M. C. & Kiers, E. T. (2013) Mycorrhizal fungal establishment in agricultural soils: Factors determining inoculation success. *New Phytologist*, 197, 1104–1109.

Wang, Y. Z., Zhang, L. W., Yang, H., Yan, G. J., Xu, Z. H., Chen, C. R. & Zhang, D. K. (2016) Biochar nutrient availability rather than its water holding capacity governs the growth of both C3 and C4 plants. *Journal of Soils and Sediments*, 16, 801–810, DOI: 10.1007/s11368-016-1357-x.

Whiting, D., Card, A., Wilson, C. & Reeder, J. (2011) *Estimating soil texture: Sandy, loamy or clayey? CMG Garden Notes #214*, Colorado Master Gardener Program, Colorado State University Extension, Fort Collins, CO, available at: http://www.ext.colostate.edu/mg/gardennotes/214.html (accessed Oct. 11, 2018).

Whiting, D., Card, A., Wilson, C. & Reeder, J. (2014) *Soil tests. CMG GardenNotes #221*. Colorado Master Gardener Program, Colorado State University Extension, Fort Collins, CO, available at: http://www.ext.colostate.edu/mg/gardennotes/221.pdf (accessed Oct. 11, 2018).

8

Water, Soils, and Plants

D.A. Cleveland, S.E. Smith, D. Soleri

Chapter 8 in a nutshell.

- Agriculture accounts for most fresh water use globally.
- Water is an essential, but increasingly scarce garden resource.
- Storage and movement of water in the soil depends on soil texture and structure, and the amount of organic matter.
- Evaporation, plant roots and gravity remove water from the root zone.
- To maximize harvest per area of land, keep water in the root zone at 50% or more of field capacity.
- To increase water use efficiency in order to get the most harvest and other benefits for the amount of water used, you can:
  - use deficit irrigation by letting water in the root zone drop below 50% of field capacity before irrigating, especially during the least sensitive parts of a crop life cycle;
  - choose crop varieties that are adapted to the current and changing water availability in your garden;
  - minimize excess evapotranspiration by using mulch, shades, wind breaks, and planting patterns;
  - minimize water lost to runoff by increasing infiltration;
  - minimize water lost to drainage below the root zone by not overwatering;
  - use harvested rainwater or greywater instead of piped water when possible to help increase water use efficiency at the community scale and beyond;
  - use the irrigation methods best for you and your garden;
  - use the types of garden beds best adapted to your climate and other gardening goals.

Water as a liquid, and a gas, is constantly being cycled among plants and animals, soil, air, and bodies of water such as streams, lakes, rivers, and oceans. Water is essential for all life, but good quality water for people and plants is becoming scarcer due to anthropogenic climate change (ACC), increasing demand, and pollution, especially in already dry areas. And scarcity is increasing disputes due to inequity in rights and access to water (Hsiang et al., 2013). Yet, as the consequences of the trend of increasing water scarcity become apparant, this could motivate cooperation at all levels to "balance competing demands and defuse potential conflict," in order to avoid the tragedy of an open-access resource (Rodell et al., 2018) (Box 3.3).

In our food gardens, water carries plant nutrients from the roots upward, and carbohydrates from photosynthesis throughout the plant. Water is also a medium for chemical reactions, cools the plant by evaporating from the leaves, and is essential for photosynthesis. The immediate goal of garden water management is to provide plants with enough water to produce a harvest and other benefits for a reasonable investment of time, money, and other resources, and without creating salinity or water-logging problems (Sections 8.2.1, 8.8.5). Another important goal of garden water management is strengthening equitable access to good quality water for people, while providing adequate water for food production and natural ecosystems, now and in the future.

In this chapter we look at some concepts of water-soil-plant relationships that are key to achieving these goals as water becomes scarcer.

8.1. Food's Water Footprint

The *water footprint* (WF) of food is broadly defined as the amount of fresh water needed to produce a given amount of food, which includes rainwater (the green WF), irrigation water (the blue WF), plus the water required to ensure that nutrient and other agrochemical runoff leaving the field is sufficiently diluted to be safe (the grey WF, not related to household greywater, Section 8.7.4).

Water footprints for crops are the reciprocal of agronomic water use efficiency (WUE, the amount of harvest per unit of water) described in Box 5.1 and discussed below (Section 8.3). Whereas WF is defined as g of water g^{-1} of harvest, agronomic WUE is usually defined as kg of harvest m^{-3} of water (Katerji and Mastrorilli, 2014).

Globally, food production accounts for 99% of our blue and green WF (Mekonnen and Hoekstra, 2014). One study estimated the WF for industrial fresh tomatoes in Spain to be 82 g of water to produce 1 g of tomato, with 17% rainwater (14 g), 74% irrigation (61 g), and an additional 9% (7 g) needed to dilute irrigation drainage water to an acceptable quality due to pollution in the field (Chapagain and Orr, 2009).

Fruits and vegetables have a wide range of WF, but on average they are much lower than the WF of animal foods. For example, the combined blue, green, and grey WF kg^{-1} of beef, chicken, eggs, and milk are 48, 13, 10 and 3 times that of vegetables, and even the WF of the protein content of these foods is 4.3, 1.3, 1.1 and 1.2 times that of vegetable protein (calculated using data in Mekonnen and Hoekstra, 2012). Gardens can help reduce our total food WF by encouraging more fruit and vegetable consumption to replace animal food consumption, which will reduce greenhouse gas emissions (GHGE) and can improve our nutrition and health (Section 1.1.2).

8.2. Water, Soil and Plant Interactions

Plants get the water they need through their roots, and many soil characteristics affect the availability of water to plants, and help determine how much water the garden will need.

8.2.1. Basic water soil dynamics

Water in the soil is present in soil pores, the spaces between particles of soil (Section 7.2.2). If water is added until all the pores in the soil are filled and there is no room for any more, the soil is *saturated* (Fig. 8.1). If a saturated soil is allowed to stand with no additional water being added, the *free* or *gravitational water*—water that is not held by adhesion to soil particles or cohesion to that adhered water—will drain down and out of the root zone due to gravity, and air will take its place. Since most soils drain gravitational water from the root zone relatively rapidly, it is not available to plants.

When the point is reached where no more water drains out by gravity, the water content of the soil is at *field capacity* (FC), and water movement in the soil is by *capillary action* in the soil pores (Section 5.1) due to cohesion of water molecules to each other, and adhesion to soil particles in the small capillary spaces of the soil (Brady and Weil, 2010, 137–140, 154–162). Water moves by capillary action from wetter areas where it has higher energy, to drier areas where it has lower energy. The energy of water in the soil, the *soil water potential*—a property of the water—is commonly measured in units of pressure called *bars*. The more negative the value, the lower its energy potential, the more tightly the water is held, and the more external energy required to remove it, for example by the roots. That is, water in drier soil is closer to soil particles and held more tightly to them, and therefore that water has lower energy, compared with water held less tightly in wetter soil (Brady and Weil, 2010, 135–144).

When the soil water content is at FC, conditions are optimal for crops because the water available to plants is maximized, and there is still ample pore space for the air needed for root respiration. Soil porosity is determined by soil structure, which in turn is the result of soil texture, organic matter, living organisms, cultivation, and other factors (Section 7.2.2). As water in the soil is taken up by the plant roots, the remaining water is closer and closer to soil particles and more and more tightly held by them—soil water potential is decreasing, becoming more negative. For example, the water in garden soil with a water potential of −0.5 bars is more easily available to plants than water in soil with a water potential of −0.7 bars. When the amount of water in the soil is reduced to the point where the plant cannot absorb it fast enough to stay alive, because it is held so tightly in the soil matrix, soil water content is at the *permanent wilting point* (PWP). Both FC and PWP are characteristics of soils in terms of their water content. Field capacity is often considered to be −0.1 to −0.3 bars and PWP −15 bars. In practice, however, the PWP depends on the crop and variety, and on the soil's properties including texture, structure, and organic matter.

The water held in the soil between FC and the PWP is water that can be absorbed fast enough by plants to grow and produce. This is called *plant available soil water* content (AW):

$$AW = FC - PWP$$

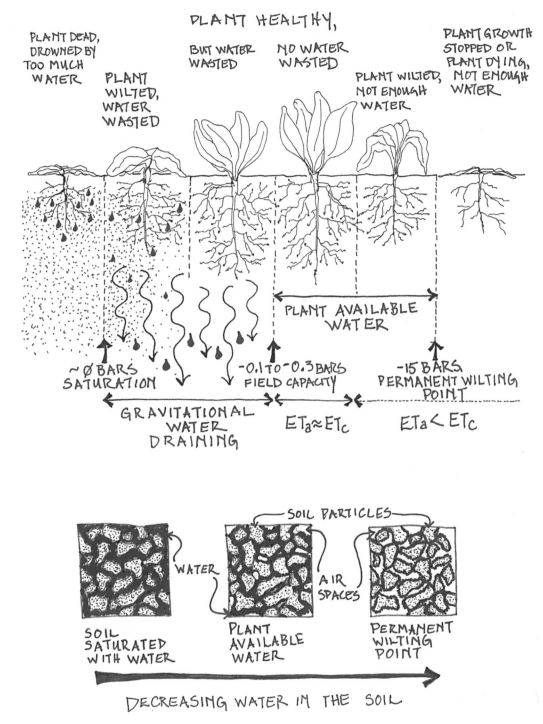

Figure 8.1. Soil water and plant growth

Adapted from Cleveland and Soleri (1991) with permission

To maximize yield (in terms of harvest per unit area), most garden crops should be watered when about 50% of plant AW has been depleted (Section 8.6). Even though a plant will be able to survive on the remaining 50%, it may have a water deficit for part of the day. This means water will be transpired faster than it can be taken up, which can cause wilting and decreased total yield (Section 8.3.1). However, as discussed in Section 8.3.1, experimenting with deficit irrigation may be a good strategy when water is the main limitation to garden production.

Too much water can also harm or even kill plants. If a soil remains saturated because water is being added faster than it can drain, or because of a high water table, the soil becomes *waterlogged*. Most plants cannot grow in waterlogged soils because there is no air for the roots, it has been squeezed out by the water (Fig. 8.1). Only a few food plants, like wild rice, are *hydrophytes*, which can tolerate waterlogged soils due to adaptations like air spaces in their stalks and roots that allow movement of oxygen from aboveground to belowground plant tissues.

8.2.2. Water movement in the soil

When irrigating, the soil is saturated, and most water movement is due to the downward force of gravity, and the pattern and rate of water movement depends on soil texture, structure, depth, and organic matter content. However, because of capillary action (Section 8.2.1) some movement takes place in all directions toward areas with less water. In clayey soils with many fine pore spaces, water moves further horizontally by capillary action than in sandy soils that have much larger particles and fewer and larger pores (c.f. Figs 8.2a, b).

Water movement in soils with distinct layers of different textures, for example layers of sand or clay in a loam soil (Fig. 8.2c), will be slowed because of the change in pore size. An abrupt change from sandy to clayey texture will slow the movement of water because it moves more slowly in the much smaller pores of the clay. Downward movement will also be slowed by a transition from clayey to sandy texture because the water potential in the clayey soil is much lower due to being held more tightly in the small pores (Fig. 8.2d, e), and the bottom of the clayey soil needs to become saturated before water will move into the sand below it. A dense layer of caliche, ironstone, or rock will practically stop downward water flow, causing the soil above it to become saturated (Fig. 8.2f). A separate layer of organic material will also slow the downward movement of

water for the same reason as a transition from clayey to sandy texture. However, when mixed in with the soil in the root zone, organic material speeds downward water movement (Fig. 8.2g). Organic material also improves soil structure as it breaks down to form humus (Section 7.5.1), creating soil aggregates and larger pore spaces that water can penetrate faster and deeper. Vertical organic mulches open to the surface allow water to move quickly into the root zone, and organic surface mulches protect the soil surface from compaction by raindrops which reduces infiltration (Fig. 8.2h; Sections 8.3.2, 8.8.3).

8.2.3. Water uptake and transport by plants

Plants contain a much greater proportion of water than animals of the same size and weight. Most water is recycled internally in animals, whereas in plants water flows through and is being lost through transpiration when stomata are open to obtain the CO_2 required for photosynthesis, to cool the plant, and to drive nutrient uptake (Section 5.5). Water and dissolved nutrients move from the roots to the stems and leaves through the vascular system (Section 5.1). Plant roots actively grow in moist soil toward areas of higher water content, and absorb water and nutrients through the large surface area of their root hairs (Section 5.3). As water is lost through transpiration it must be replaced with more water absorbed from the soil by the roots in a continuous process.

The rate of transpiration from garden crops is affected by many factors, including the garden microclimate and the types of crops being grown. Transpiration is commonly measured in combination with evaporation, referred to as *evapotranspiration* (ET) because transpiration is difficult to measure separately, and both are relevant for agronomic WUE. *Crop evapotranspiration* rate (ETc) is the maximum ET rate for a specific crop when the crop requirements for water are fully met, i.e., growth and production are not limited by water availability or plant stress (Snyder, 2014). That's why an important goal of garden management when thinking of getting the most productivity for an area is to maintain *actual evapotranspiration* (ETa) close to ETc by keeping water in the root zone at no less than 50% of field capacity. Deficit irrigation is an alternative strategy to maximize WUE when water is more limited than land (Section 8.3.1).

It's also important to optimize WUE by minimizing excess transpiration from stressful conditions such as high temperature, low humidity, or high wind, and minimizing evaporation, runoff and deep percolation

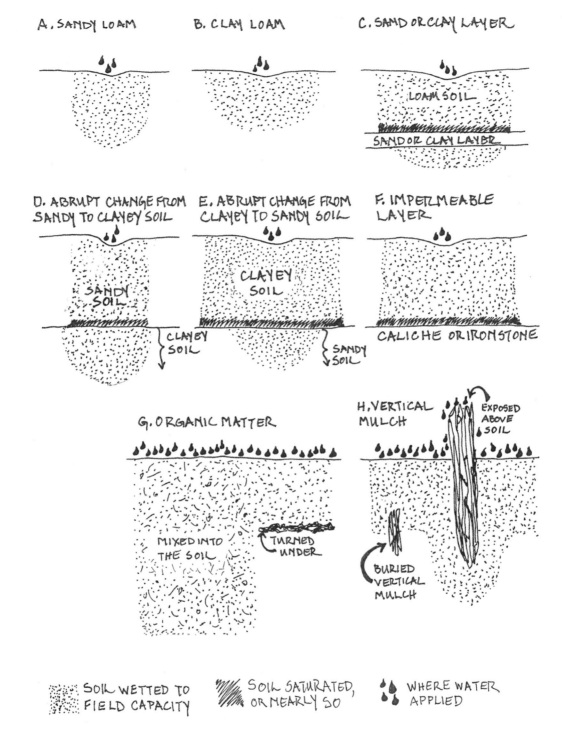

Figure 8.2. Water movement in the soil

Based on Gardner (1962, reprinted 1979); Gardner (1988)

(Figs 8.1, 8.3), and we'll see how to do this in the next section. In Section 2.3.1 and Table 2.4 we discuss the possible effects of ACC on transpiration rates.

8.3. Dryland Garden Water Management

Agronomic WUE—the amount of harvest per unit of water applied (irrigation plus precipitation)—is a key concept for understanding how to respond to increasing water scarcity (Box 5.1).

8.3.1. Water management for water use efficiency

Conserving scarce water resources by reducing unnecessary losses is key for increasing agronomic WUE, by maximizing the amount of the available water in the root zone where plants can use it (Fig. 8.3). To increase agronomic WUE, gardeners can:

- use an appropriate source of water and irrigation method (Sections 8.7, 8.8);
- minimize loss of water below the root zone by not overwatering, and by increasing the soil's water-holding ability with organic matter (Section 7.5);
- minimize runoff and maximize infiltration of rain or irrigation water into the garden soil by improving soil structure and by using mulch (Sections 7.2.2, 8.3.2), and efficient irrigation methods (Section 8.8.1);
- minimize evaporation and excess transpiration by mulching, dense plantings, multiple plant levels (Section 6.2.3), shading, windbreaks (Section 8.3.2), and careful selection of planting dates;
- plant crops and varieties that produce large harvests per unit of water, and are heat and drought adapted (Section 5.6).

As water in many areas becomes scarcer and more expensive due to rising demand, warmer temperatures, and increasing ET, and in some places decreasing precipitation with ACC, there is also more interest in increasing agronomic WUE through deficit irrigation. *Deficit irrigation* maximizes harvest per unit of water, by applying less water than crops need to maximize yield (harvest per unit land area) (Fig. 8.4). That is, it aims for a water application rate where the response (harvested food) per unit of water will be greatest, recognizing the diminishing rate of response for each additional unit of water past that point. This makes water available to grow additional crops, which makes sense when water is scarcer than land. In other words, using less water than the optimum per unit of garden area so that more area can be planted, rather than irrigating a smaller area to maintain soil water at 50–100% of FC, can result in overall greater harvest for the amount of water available. Deficit irrigation can also support more equitable water use when water is scarce, if it means using less water in each garden, so that more people can have productive gardens.

Using deficit irrigation to optimize WUE means reducing the amount of water applied so plants experience some water stress, but keeping yield reductions relatively small. In fact, a number of studies have found that using deficit irrigation not only increases WUE, but may even maintain yield, and improve quality, as found for example in research with oranges in Spain (Zapata-Sierra and Manzano-Agugliaro, 2017). One study that modeled different combinations of irrigation method (full or deficit irrigation), irrigation technique (furrow, sprinkler, drip, or subsurface drip), and mulches (none, organic, or plastic), concluded that to increase WUE, deficit irrigation should be implemented first because it reduces water use while reducing cost, and with very little reduction in yield (Chukalla et al., 2017). Mulch and drip irrigation could then be added, which would increase WUE much more than deficit

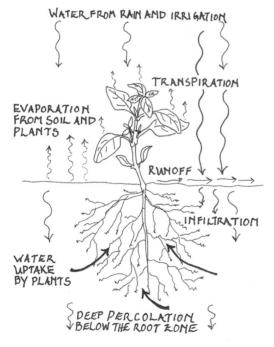

Figure 8.3. The water cycle in the garden

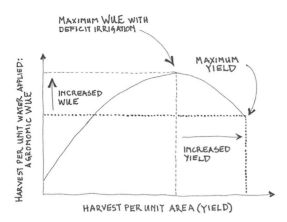

Figure 8.4. Deficit irrigation

Based on Solh (2010)

irrigation, but at a cost, with the total cost highest for drip irrigation with synthetic mulch—yet this was also the most cost effective (highest WUE increase per dollar), so there was a tradeoff between cost and cost effectiveness, although this did not include manufacture and disposal of these plastics.

Using deficit irrigation optimally involves making precise measurements of soil and plant water status that are beyond the scope of gardeners. But understanding the concept of deficit irrigation makes it easier to intentionally choose what criterion for efficiency is most valuable—land or water—and can help in designing informal and formal experiments to increase WUE in garden management. Deficit irrigation underscores the importance of selecting appropriate crops and varieties (Box 5.1, Section 5.6) for the location, season, and amount of water available, and having an understanding of crop reproductive cycles (Section 8.4, Fig. 8.5).

8.3.2. Reducing excess evapotranspiration

Reducing excess ET in your garden means less time spent watering, more efficient water use, and lower water bills. Simple methods to do this include using mulches, windbreaks, shades, and cropping patterns.

Surface mulches. Plant debris like leaves, straw, and cut or pulled weeds can be used as a mulch to cover the soil surface, shading and cooling it, and helping to reduce evaporation and discourage weeds that compete with crops for water. Mulch also protects the soil from the impact of raindrops, which can compact the soil surface, reducing infiltration and increasing runoff. However, if you water on top of thick surface mulches, much of the water can remain in the mulch, and is then lost to evaporation. This can be avoided by watering under the mulch, including with drip irrigation, or using subsurface irrigation (Section 8.8.3). When mulching around newly planted seeds (Section 6.2.4), keep checking for and removing garden pests such as cutworms or sow bugs that can quickly destroy young seedlings. If these pests become a problem, remove the mulch from the area immediately around the seedlings, and try collars such as cans with top and bottom removed to exclude the pests (Fig. 9.4).

For fruit trees and other perennial crops, stones and rocks with a higher albedo than the bare soil (Section 4.1.3) can make good surface mulch, keeping soil cool in hot summers (Jafari et al., 2012). They can also reduce evaporation by condensing moisture that evaporates from the soil during the night on the cooler lower surface of the stones or rocks, keeping it from being lost to the air.

Windbreaks and shades. Wind increases ET rates by disrupting the protective boundary layer of moist air near the surface of leaves and soil (Section 4.1.4), and direct sunlight raises leaf and soil temperatures. Section 4.1 has more on orientation in relation to the sun. Windbreaks and shades reduce water use and can be created from trellises, fences, or plantings of trees, shrubs or other crops, or by making use of existing buildings, fences, walls, or other landscape features. You can push branches, stalks, and other plant parts into the soil or tie them to frames of branches or bamboo. We often use palm fronds in our garden for shades and windbreaks for both individual plants and beds.

A 50% decrease in wind speed can reduce ET rates by 33% (Shaxson and Barber, 2003, 59). Windbreak effects are typically seen downwind for a distance of about 10 times the height of the windbreak, but this distance should be longer than the area needing protection, in order to prevent strong winds from creating eddies on the edges of the planted area. A study in southern Italy found that a windbreak blocking 80% of the wind reduced excess ETa 20% for rain-fed wheat, and 31% for irrigated beans, up to a distance of 15 times the windbreak's height (Campi et al., 2012). Windbreaks can also prevent the warmer air around plants and soil from being blown away when low temperatures are a problem. The same Italian study found the air temperature was warmer within an area extending five times the height of the windbreak out into the field. Windbreaks should only block about 50–80% of the wind.

Cropping patterns. Polycultures in the garden, with a dense planting of crops that have different

---

Water, Soils, and Plants

aboveground heights, shades the soil and shorter plants (Section 6.2.3). Similarly, different rooting depths make the best use of water in the soil. Where winters are cold, deciduous trees such as fig, jujube, peach, or pomegranate that lose their leaves to let in the winter sunlight, but leaf out in the summer to provide shade and wind protection for other garden crops, are especially good. Plants like melon, squash, and *verdolaga* provide a living mulch by spreading out horizontally, shading the soil surface.

8.4. Soil Water and Garden Yield

We use the term *drought* to describe a condition in the environment when below normal precipitation results in unusually dry conditions, sometimes called *meteorological drought* (UNL, 2016). As temperatures rise and water availability changes due to ACC, some conditions that have been defined as drought are becoming "normal," and droughts are becoming more severe, requiring new responses (Chapters 2 and 3).

When a plant's physiological condition is affected by drought, it experiences water stress (Section 5.6). A plant is *water stressed* when ETa falls below its optimal rate (ETc), that is, its optimal transpiration water requirements are not met and it wilts (Fig. 8.1).

Often this is the result of insufficient water in the soil, but wilting may also be caused by disease or pests that reduce a plant's ability to take water from the soil, or to use it (Section 9.8.4). When a crop experiences water stress, the *yield* (as amount harvested per unit area planted) is often reduced. How much the yield is reduced depends on when in the crop's lifespan and reproductive cycle the water stress occurs, the crop and crop variety, the amount of water deficit, and other growth conditions, such as temperature. Of course, some crop species and varieties have drought adaptation in the form of escaping drought entirely, for example through very short lifespans, and so may never be exposed to water stress (Section 5.6).

Crops are very vulnerable during germination and seedling establishment; then, as they develop through vegetative, flowering, yield formation (typically development of fruit, seed, or tuber), and ripening stages of their reproductive cycle, their sensitivity to water stress changes (Steduto et al., 2012). Generally, established garden plants are most sensitive to water stress during flowering, followed by fruit and seed formation, with vegetative growth and fruit and seed maturation (ripening) being the least sensitive (Smith, 2012). Figure 8.5 summarizes these general findings and shows how, compared to no water stress (0), water stress (WS)

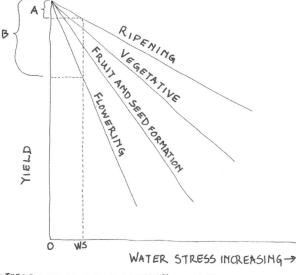

A = THE DROP IN YIELD WITH WATER STRESS WS OCCURRING DURING RIPENING
B = THE DROP IN YIELD WITH WATER STRESS WS OCCURRING DURING FLOWERING
WS = $ET_a < ET_c$

Figure 8.5. Yield response to water stress and crop reproductive cycle

Source: Food and Agriculture Organization of the United Nations [Smith (2012) and Doorenbos & Kassam (1979, 38)]. Adapted with permission.

affects yield depending on when in the reproductive cycle it occurs, with the examples of ripening and flowering (yield reductions of A and B, respectively). So it's generally important to make sure garden crops have enough water when they are flowering and early in fruit development. However, there can be variation for specific crops and conditions, as found in a two-year field experiment with eggplant in Lebanon, which has a Mediterranean climate similar to that of many other areas that will experience increasing drought with ACC (IPCC, 2013). Researchers found that yield reduction due to drought (resulting from deficit irrigation) was least when the drought occurred two weeks prior to flowering, and greatest when it occurred at the vegetative and fruit ripening stages (Karam et al., 2011). Still, deficit irrigation not only increased agronomic WUE, but also dry matter content, which improved flavor, an example of the potential benefits of deficit irrigation (Section 8.3.1).

The specific crop and variety also affect the amount of reduction in yield as water stress increases. The more water stress-resistant a variety or crop species is, either by avoiding or tolerating water stress (Fig. 5.8), the less its yield will be reduced by the stress, and the more stable its production will be over time in a drought-prone environment. Fig. 8.6 shows that for a more stress resistant variety (Y), the water stress caused by drought results in relatively less yield reduction than for a stress sensitive variety (W). Varieties like W are often bred specifically for high production in optimal environments, and so have higher yields than

drought adapted varieties when there is no drought (S_0), but have a greater reduction in yield when drought results in stress. When stress is relatively mild (S_1), the yield of variety W may still be greater (B) than the stress resistant variety (D), but as stress becomes more severe (S_2), there may be a qualitative genotype-by-environment interaction (Box 10.3), with the yield of variety W dropping below that of a resistant variety (compare yield E and F).

This means that where water stress is likely because of lack of rain or irrigation water during the growing season, adaptation to water stress is an important criterion for choosing crops (Section 5.6), and growing a highly responsive, high-yielding variety unadapted to stress can make gardeners more vulnerable.

The chance that garden plants will be water stressed, and yields reduced, can be minimized by:

- using more drought-adapted crops, varieties, and crop mixtures (Section 6.2.3), including those adapted to new and projected climate changes;
- changing planting times to improve the fit between availability of water in the soil and crop demand, especially as temperatures and precipitation patterns or extremes are affected by ACC;
- increasing available soil water by applying enough water to keep soil in the root zone at 50–100% of field capacity; by decreasing water lost to runoff and excess deep percolation; and by using mulch, shade, and windbreaks to decrease water lost to evaporation and excess transpiration (Section 8.3.2);

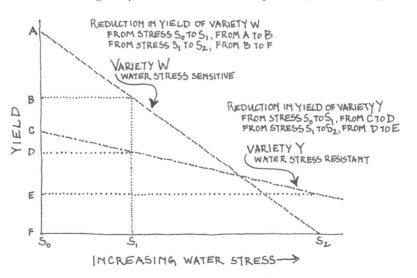

Figure 8.6. Crop varietal differences in response to water stress

- reducing irrigation in the least sensitive stages when experimenting with deficit irrigation (Section 8.3.1), to minimize any reduction in yield while increasing WUE.

Trying any of these approaches can involve experimenting with new plants or methods, which always includes some uncertainty as well as investment of time, and often more resources for mulching or shading. This means thinking in terms of benefit:cost, and comparing the value of different benefits and costs, and tradeoffs among them (e.g., comparing the increased yields that you may get with the increased investments, and also with the other benefits you might get by investing the time and resources in different ways). It also means considering how to advance our larger social goals, for example by investigating policies supporting more equitable access to water and land for food gardens (Section 1.6.1).

8.5. How much Water?

Observing how plants respond (e.g., growth, yield, quality) to different watering methods, and talking with experienced gardeners in your area, will help you learn to judge when and how much to water your garden. There is very little information about the water requirements of mixed dryland gardens. In our case study of two urban desert gardens in Tucson, Arizona (Section 1.5.2), we applied an annual average of 48 L m^{-2} week^{-1} (1.2 gal ft^{-2} week^{-1}) (Cleveland et al., 1985). The gardens were mixtures of annual vegetables, producing food throughout the year. The amount of irrigation water needed differed greatly during the year depending on temperature, rainfall, and humidity. For example, even with some shading and mulching the amount of rainwater plus irrigation water applied in one of our gardens during the hot, dry season (April–June) was more than four times that in the cool, wet season (January–March), and the rainfall only supplied 5% of hot, dry season water compared with 53% in the cool, wet season.

The WUE (kg of harvest m^{-3} of rainfall plus irrigation water applied, 1 m^3 of water = 1000 L = 264.2 gal) also varied a lot between seasons, as shown in Fig. 8.7 for one of those gardens. In the cool, rainy first

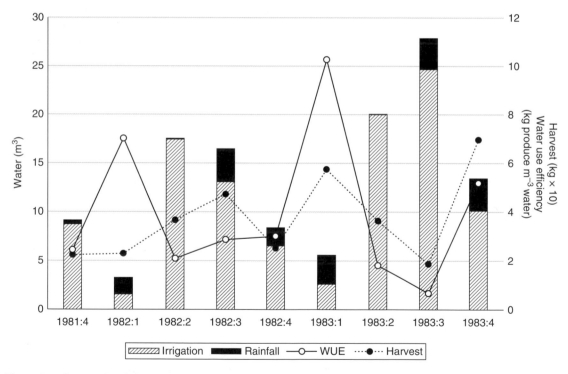

Figure 8.7. Seasonal variation in water use and food harvest in a desert garden

Based on Cleveland and Soleri (n.d.)

Year: quarter, 1 = Jan, Feb, Mar; 2 = Apr, May, Jun; 3 = Jul, Aug Sep; 4 = Oct, Nov, Dec

quarters (January–March) we harvested 22 and 57 kg, while using only 3.3 and 5.6 m³ of water, for WUE of 7.0 and 10.3 (Cleveland and Soleri, n.d.). However, in the hot, dry second quarters and the hot, wet third quarters, the WUE was much lower, varying from 0.7 (27.9 m³ of water and 18.7 kg harvested) to 2.9 (16.5 m³ of water and 47.2 kg harvested). This figure also illustrates the short-term variability in rainfall, irrigation and harvest, and WUE, for a year-round garden over a 27-month period.

Having good estimates of how much garden produce you are harvesting and how much water you are using in different seasons, or for different crops, can increase your awareness, leading to ideas for improving WUE or for doing simple experiments, and can be valuable when applying for grants or other support for a garden project (Box 8.1). For example, because of very low WUE in some seasons, it may make sense to minimize planted area then, and shift resources to seasons with higher WUE. But if the harvest per unit area is also much greater when there is low WUE, as in our Tucson garden study, there will be a tradeoff. However, in a year-round mixed garden with continuous harvesting, some of the harvest in one month or season will of course be attributable in part to the water used in the previous month or season.

The objective in watering is to wet the soil where the roots are growing, and a short distance beyond to encourage deeper rooting (Section 5.3), and to flush salts out. All water, even rainwater, contains some salts, and some greywater and piped water can contain a relatively high level of salts. The amount of water needed will increase with the saltiness of the water (Section 8.8.5). Too little water can restrict root growth, increase salt in the root zone, and lead to a larger proportion of water lost to evaporation from the soil. Too much water means a waste of resources, and can cause waterlogging.

While most successful gardeners don't make numerical calculations about water requirements or use, if you are planning a new or expanded garden, a community garden, or writing a proposal for a garden project, having an estimate of future water needs could be very useful. For calculating how much water to apply, see resources in Section 8.10.4, and consult local Cooperative Extension and other experts.

8.6. When to Water?

When you see plants in your garden wilting, with leaves drooping, that's a good sign that it's time, or past time, to water. However, a few crops such as cucurbits like squash and melon can wilt during hot afternoons before they need to be watered. If they recover quickly as the temperature drops in the evening, then they probably do not need watering, but may benefit from shading during the day. For most crops, wilting when there is adequate soil moisture is likely caused by disease or pest damage, for example, beetle larvae feeding on roots, or fungi blocking the movement of water in the plant's vascular system (Section 9.8.4).

Box 8.1. Measuring water applied to the garden

If you are going to calculate the amount of water used in the garden, then you will need some way of measuring the water applied. Even if you aren't calculating water requirements, measuring the water gives you an estimate of amount and costs, which along with information on harvest and other garden benefits can be used to estimate benefit:cost ratios to help in deciding if it's worth growing water-demanding crops, or any crops during hot, dry seasons. These data will also be useful documentation for project proposals.

If water is being applied using containers such as buckets, the volume of the container can simply be multiplied by the number used. Water applied by a sprinkler can be estimated the same way as rainfall (Section 8.7.2), by averaging the water collected at several places within the garden, and converting mm of water on the area under the sprinkler to liters (1 mm depth on 1 m² = 1 liter). When water is delivered by pipe or hose, the average of three or more measures of the time taken to fill a container of known volume can be divided into the total time taken for watering the garden. For example, if it takes 25 minutes to water the garden and an average of 1.75 minutes to fill a 19-L (5-gal) container, then the total water used = (25/1.75) x 19 = 271.4 L (71.4 gal). Alternatively, inline (at the faucet or on the hose) water meters make measuring applied water very easy, and are the best way to measure water with a drip or similar system. With drip irrigation, emitters are described in liters (gallons) applied per hour, so water application can also be estimated by multiplying the number of emitters, their application rate per hour, and the number of hours.

You can also look at the soil for signs of when it's time to water. As already discussed, most crops should be watered to maintain plant available water in the root zone at 50% or more, unless you are using deficit irrigation (Section 8.3.1). The soil in the root zone where most of the roots are growing can be sampled by digging down in several places. The descriptions in Table 8.1 are a guide for making rough estimates of soil water deficit based on handling soil samples; Section 8.10.4 lists other resources. Experimenting with crops and crop mixes, time of planting, time of day water is applied, mulching, and shading, can increase the length of time between irrigations, reduce overall water use, and increase yield and agronomic WUE. For calculating irrigation timing for larger projects, see resources in Section 8.10.4, and again, consult local experts.

8.7. Sources of Water for your Garden in a Time of Climate Change

For most food gardens in the global north there are three main sources of water: piped water through a centralized distribution system, rainfall, and recycled greywater. Most gardens in dry areas of the global north depend for some or all of their irrigation requirements on piped water supplied by the local water agency or company.

8.7.1. Piped water

Piped water is sourced almost entirely from either surface water in rivers and lakes that's often stored in reservoirs, or groundwater pumped from aquifers. As aquifers are over-drafted over the longer term in many drylands due to reduced precipitation with ACC and increasing demand, aquifer sediments may compact, permanently reducing capacity (Rodell et al., 2018). With increasing water scarcity, treated wastewater is becoming a major potential source of water worldwide (WWAP, 2017). Regional and local agencies are increasingly reclaiming treated water for agricultural use and for recharging the aquifers that are a source of residential piped water (for example in California, CDWR, 2016). Depending on the level of treatment, reclaimed water can be used on all food crops, with disinfected tertiary reclaimed water (wastewater that has been both filtered and disinfected according to government standards) allowed to be in contact with edible parts of crops (CAWSI, 2018). In Los Angeles County over a third (625 million L or 165 million gal day^{-1}) of treated water is reclaimed (SDLAC, 2018), and the city of Los Angeles is making reclaimed water available for pick-up by homeowners for use in their gardens. *Conjunctive water use*, that is coordinating the use of surface and groundwater, for example by using reclaimed water, and surface water in wet years, to

Table 8.1. Soil texture and water deficit estimates[a]

Texture	Field capacity (0% deficiency, and no free water when soil squeezed)	Amount of water needed to bring soil to field capacity (mm m^{-1} soil depth)	
		Time to water (50–75% deficiency)[b]	Permanent wilting point (100% deficiency)[b]
Sand (coarse)	Will not form ball, but wet outline left on hand	Appears dry, will not form ball with pressure, but still some clumping (40–65 mm m^{-1})	Dry, loose, single grains flow through fingers (85 mm m^{-1})
Sandy loam (moderately coarse)	Makes weak ball, outline left on hand	Appears dry, will not form a ball, sticks together slightly (65–100 mm m^{-1})	Dry, loose, flows through fingers (125 mm m^{-1})
Loam (medium)	Can form cylinder, wet outline left on hand	Crumbly but makes weak to good ball when squeezed (85–125 mm m^{-1})	Powdery, dry, small clods easily broken into powder (170 mm m^{-1})
Clay loam (fine)	Can form ring, wet outline left on hand	Pliable, forms ball but not cylinder (100–160 mm m^{-1})	Hard, cracked (200 mm m^{-1})

[a]Based on information in Merriam et al. (1980, 759); Stegman et al. (1980, 798); Doneen and Westcot (1984, first published 1971, 9, 11); USDA NRCS (1997, 9-8)

[b]To maintain ETc, some crops may need to be watered at less than 50% deficiency, that is, before 50% of AW has been removed. The depth of water in mm (equal to L per m^2, since one L of water on one m^2 is 1 mm deep) that would have to be applied to bring the soil to field capacity. See Section 7.2.1 for determining soil texture.

recharge groundwater, is also increasingly important for piped water supplies.

8.7.2. Rainfall

Ultimately all fresh water on Earth comes from condensation and rainfall or other forms of precipitation like snow and hail. Fresh water sources include streams, rivers, lakes, and many aquifers that can be tapped with wells. While we have a very limited capacity to change precipitation intentionally, we are having large unintentional effects via ACC, resulting in accelerating trends in precipitation patterns that are increasing the potential for negative impact and vulnerability (Chapters 2, 3).

Rainfall varies in total amount (mm per year), intensity (mm per hour), and distribution between and within locations and seasons. Where the rainfall is adequate, it might be possible to have a successful garden in the rainy season with little or no additional water. This is especially true when the rainy season is also the cool season, as with the Mediterranean climate including the west coast of North America, southeastern Australia, and southwestern South Africa. On the central coast of California we let many winter crops go to seed in the spring, so the first rains in the fall provide the moisture to germinate seeds of arugula, chard, cilantro, collard, kale, lettuce, and mustard in our garden, and the mulched areas of our yard. In years with about 46 cm (18 in) of winter rain (our historical average, not accounting for ACC trends), they produce an abundant harvest through late spring with no irrigation.

Garden decisions such as when to plant and what crop mixtures to grow have often depended on what we think will be the amount and timing of rainfall, based on our past experience. We might think, "to take advantage of summer rainfall based on past years' timing of rainfall, I should plant no later than ..." These are our informal estimates of the probability of receiving a certain amount of rain during a given period, but probabilities can also be quantified (Box 8.2). *Probability* is the likelihood that an event will occur based on the information available (Box 2.3). A higher probability of adequate rainfall means a lower risk (a lower probability of vulnerability, Section 3.1) for gardeners who depend on that rainfall.

When gardens depend at least in part on rainfall, there will always be some risk due to lack of rain. The level of vulnerability (Section 3.1) you are willing to accept will depend on how you value your investment in labor, water, and other resources, and the harvest and other outputs from your garden, like community cohesion, a more pleasing diet, or improved nutrition. Harvesting and storing rainfall are ways to reduce, though not eliminate, vulnerability when your garden depends at least in part on rainfall.

In the past, rainfall data from the previous 20–30 years at a location could be used to calculate descriptive statistics (mean, standard deviation) by month or even week, which could be used for planning. But because our climate is changing, there is less reason to assume that rainfall can be described strictly by familiar variation around an established mean, because new trends, including increased variability, are emerging in most locations (Milly et al., 2008) (Box 2.1, Section 2.3.2), as discussed in Box 8.2. Rainfall and other weather projections for particular locations can be found online, which are useful for planning large-scale community garden projects, especially those depending on rainfall, including rainwater harvesting. The utility of official weather information, including projections, is improving, for example, Climate Explorer (Section 8.10.2) provides county-level projections for the US. But weather can be very localized, and local landscape features create consistent patterns that may not be reflected in official data. For example, a location may be at a higher elevation than the reporting station, and so usually have more rainfall. Collecting local weather data is a way to begin to understand locally-specific characteristics of ACC, eventually providing a better basis for future planning.

Measuring and recording rainfall at your garden is a great way to increase your understanding of the relationship between water and garden management. Rainfall can be quite easily measured with a large container, like a 19-L (5-gal) plastic bucket with fairly straight sides, and the depth of rainwater measured with a ruler. For more accurate measurements, inexpensive, simple plastic rain gauges can be purchased from scientific, agricultural, or garden suppliers; the tapered type is better than the flat-bottomed kind, because it makes reading small amounts of rainfall easier. Local weather documentation is now the focus of a number of very active "citizen" science projects (Section 8.10.2). For example, gardeners in the US or Canada can join the Community Collaborative Rain, Hail and Snow Network (CoCoRaHS), use their standardized 4-in

Box 8.2. Calculating rainfall probabilities from weather records

Probabilities can be calculated from rainfall records. For example, in the desert areas of the southwestern US, the beginning of the summer rains in July is an important time for planting summer crops including beans, maize, and squash. We can calculate the probability that a given year will have enough rainfall for July planting. Given data on the amount of rainfall in the month of July for the past 30 years, the amounts of rainfall from most to least are listed (Table 8.2). Using a simple formula, the probability of any amount can then be estimated, as shown in the table. In example #1, if a successful harvest from a July planting depends on at least 40 mm (1.6 in) of July rainfall, then it will be successful in 68% of years, on average. This means that, on average, in 32% of the years you would need to replant in August, or supplemental irrigation will have to be provided. It does not mean that in any given 10-year period there will be at least 6 years of successful harvests. Probability is a calculation based on long-term, average observations; it is a useful guideline but should never be taken as an absolute predictor of growing conditions and resulting harvests.

The greater number of years the calculation is based on, up to about 30, the greater the reliability.

You could begin by recording rainfall in your own garden, and comparing these data with the 30-year mean for the closest weather station to get some sense of where you stand in comparison.

Doing this calculation assumes there are no trends, which is no longer realistic in many areas due to ACC (Chapter 2). However, for most gardeners planting annual or biennial crops, predicting precipitation based on accumulated experience with rainfall and locally relevant indicators such as air temperature, unwatered soil moisture, and the activity of other organisms, together with long-term historical data such as in Table 8.2 can work pretty well. In addition, there are now tools such as those provided by the US National Weather Service (NWS NOAA, 2018), with weather predictions for the upcoming week to three-month periods for the coming year that could be very useful for certain gardening decisions. These predictions take ACC into account in real time as they are based on current atmospheric and oceanic conditions and also identify anomalies, which are predictions significantly deviating from the average at a location. For example, mid- to late-summer rainfall is important for many gardens in southeastern Arizona, and can make a big difference in irrigation demands

Table 8.2. Calculating rainfall probabilities. An example of July rainfall for Tucson, Arizona, 1985–2017

m	Year	Rainfall (mm)	m	Year	Rainfall (mm)	m	Year	Rainfall (mm)
1	2017	172.7	11	2010	68.8	21	2000	40.4
2	1990	138.4	12	2013	66.0	22	2014	36.3
3	2006	137.2	13	2003	63.5	23	1989	36.1
4	2007	132.6	14	2002	62.7	24	2001	27.7
5	1999	105.4	15	2015	52.8	25	1992	23.6
6	2012	104.9	16	1996	47.8	26	2004	21.8
7	1998	103.1	17	1986	46.2	27	2005	18.3
8	2008	86.9	18	2009	45.2	28	1997	13.0
9	2016	84.3	19	1988	42.9	29	1991	11.2
10	1985	79.8	20	2011	41.7	30	1994	10.4

$p = m/(n + 1)$
p = estimated probability of equal or greater rainfall
m = position in the series, ranked from highest to lowest rainfall
n = number of years of measurements = 30

Example #1: To find the probability of a July with over 40 mm of rainfall, find the value closest to 40 but greater than this in the table above and note that its position is m = 21, then p = 21/(30 + 1) = 0.68. This means that there is about a 68% chance (about 2 out of 3) of having a July with more than 40 mm of rain, and a 32% chance of a July with less than 40 mm of rain.

Example #2: To find the amount of rainfall that can be expected with a 60% probability (p = 0.60), m = p(n + 1), so m = 0.60(30 +1) = 18.6. To be conservative we round "down" (to the next entry with a lower amount of rainfall) to m = 19, and find in the table that rainfall for m = 19 was 42.9 mm. Therefore, in 60% of the years the rainfall in July will be 42.9 mm or more.

SCENIC (2018, Station Tucson International Airport, 23160)

Continued

Box 8.2. Continued.

during important times in crop reproductive cycles and lifespans, such as fertilization and ear-filling for maize (Fig. 8.5). With a baseline of 1981–2010, the NWS air temperature and precipitation probability predictions identify probability of "normal" (represented by the median), and above- and below-normal temperatures and amounts of precipitation, with each of these three classes comprising 33.3% of the baseline disribution, in the absence of any other information. When current conditions indicate a statistically significant deviation, the probability distribution of the predictions is adjusted accordingly. Knowing there was a >50% chance of a "normal" amount of rainfall or better over mid- to late-summer might encourage larger plantings of maize, or other crops that would benefit from this.

For gardeners considering larger investments such as planting perennials, building terraces, or establishing school and community gardens, even longer-term local climate information is desirable. While trying to account for trends beyond several months will be difficult, where available the same sorts of calculation as outlined above could be done using projected precipitation (and temperature) based on climate models under a particular IPCC scenario (Box 2.3). The Climate Explorer, which is part of the US Climate Resilience Toolkit, could be very useful for this. This might be worthwhile including if you are planning perennial plantings or garden layout for water harvesting, or searching for crop varieties more appropriate for the projected

conditions. However, because it is based on model projections, that estimate would have even greater uncertainty, and so to inform decisions it may be best to start with the conservative (in terms of impacts of ACC trends) moderate emissions scenario (RCP4.5) and model projections closest to the trendline based on all models used. For example, Climate Explorer uses up to 32 models for developing temperature projections, and provides minimum, maximum, and mean (represented by the trendline) projections, all for either RCP4.5 or 8.5 scenarios. Using tools such as this may be your best option for objectively anticipating the outcomes of current trends.

Overall, there is no getting around the enormous complexity, uncertainty, and local specificity in the future climate. The tools we've described here will be helpful for gardeners, but limited. This challenge is much better addressed at scales larger than individual or community gardens. We believe the most valuable approach is for gardeners and their communities to organize to identify their needs, and then partner with respectful scientists to develop locally relevant projections of temperature and precipitation that have broad relevance for food gardens, public health, urban planning, and more. This is already starting to occur. For example, the American Geophysical Union's Thriving Earth Exchange is devoted to facilitating and supporting just such partnerships, and promising examples of community risk assessment for the increasing heat and drought due to ACC can be found on their website (Section 8.10.2).

---

diameter rain gauge, post measurements on the CoCoRaHS website, and review the data from their area and elsewhere. At the same time, these data enable scientists to create more precise descriptions and forecasts of current and future weather.

8.7.3. Harvested rainwater

Rain falling directly on your garden is a free resource, and can produce a good harvest in rainy, cool seasons, but probably won't be enough in hot, dry seasons. However, rainfall can be concentrated by harvesting the rainfall runoff from adjacent areas, including rooftops, and directing it to the garden, or storing it for later use. Harvesting rainwater involves collecting it in a *catchment* and applying it to the garden (or a garden plot, or single tree), and is a simple way to increase the amount of water available for your garden, without the expense or trouble of importing it. The catchment

can range from a walkway, with water directed into the basin around an adjacent fruit tree, to the roof of a large building, with water stored in containers for watering a community or school garden. Runoff from roofs can be easily stored in aboveground containers, and runoff from the ground in belowground cisterns.

Rainwater harvesting for food production has been practiced in dry areas for thousands of years using a diversity of techniques, and is increasingly popular worldwide, because water is becoming scarcer and more expensive, due in part to ACC (Mekdaschi Studer and Liniger, 2013). In the western US, surface water is increasingly scarce, groundwater tables are falling, and the cost of energy to pump water to the surface has risen. For example, the city of Tucson, Arizona, is offering rebates to homeowners to install a rainwater harvesting system (City of Tucson, 2017), the San Francisco Public Utilities Commission has created a resource page that includes a rainwater harvesting manual and an Excel spreadsheet for calculating irrigation demand and cistern storage capacities (SFPUC, 2015), and the Texas Water Development Board has published a water harvesting manual (TWDB, 2005).

Small microcatchments that direct runoff to one or two tree basins or garden beds are much more efficient at harvesting rainwater than larger catchments, because infiltration and evaporation in the catchment are minimized, and water doesn't need to be transported to the garden. Because of the small volumes involved, only simple structures are needed to control it (Cleveland and Soleri, 1991, 212–217). Few if any engineering or construction skills are needed for design and construction of small microcatchments, and gardeners can experiment and adjust them based on performance.

Local experience and knowledge is a starting point for experimentation without the need for an expensive design process or detailed rainfall data. For large community or school gardens, designing the runoff system and consulting experts can save time and resources, and avoid problems, and many resources are available (Section 8.10.3).

8.7.4. Greywater

Instead of sending your household greywater to a treatment plant via the sewer system, it can be a source of water for your garden. Greywater is classified into light and heavy portions: *light greywater* (hereafter simply greywater) is from the shower and laundry, and has less suspended organic material than *heavy greywater* from the kitchen (Cohen, 2009). Most households create enough light greywater for a food garden that would produce much of their fruits and vegetables. For example, the average wastewater generated in the US is about 379 L (100 gal) per person per day (EPA, 2013, 8–18), and about 46% of a single-family's wastewater is light greywater (Cohen, 2009), which means on average each person in the US produces 63,000 L (17,000 gal) of light greywater annually.

How much garden produce could you grow using greywater as the source of irrigation water? Based on Jeavons' irrigation rate for intensive vegetable gardens of 1487 L m^{-2} year^{-1} (36 gal ft^{-2} year^{-1}) (Jeavons, 2006), just 38% (55,000 L) of the amount of greywater generated by the average California household of 2.9 persons, 145,000 L year^{-1} (38,000 gal year^{-1}), would be enough to irrigate a 38 m^2 (400 ft^2) garden, supplying 100% of the average US vegetable consumption for the household, assuming a yield of 5.7 kg m^{-2} (1.2 lb ft^{-2}) of vegetables per year (Cleveland et al., 2017).

Light greywater does contain some organic material and microorganisms (some of which can be pathogenic), and so it is generally considered safe for fruit trees and vegetables that do not come into contact with it, like artichoke or trellised tomato, but can also be used with caution for other vegetables (Ludwig, 2015) (e.g., by directing it to the root zone with vertical mulch). Ludwig's book is a great resource if you want to install a simple, low-maintenance greywater system, for example a laundry-to-landscape system that Ludwig invented, using the washing machine pump to move greywater, including uphill. Greywater systems allow diversion of greywater to the sewer or septic if there are contaminants, for example when washing dirty diapers, or when someone in the household with an infectious disease takes a shower. Check with your local government about rebates for greywater systems (e.g., City of Tucson, 2018) and regulations governing recycling of greywater (Section 8.10.3).

8.8. Irrigation

Irrigation is the directed application of harvested rainwater, piped water, or greywater to the garden. *Surface* irrigation is the application of water using basin, furrow, or surface drip systems; *overhead* or "sprinkler" irrigation is done by hand or using a sprinkler attached to a hose; *subsurface* irrigation is applying water directly to the root zone (e.g., using pots, vertical mulch, or buried drip lines). In this section we discuss irrigation efficiency, and surface, subsurface, and drip irrigation.

8.8.1. Irrigation efficiency

Irrigation efficiency (Ei), a measure of how much water actually ends up in the root zone of growing crops compared with the amount of water at its source (Howell, 2003). Ei is an important component of WUE, and can be calculated by multiplying conveyance efficiency (Ec) times application efficiency (Ea):

$$Ei = Ec \times Ea$$

Conveyance efficiency is the amount of water applied to the garden, as a percentage of the amount of water extracted from the source, such as an aquifer or lake. The immediate source for gardeners is a hose end or microcatchment outlet, so Ec is assumed to be 100%, although ecological thinking recognizes conveyance costs such as evaporation from water system delivery canals. Therefore for gardens, Ei is determined by *application efficiency*, the amount of water applied that reaches and stays in the root zone, divided by the amount of water applied to the garden. An Ea of 100% would mean that all of the water applied reaches and stays in the root zone with sufficient excess to leach salts below the root zone. But in reality Ea is always less than 100% because of evaporation from the surface, runoff of excess water out of the garden, or excess percolation below the root zone (Fig. 8.3). Salinity (Section 8.8.5) and waterlogging are often the result of inefficient irrigation in dry areas, but managing water for optimum application efficiency can avoid these problems. You can increase application efficiency in your garden by reducing excess evapotranspiration (Section 8.3.2), and also by ensuring equal time for water infiltration in all areas of the garden being irrigated, minimizing water lost in runoff, or deep percolation below the root zone, except as needed to leach out salt.

8.8.2. Surface irrigation

Surface irrigation is commonly used with furrows and sunken beds (Section 8.9.1). Low Ea due to evaporation and excess deep percolation (i.e., more water than needed to leach excess salts) are common problems with surface irrigation. Deep percolation can also reduce production because it leaches nutrients below the root zone. In areas with high water tables or soil layers that impede percolation, excess water can raise the water table into the root zone over the long run, causing waterlogging and salt concentration in the soil there.

Water can be lost below the root zone when too much water is applied, even when the surface is level, or it can be the result of uneven application due to low and high spots or to long irrigation "runs" in furrows, resulting in excess deep percolation in the low spots or at the beginning of the run. It can also be due to differences in soil texture, with water infiltrating more rapidly in sandy areas where it moves well beyond the root zone while in the same amount of time water may not even reach the entire root zone in more clayey areas of the garden. If this is a problem, locate garden beds or furrows so that they have a consistent soil texture—that is,

stratify them by soil texture (see Appendix 3A, Fig. 3A.3). Gopher or other animal burrows can move lots of water rapidly away from the root zone, and it can be startling to see water gushing out of sight as the soil on the surface collapses into a hidden burrow.

If you have low Ea because of loss of water below the root zone you can apply less water to increase efficiency, but if this results in insufficient water reaching the root zone in some areas of the garden, it's not a good strategy. Again, when too little water reaches the root zone, you can get salt build-up in the soil as a result of evaporation increasing salt concentration, and not enough water to leach dissolved salts below the root zone.

8.8.3. Subsurface irrigation

Subsurface or *root zone irrigation* is the delivery of water below the surface of the soil, directly to the root zone. It reduces evaporation by minimizing wetting of the soil surface and upper layers, and can dramatically increase Ei and agronomic WUE.

Pot (or *pitcher*) *irrigation* is a traditional method developed in Iran and northern Africa, and long used in China and elsewhere. Pot irrigation has recently become more popular, including in major development projects, for example in Ethiopia (Woldu, 2015) and Pakistan (Shaikh and Tunio, 2012), and with gardeners in industrialized countries including the US (Bainbridge, 2015). It uses an unglazed ceramic pot, for example a traditional *olla* (Fig. 8.8), or even a standard flower pot with the hole in the bottom plugged, that is filled with water which slowly seeps into the adjacent soil, especially near the bottom of the pot. Covering the top of the pot reduces evaporation. Because of hydrotropism (Section 5.3), plant roots will grow in this zone, close to the pot. The rate of water seepage through the pot is self regulating, because it decreases as the soil near the pot becomes wetter and the water potential in that soil increases (Abu-Zreig et al., 2006). Water flow is also affected by the size and porosity of the pot. One study compared small and large pots (11- vs 20-L [2.9- vs. 5.3-gal]), with the smaller pots made with twice the hydraulic conductivity of larger pots (rate of water movement through the pot walls; 0.14 vs. 0.07 cm [0.06 vs. 0.03 in] day^{-1}) by incorporating organic material into the clay (Siyal et al., 2009). The two pot sizes produced similar patterns of wetting in the soil, but the smaller pots resulted in higher soil water content at all depths and distances measured—up to 40% greater water content 60 cm (23.6 in) from the pot center at 40 cm (15.7 in) depth. Relatively low firing temperatures also increase the hydraulic conductivity of clay pots, peaking at about 482°C (900°F), and declining rapidly after 538°C (1,000°F) (Bainbridge, 2001), useful information if you or someone you know is interested in making pots for irrigation.

A problem that may occur with pot irrigation is salt accumulation in the soil, since not enough water may be applied to flush the salt below the root zone. In addition, as with drip irrigation, clean water must be used because clay, silt, or organic material in the water can clog the pores of the pot and prevent water from seeping out. Dissolved solids can also clog the pores.

Another method of root zone irrigation is *vertical mulching*. Holes or narrow trenches in garden beds can be filled with stalks of plants like amaranth, maize, millet, sunflower, or wheat, that provide a pathway for water movement. The large air spaces between and within these stalks allow gravity to move the water quickly down to the root zone (Fig. 8.9). In addition, these organic mulches will improve soil water-holding capacity as they decompose. Sand can also be used to fill the trenches or holes, and topped with gravel to slow

evaporation. Compared with pot irrigation, there is no problem with water quality with vertical mulch, but water movement into the soil is not self regulating. Bainbridge describes several other methods of subsurface irrigation which gardeners can try (Bainbridge, 2015).

8.8.4. Drip irrigation

Drip irrigation is a great way to increase agronomic WUE, and can be used for either surface or subsurface irrigation. It works very well for trees or for rows of one species, but is not as appropriate in mixed gardens (Section 6.2.3). Drip irrigation typically involves a system of valves, and plastic pipes with water emitters or perforated flexible pipe ("drip tape"), that delivers low volumes of water per unit time. The system can be automated using electronic controllers that regulate the valves feeding the irrigation pipes or tapes, so that you can irrigate in the very early morning (midnight–4:00 a.m.) when you are sleeping, and evaporation rates are lowest, or when you are away from the garden. Controllers are now available that receive local weather data and include this in irrigation scheduling. Designing, constructing, and maintaining a drip irrigation system is more complicated than most other types of irrigation, and the initial equipment expenses are higher, but the

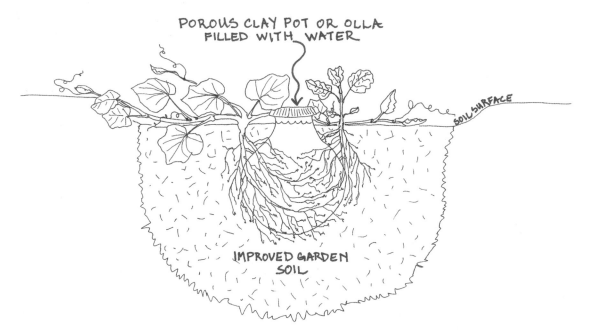

Figure 8.8. Pot irrigation

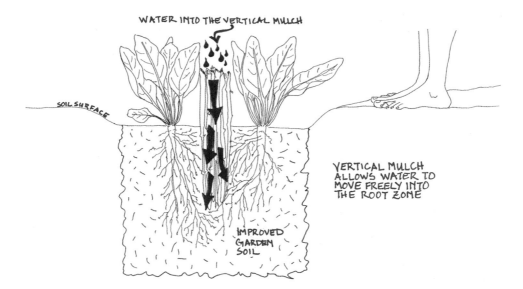

Figure 8.9. Vertical mulch

large increase in WUE may mean it is worth it. Water is typically distributed more uniformly within the root zone with drip irrigation than with other irrigation systems, and because there is less water loss due to evaporation or runoff, Ea with drip irrigation can be very high.

Subsurface drip irrigation has both advantages and disadvantages compared with surface drip irrigation. It can potentially supply water more uniformly with higher WUE, but may be more affected by emitter clogging, root intrusion and pinching of the pipes, mechanical and pest damage, compression by overlying soil, soil water parameters, and pipe deterioration with age (Lamm and Camp, 2007).

Drip irrigation pipes, tape, and emitters are all made of plastic and will eventually need replacing, and the resulting waste, along with their production, have substantial environmental costs. The extensive use of drip irrigation for water conservation in agriculture is one part of the growing challenge of plastic waste in farming. One response is to develop biodegradable drip systems, work that is ongoing (Hiskakis et al., 2011).

8.8.5. Salinity

Salty soils occur naturally in dry areas where not enough rain falls to wash soluble salts down and out of the root zone. Irrigation can make the situation worse, since surface water and groundwater contain more salt than rainwater, and salt tends to build up in the soil as water is continually added with irrigation. This is especially true when air temperatures are high and humidity is low, which increases ET. As water is used by plants and evaporates from the soil surface, salts are carried to high points in the garden bed, and the concentration of salt in soil water increases. So reducing excess ET to reduce the amount of irrigation required is an important way to combat salinization.

To avoid salinity problems, it's also important to water to the bottom of the root zone and slightly beyond. *Leaching* is washing salt from the root zone by adding water beyond what's needed by the plants, and is commonly used to prevent salt build-up. Where high water tables are not a problem, leaching with irrigation water can be used periodically to rid the root zone of salt. The amount of water needed, the leaching requirement, will depend on the salt content of irrigation water, salt sensitivity of the garden crops grown, and the amount of yield reduction that you can accept, and we list some resources for estimating this in Section 8.10.4.

Most gardeners can identify the need for leaching by checking high points in the garden like the sides of sunken beds for white salt deposits, or looking for leaf margin burning in young plants that are adequately watered (Section 9.8). In some cases, irrigation water may be so salty that no amount of water applied for leaching will permit acceptable garden yields. In this case, selecting more salt-tolerant crops (Section 5.7), or less salty water, or gardening elsewhere are the options.

8.9. Building Garden Beds

Garden beds separated by walkways help contain and concentrate soil nutrients and water for plant growth. They separate soil that has not been improved, or may even be contaminated, from the soil where garden plants are grown (Section 4.2.1). Using beds also makes it easy to reach and work in the garden without walking on and compacting the soil in the growing area. Using types of garden beds appropriate for different and changing local climates can make watering easier, and help to increase irrigation efficiency. For these reasons we focus on garden beds, even though some gardeners plant crops in rows at ground level without beds. The two main types of garden beds are sunken and raised.

8.9.1. Sunken beds

Sunken beds are deep pockets of improved soil that make good use of water, and have a long tradition in dry areas. In the US Southwest, the Zuni people have traditionally used very small rectangular beds bordered with clay berms for growing crops such as chile, herbs, melon, and onion. These gardens, near the main villages and surrounded with juniper stick fences, were gently watered by hand from large containers using small dippers (Ladd, 1979). Today many Zuni still garden in sunken beds. The nearby Hopi people also continue their tradition of sunken beds in their terraced gardens, using a small, gravity-fed irrigation system to water them (Soleri, 1989). In Oaxaca, Mexico, we worked with farmers making sunken nursery beds to start vegetable, herb, and flower seedlings, using palm fronds to protect the seedlings from sun and wind (Fig. 6.9). We have had very good results with sunken beds, including in Tucson, Arizona, where we had to excavate a lot of caliche and replace it with compost and soil from the A horizon (topsoil) (Section 7.1.1) (Soleri and Cleveland, 1988).

Surface irrigation is traditionally used with sunken beds, which are also known as *basin beds* because the berms form a basin that holds water. However, you can also use subsurface irrigation with sunken beds. Given the goal of efficient application, planning the area and depth of your basins will depend on a combination of crop water needs, soil texture and structure, water quality and speed of delivery, and rooting depth. The deeper the rooting depth of the crop, the more sunken the basin needs to be, that is, the greater the distance between the top of the berm and the soil surface inside the basin. To increase efficiency, have level basins that can be filled as quickly as possible with enough water so that soil in all parts of the basin is wetted throughout the root zone, minimizing loss of water below the root zone, except for deep irrigations to flush out salts. As mentioned, efficiency is decreased by low spots that get too much water, or when there are different infiltration rates, because of differences in soil texture or structure.

Basins should be sized and located so that the soil within a bed is the same texture and structure. The faster the rate of infiltration, the smaller its surface area should be so that too much water is not added and lost to deep percolation in one area of the basin before other areas are wet. Porous soils with sandy texture or strong structure have higher rates of infiltration than soils with clayey texture or weak structure, but porous soils hold less water, so basin depth and surface area need to be smaller for these compared to clayey soils. Infiltration is also slowed when there are large amounts of suspended matter in the water that clog the pores in the surface layer of the soil.

The faster water can be applied, the larger the basins can be; the slower the application, the smaller the basins should be. If the basin is too big, or the water is applied too slowly, too much water will have infiltrated at the point of application by the time all soil is at field capacity. In this case the basins will have to be watered at several different places or, preferably, made smaller. You can use bubblers or other methods to slow and spread the flow of water so that it doesn't erode the soil. We have worked with Hopi gardeners who use tin cans they have punched with holes, and fastened to the end of a hose with wire as a homemade bubbler.

In hot, dry areas and during dry seasons, sinking the beds can be better than raising them for several reasons:

- they are easier to water quickly and efficiently by surface irrigation;
- the berms give the moist soil, young seedlings, and transplants some protection from drying winds;

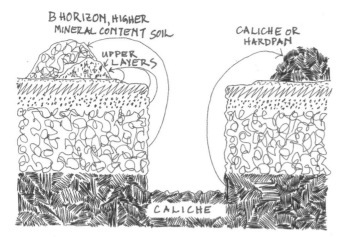

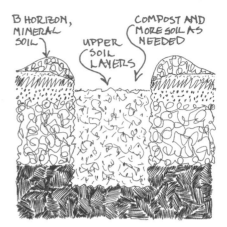

THE BED IS DUG ~45CM (18IN) DEEP. UPPER LAYERS AND HIGHER MINERAL CONTENT PART OF B HORIZON ARE SET ASIDE, CALICHE OR HARDPAN ARE DISCARDED.

HIGHER MINERAL CONTENT SOIL IS USED TO BUILD BERMS AROUND THE BED. THEN THE BED IS REFILLED WITH SOIL FROM THE UPPER LAYERS MIXED WITH COMPOST AND SOME B HORIZON SOIL.

Figure 8.10. Making a sunken bed

- salt residues due to evaporation are drawn up the walls of the berms, away from the growing area;
- young plants can easily be protected by laying palm fronds or other material across the beds;
- when rains do come, the rainwater is captured by the beds, and the garden soil is not eroded.

In places with marked wet and dry growing seasons, it may be better to switch to raised beds during the wet season. Raised-sunken beds may also be worth trying. These are raised beds with berms or other barriers around the edges that hold irrigation water during dry periods and can be broken in places or removed during the rainy season. In drier areas, like much of the southwestern US, this may not be necessary because only rarely does enough rain fall to cause damaging waterlogging. However, one of the consequences of ACC is an increase in short-duration, high intensity rainfall in some areas (Section 2.3.2, Table 2.4), including those where total annual precipitation will be decreasing, so you will need to observe the effects of those changes, experiment, and adapt. Figure 8.10 outlines the steps for making a sunken garden bed in desert soils with caliche in the B horizon, the method we used for our Tucson, Arizona, gardens (Soleri and Cleveland, 1988).

8.9.2. Raised beds

Raised beds are most commonly used in regions with higher precipitation, where they can improve drainage of soil that is seasonally flooded or water-logged. Like sunken beds, they also create a clearly defined area of improved soil. Raised beds used in the US are a gardening model originally introduced by English master gardener Alan Chadwick in the 1960s, and were based on nineteenth-century "French intensive" gardening developed by market gardeners around Paris, and influenced by German philosopher Rudolf Steiner's biodynamic principles (Martin, 2007). Since the 1960s, very influential garden educators including Chadwick and John Jeavons have promoted raised beds in the US, based on their experience in areas where raised bed gardening is an effective alternative to rows of plants on flat ground, a style modeled on farm fields. Jeavons' classic book, *How to grow more vegetables*, now in its ninth edition, not only provides detailed information on how to construct raised beds, but also how to double dig, and how to plant in patterns that he found to maximize garden production (Jeavons, 2017).

As a result, raised beds have become part of the alternative and organic gardening model in North America, Europe, and Australia and as such have been promoted as the model for gardens everywhere as part of an alternative to industrial gardens (Cleveland and Soleri, 1987). While raised beds work really well where there is plenty of rain and cooler temperatures, and where providing adequate drainage is an issue, they may not be the best choice under hot, dry conditions, because raised beds on open ground expose a lot of soil surface to the sun and air which raises soil temperatures in the warm season, increasing evaporation, leading to salt build-up in the growing area. However, using drip or pot irrigation can eliminate many problems with watering raised beds.

One major advantage of raised beds is that they can make it easier for people with back pain or other physical challenges to garden. Raised beds also make protecting the garden from gophers and other burrowing, root-eating animals easier because you will not need to dig down as far in order to line the base and sides of the bed with aviary wire or hardware cloth to exclude those animals.

A great alternative to raised beds built with only soil is to use borders of wood, stone, concrete blocks or other material to protect the sides of the beds, reduce evaporation, and moderate temperature fluctuations, providing many of the advantages of sunken beds. Bordered raised beds also make it easier to keep contaminated soil out of the growing area and so are a good choice for many urban gardens. Borders also facilitate making raised sunken beds. Using cinder blocks filled with soil for the border as the Community Seed Exchange in Northern California does in its seed garden, is a great idea because compared to wood, concrete blocks are relatively inexpensive, more durable, and provide much better insulation.

Stand-alone containers are another option for garden beds. A friend of ours in Arizona who is a longtime gardener has half a dozen old bathtubs in his backyard filled with thriving tomato, pepper, and other crops. He has converted his entire garden into bathtub "beds." Stand-alone containers also eliminate gopher problems, and can help deter other large pests. The main challenge is finding suitable containers. To prevent the soil from overheating in the summer, you can paint the outside of containers white, insulate them with mulch, and provide shade. A non-profit project in Austin, Texas, has instructions for building raised beds using discarded pallets and including water wicking systems (Section 4.3.4).

8.10. Resources

All the websites listed below were verified on June 6, 2018.

8.10.1. Water footprints

See the Water Footprint Network, Section 2.7.3.

The Pacific Institute is a nonprofit organization in Oakland, California that "is a global water think tank" conducting scientific research supporting policy and education for sustainable water policies. The website is a good source for publications, tools, including for community action on water issues, research on salinization, urban water conservation, water-based conflicts worldwide, and more.
http://pacinst.org/

8.10.2. Climate and weather

A site about climate in the southwestern US (Arizona, New Mexico) run by the University of Arizona, includes explanations and links useful for finding and understanding climate projections for that region.
http://www.climas.arizona.edu/

For California climate history and projections by year based on different IPCC scenarios see Cal-Adapt, Section 2.7.2.

The Climate Explorer is a map-based tool that is part of the US Climate Resilience Tookit and is useful for examining past climate and long-term climate projections generated from climate models for the continental US.
https://crt-climate-explorer.nemac.org/

The Community Collaborative Rain, Hail and Snow (CoCoRaHS) Network is a public science project started by the Colorado Climate Center at Colorado State University and supported by NOAA and the NSF. Thousands of members of the public are collecting precipitation data in their backyards or other locations and entering them into the CoCoRaHS public website. When coverage is good it gives an idea of how spatially variable precipitation can be. They also link to historical precipitation databases.
http://www.cocorahs.org/

The American Geophysical Union (AGU) "is an international non-profit scientific association with 60,000 members in 137 countries" who are earth and space scientists conducting and communicating research. The AGU's Thriving Earth Exchange (TEX) is developing a framework and process for communities and scientists to work directly with each other to solve local problems, many associated with ACC. TEX supports these partnerships in multiple ways.
https://thrivingearthexchange.org/

See the online community weather and climate journal *ISeeChange* in Section 2.7.2.

8.10.3. Greywater and rainwater harvesting

The US National Conference of State Legislatures has a website with up-to-date information on rainwater harvesting laws and legislation.
http://www.ncsl.org/research/environment-and-natural-resources/rainwater-harvesting.aspx

HarvestH2o is a commercial website with information on rainwater harvesting regulations and financial incentives for states and cities, although these may not all be current.
http://www.harvesth2o.com/statues_regulations.shtml

Water harvesting. Guidelines to good practice by Mekdaschi Studer and Liniger (2013) has much information about a variety of water harvesting techniques from around the world, including technical details, guidelines for practice, and photos; available gratis online.

Brad Lancaster's website and books are a good resource for practical information about rainwater harvesting, based on his extensive personal experience and research (Lancaster, 2008, 2013).
http://www.harvestingrainwater.com/

Greywater Action has a website with lots of resources on greywater and rainwater harvesting, as well as composting toilets, and includes a page on greywater regulations and policies.
http://greywateraction.org/

Oasis Design is a website by Art Ludwig, developer of the laundry-to-landscape method for recycling greywater. Ludwig is the author of a book on greywater useful for gardeners (Ludwig, 2015), and a companion book for professional installers.
http://oasisdesign.net/

Many cities and states have programs supporting greywater recycling and rainwater harvesting, and you can check to see what applies in your location. San Francisco's resources include a rainwater harvesting calculator.
http://sfwater.org/index.aspx?page=178

8.10.4. Irrigation

The University of California's Master Gardener program has a number of online resources about installing and using drip irrigation systems. Here, for example, are ones from the Sonoma County program.
http://sonomamg.ucanr.edu/Drip_Irrigation/

The USDA's 1997 *Irrigation guide* is a comprehensive guide of over 700 pages, with many good graphs and illustrations, although oriented toward large-scale production, and available free online (USDA NRCS, 1997).

The UN FAO has published a series of irrigation and drainage papers, many of which have information on low-tech ideas for irrigation and solving problems.
http://www.fao.org/publications

Brouwer and Heibloem's FAO manual gives simple, detailed instructions for calculating timing of irrigation, and is available free online (Brouwer and Heibloem, 1986).

For discussion of salinity management and estimating the leaching requirement, the FAO publication *Water quality for agriculture* is a main source (Ayers and Westcot, 1994). The University of Arizona and the University of California also have brief, practical guides.
https://extension.arizona.edu/sites/extension.arizona.edu/files/pubs/az1107.pdf
https://anrcatalog.ucanr.edu/pdf/8550.pdf, and
http://ucanr.edu/sites/UrbanHort/Water_Use_of_Turfgrass_and_Landscape_Plant_Materials/SLIDE__Simplified_Irrigation_Demand_Estimation/

The USDA has published a detailed guide with color illustrations for estimating soil water content by feel and appearance, with corresponding irrigation requirements in inches of water ft^{-1} of soil (USDA NRCS, 1998).

David Bainbridge shares his knowledge from years of experimenting in *Gardening with less water: Low-tech, low-cost techniques; use up to 90% less water in your garden*, an excellent resource for just what the title states, including pot irrigation (Bainbridge, 2015).

Robert Kourik's book (2009) is a very accessible how-to manual for installing a home drip irrigation system, with California examples.

A simple introduction to irrigation scheduling with adjustable controllers has been developed by the Alliance for Water Efficiency. http://www.allianceforwaterefficiency.org/Irrigation_Scheduling_Introduction.aspx

8.11. References

Abu-Zreig, M. M., Abe, Y. & Isoda, H. (2006) The auto-regulative capability of pitcher irrigation system. *Agricultural Water Management*, 85, 272–278, DOI: http://dx.doi.org/10.1016/j.agwat.2006.05.002.

Ayers, R. S. & Westcot, D. W. (1994) *Water quality for agriculture*. UN Food and Agriculture Organization, available at: http://www.fao.org/docrep/003/t0234e/T0234E00.htm#TOC (accessed Oct. 15, 2018).

Bainbridge, D. A. (2001) Buried clay pot irrigation: A little known but very efficient traditional method of irrigation. *Agricultural Water Management*, 48, 79–88, DOI: http://dx.doi.org/10.1016/S0378-3774(00)00119-0.

Bainbridge, D. A. (2015) *Gardening with less water: Low-tech, low-cost techniques; use up to 90% less water in your garden*. Storey Publishing, North Adams, MA.

Brady, N. C. & Weil, R. R. (2010) *Elements of the nature and properties of soil*, 3rd edn. Prentice Hall, Upper Saddle River, NJ.

Brouwer, J. & Heibloem, M. (1986) *Irrigation water management: Irrigation water needs*. Irrigation water management training manual no. 3. Food and Agriculture Organization of the United Nations, Rome, Italy, available at: http://www.fao.org/docrep/S2022E/S2022E00.htm#Contents (accessed May 28, 2018).

Campi, P., Palumbo, A. D. & Mastrorilli, M. (2012) Evapotranspiration estimation of crops protected by windbreak in a Mediterranean region. *Agricultural Water Management*, 104, 153–162.

CAWSI (California Agricultural Water Stewardship Initiative) (2018) Use of municipal recycled water, available at: http://agwaterstewards.org/practices/use_of_municipal_recycled_water/ (accessed May 21, 2018).

CDWR (California Department of Water Resources) (2016) *Municipal recycled water. A resource management strategy of the California Water Plan*. California Natural Resources Agency, available at: http://www.water.ca.gov/waterplan/docs/rms/2016/11_Municpal_Recycled_Water_July2016.pdf(accssed Oct. 15, 2018)

Chapagain, A. K. & Orr, S. (2009) An improved water footprint methodology linking global consumption to local water resources: A case of Spanish tomatoes. *Journal of Environmental Management*, 90, 1219–1228, DOI: http://dx.doi.org/10.1016/j.jenvman.2008.06.006.

Chukalla, A. D., Krol, M. S. & Hoekstra, A. Y. (2017) Marginal cost curves for water footprint reduction in irrigated agriculture: Guiding a cost-effective reduction of crop water consumption to a permit or benchmark level. *Hydrology and Earth System Sciences*, 21, 3507–3524, DOI: 10.5194/hess-21-3507-2017.

City of Tucson (2017) Rainwater harvesting rebate, available at: https://www.tucsonaz.gov/water/rainwater-harvesting-rebate (accessed June 21, 2018).

City of Tucson (2018) Gray water rebate, available at: https://www.tucsonaz.gov/water/gray-water-rebate (accessed June 13, 2018).

Cleveland, D. A. & Soleri, D. (n.d.) Quantifying physical benefits and costs in an urban desert garden.

Cleveland, D. A. & Soleri, D. (1987) Household gardens as a development strategy. *Human Organization*, 46, 259–270.

Cleveland, D. A. & Soleri, D. (1991) *Food from dryland gardens: An ecological, nutritional, and social approach to small-scale household food production*. Center for People, Food and Environment (with UNICEF), Tucson, AZ. https://tinyurl.com/FFDG-1991

Cleveland, D. A., Orum, T. V. & Ferguson, N. F. (1985) Economic value of home vegetable gardens in an urban desert environment. *Hortscience*, 20, 694–696.

Cleveland, D. A., Phares, N., Nightingale, K. D., Weatherby, R. L., Radis, W., Ballard, J., Campagna, M., Kurtz, D., Livingston, K., Riechers, G., et al. (2017) The potential for urban household vegetable gardens to reduce greenhouse gas emissions. *Landscape and Urban Planning*, 157, 365–374, DOI: http://dx.doi.org/10.1016/j.landurbplan.2016.07.008.

Cohen, Y. (2009) Graywater—a potential source of water. UCLA Institute of Environment and Sustainability, Los Angeles, CA, available at: http://www.environment.ucla.edu/reportcard/article4870.html (accessed Sept. 24, 2014).

Doneen, L. D. & Westcot, D. W. (1984 [first published 1971]) Irrigation practice and water management.

Revision 1. *FAO Irrigation and Drainage Paper Number 1*, FAO, Rome.

Doorenbos, J. & Kassam, A. H. (1979) *Yield response to water. FAO irrigation and drainage paper No. 33*. FAO, Rome.

EPA (US Environmental Protection Agency) (2013) *Inventory of U.S. greenhouse gas emissions and sinks: 1990–2011*. EPA, Washington, D.C, available at: http://www.epa.gov/climatechange/ghgemissions/usinventoryreport.html (accessed April 20, 2013).

Gardner, W. H. (1962 [reprint 1979]) How water moves in the soil. *Crops and Soils Magazine*, 1979, 13–18, available at: http://hydrogold.org/jgp/pdf/lib.walter_h_gardner.water_movement_in_soils_1962.pdf (accessed June 13, 2018).

Gardner, W. H. (1988) Water movement in soils. *USGA Green Section Record*, March/April 1988, 23–27, available at: http://gsrpdf.lib.msu.edu/ticpdf.py?file=/1980s/1988/880323.pdf (accessed June 13, 2018).

Hiskakis, M., Babou, E. & Briassoulis, D. (2011) Experimental processing of biodegradable drip irrigation systems—possibilities and limitations. *Journal of Polymers and the Environment*, 19, 887–907, DOI: 10.1007/s10924-011-0341-1.

Howell, T. A. (2003) Irrigation efficiency. In Stewart, B. A. & Howell, T. A. (eds) *Encyclopedia of water science*, 467–472, DOI: 10.1081/E-EWS 120010252.

Hsiang, S. M., Burke, M. & Miguel, E. (2013) Quantifying the influence of climate on human conflict. *Science*, 341, DOI: 10.1126/science.1235367.

IPCC (Intergovernmental Panel on Climate Change) (2013) *Climate change 2013: The physical science basis*. Working Group I contribution to the Fifth Assessment Report of the Intergovernmental Panel on Climate Change, IPCC, Geneva, available at: http://www.climatechange2013.org/images/report/WG1AR5_SPM_FINAL.pdf (accessed Oct. 15, 2018)

Jafari, M., Abdolahi, J., Haghighi, P. & Zare, H. (2012) Mulching impact on plant growth and production of rainfed fig orchards under drought conditions. *Journal of Food, Agriculture & Environment*, 10, 42–433.

Jeavons, J. (2006) *How to grow more vegetables (and fruits, nuts, berries, grains, and other crops): Than you ever thought possible on less land with less water than you can imagine*, 6th edn, Ten Speed Press, Berkeley, CA.

Jeavons, J. (2017) *How to grow more vegetables (and fruits, nuts, berries, grains, and other crops): Than you ever thought possible on less land with less water than you can imagine*, 9th edn. Ten Speed Press, Berkeley, CA.

Karam, F., Saliba, R., Skaf, S., Breidy, J., Rouphael, Y. & Balendonck, J. (2011) Yield and water use of eggplants (*Solanum melongena* L.) under full and deficit irrigation regimes. *Agricultural Water Management*, 98, 1307–1316.

Katerji, N. & Mastrorilli, M. (2014) Water use efficiency of cultivated crops. In *eLS*, John Wiley & Son, Ltd, DOI: 10.1002/9780470015902.a0025268.

Kourik, R. (2009) *Drip irrigation for every landscape and all climates*, 2nd edn. Chelsea Green Publication, White River, VT.

Ladd, E. J. (1979) Zuni social and political organization. In Ortiz, A. (vol. ed.) *Southwest*, Vol. 9, in Sturtevant, W. C. (general ed.) *Handbook of North American Indians*, 482–491. Smithsonian Institution, Washington, DC.

Lamm, F. R. & Camp, C. R. (2007) Subsurface drip irrigation. In Lamm, F. R., Ayars, J. E. & Nakayama, F. S. (eds) *Developments in agricultural engineering*, 473–551. Elsevier, DOI: https://doi.org/10.1016/S0167-4137(07)80016-3.

Lancaster, B. (2008) *Rainwater harvesting for drylands and beyond*, Vol. 2, *Water-harvesting earthworks,* Rainsource Press, Tucson, AZ.

Lancaster, B. (2013) *Rainwater harvesting for drylands and beyond*, Vol. 1, 2nd edn: Guiding principles to welcome Rain into your life and landscape. Rainsource Press, Tucson, AZ.

Ludwig, A. (2015) *Create an oasis with greywater. Integrated design for water conservation: Reuse, rainwater harvesting & sustainable landscaping*, 6th edn. Oasis Design, Santa Barbara, CA.

Martin, O. (2007) French intensive gardening: A retrospective. *News & Notes of the UCSC Farm & Garden*, 1–2, 10–11.

Mekdaschi Studer, R. & Liniger, H. (2013) *Water harvesting. Guidelines to good practice*. Centre for Development and Environment (CDE) and Institute of Geography, University of Bern; Rainwater Harvesting Implementation Network (RAIN), MetaMeta; and The International Fund for Agricultural Development (IFAD), available at: https://www.sswm.info/sites/default/files/reference_attachments/MEKDASCHI%20STUDER%20and%20LINIGER%202013%20Water%20Harvesting.pdf (accessed Oct. 15, 2018).

Mekonnen, M. M. & Hoekstra, A. Y. (2012) A global assessment of the water footprint of farm animal products. *Ecosystems*, 15, 401–415, DOI: 10.1007/s10021-011-9517-8.

Mekonnen, M. M. & Hoekstra, A. Y. (2014) Water footprint benchmarks for crop production: A first global assessment. *Ecological Indicators*, 46, 214223. DOI: 10.1016/j.ecolind.2014.06.013.

Merriam, J. L., Shearer, M. N. & Burt, C. M. (1980) Evaluating irrigation systems and practices. In Jensen, M. E. (ed.) *Design and operation of farm irrigation systems*, 721–760. American Society of Agricultural Engineers, St. Joseph, MI,.

Milly, P. C. D., Betancourt, J., Falkenmark, M., Hirsch, R. M., Kundzewicz, Z. W., Lettenmaier, D. P. & Stouffer, R. J. (2008) Stationarity is dead: Whither water management? *Science*, 319, 573–574, DOI: 10.1126/science.1151915.

NWS NOAA (Climate Prediction Center, National Weather Service, National Oceanic and Atmospheric Administration) (2018) Outlook maps, graphs and tables. National Center for Weather and Climate Prediction, College Park, MD, available at: https://www.cpc.ncep.noaa.gov/ (accessed June 13, 2018).

Rodell, M., Famiglietti, J. S., Wiese, D. N., Reager, J. T., Beaudoing, H. K., Landerer, F. W. & Lo, M. H. (2018) Emerging trends in global freshwater availability. *Nature*, 557, 651–659, DOI: 10.1038/s41586-018-0123-1.

SCENIC (2018) Southwest Climate and Environmental Information Collaborative, available at: https://wrcc.dri.edu/csc/scenic/ (accessed May 24, 2018).

SDLAC (Sanitation Districts of Los Angeles County) (2018) Water reuse program, available at: http://www.lacsd.org/waterreuse/ (accessed May 21, 2018).

SFPUC (San Francisco Public Utilities Commission) (2015) Rainwater harvesting, available at: http://sfwater.org/index.aspx?page=178 (accessed Oct. 13, 2016).

Shaikh, S. & Tunio, S. (2012) Pitcher irrigation brings vegetables to Pakistani desert. *Thomson Reuters Foundation News*, available at: http://news.trust.org/item/20120813154500-vmj4x (accessed Oct. 15, 2018).

Shaxson, F. & Barber, R. (2003) *Optimizing soil moisture for plant production: The significance of soil porosity*. *FAO Soils Bulletin*, 79, Food and Agriculture Organization of the United Nations, Rome.

Siyal, A. A., van Genuchten, M. T. & Skaggs, T. H. (2009) Performance of pitcher irrigation system. *Soil Science*, 174, 312–320.

Smith, M. (2012) Yield response to water: the original FAO water production function. In Steduto, P., Hsiao, T. C., Fereres, E. & Raes, D. (eds) *Crop yield response to water. FAO irrigation and drainage paper* 66, 6–13. FAO, Rome,.

Snyder, R. L. (2014) Irrigation scheduling: Water balance method. University of California, Department of Land, Air and Water Resources, Atmospheric Science, Davis, California, available at: http://biomet.ucdavis.edu/irrigation_scheduling/bis/ISWBM.pdf (accessed June 22, 2018).

Soleri, D. (1989) Hopi gardens. *Arid Lands Newsletter*, 29, 11–14.

Soleri, D. & Cleveland, D. A. (1988) Managing water in arid gardens: Simple, efficient, economical methods. *Fine Gardening*, 14–18.

Solh, M. (2010) Research outputs and approaches to enhance food security and improve livelihoods in dry areas. IFAD/ICARDA Information Exchange Workshop, ICARDA (International Center for Agricultural Research in the Dry Areas), available at: https://www.slideshare.net/ifad/1-mahmoud-solh-ifad-workshop-pre-final-oct-2009 (accessed June 12, 2018).

Steduto, P., Hsiao, T. C., Fereres, E. & Raes, D. (2012) *Crop yield response to water. FAO irrigation and drainage paper 66*. FAO, Rome.

Stegman, E. C., Musick, J. T. & Stewart, J. I. (1980) Irrigation water management. In Jensen, M. E. (ed.) *Design and operation of farm irrigation systems*, 763–817. American Society of Agricultural Engineers, St. Joseph, MI.

TWDB (Texas Water Development Board) (2005) *The Texas manual on rainwater harvesting*, 3rd edn. Texas Water Development Board, Austin, TX.

UNL (2016) National Drought Mitigation Center. University of Nebraska, Lincoln, Lincoln, NE, available at: https://drought.unl.edu/Education/DroughtIn-depth/TypesofDrought.aspx (accessed Dec. 6, 2018).

USDA NRCS (United States Department of Agriculture, Natural Resources Conservation Service) (1997) *Irrigation guide, national engineering handbook, part 652*, available at: https://policy.nrcs.usda.gov/OpenNonWebContent.aspx?content=17837.wba (accessed Aug. 10, 2018).

USDA NRCS (United States Department of Agriculture, Natural Resources Conservation Service) (1998) *Estimating soil moisture by feel and appearance*, available at: https://www.cdpr.ca.gov/docs/county/training/inspprcd/handouts/soil_moist_feel_test.pdf (accessed June 23, 2018).

Woldu, Z. (2015) Clay pot pitcher irrigation: a sustainable and socially inclusive option for homestead fruit production under dryland environments in Ethiopia (a partial review). *Journal of Biology, Agriculture and Healthcare*, 5.

WWAP (United Nations World Water Assessment Programme) (2017) *The United Nations world water development report 2017: Wastewater, the untapped resource*. UNESCO, Paris, available at: http://www.unesco.org/new/en/natural-sciences/environment/water/wwap/wwdr/2017-wastewater-the-untapped-resource/ (accessed Oct. 15, 2018).

Zapata-Sierra, A. J. & Manzano-Agugliaro, F. (2017) Controlled deficit irrigation for orange trees in Mediterranean countries. *Journal of Cleaner Production*, 162, 130–140, DOI: https://doi.org/10.1016/j.jclepro.2017.05.208.

9 Managing Pests, Pathogens, and Beneficial Organisms

T.V. ORUM, D. SOLERI, D.A. CLEVELAND, S.E. SMITH

Chapter 9 in a nutshell.

- There are many organisms living in the garden besides crop plants—other plants, bacteria, protozoa, fungi, viruses, insects and other arthropods, nematodes, and large animals—most are beneficial, but some are harmful.
- Harmful organisms are called pests or pathogens.
- Pests affect the plant physically (e.g., grasshoppers eating leaves, gophers eating roots, or weeds crowding out young crop seedlings).
- Pathogens are infectious organisms that directly attack the plant physiologically, causing disease by disrupting plant functions (e.g., fungi that cause powdery mildew).
- An ecological approach encourages beneficial organisms, and emphasizes long-term prevention of pest and pathogen problems through:
  - environmental management of soil, water, and air;
  - choice of crop species and varieties adapted to the garden environment and its organisms;
  - biological control using living organisms;
  - physical control using mechanical barriers or removal; and
  - chemical control using substances that repel, sicken, or kill pests or pathogens.
- When major short-term problems develop in spite of this preventive long-term management strategy, the last three components of an ecological approach can also be used for short-term, problem-specific, curative management.
- Diagnosing pest, pathogen, and other plant problems helps to manage them, although gardeners can often manage effectively without identifying the specific causal agent.
- Diagnosis is based on the type of damage observed, and a general knowledge of the kinds of organisms responsible.

Especially when it's hot and dry, gardens are oases of green vegetation that people find enjoyable, and that also attract insects, worms, birds, rodents, and other wild and domestic animals. Gardens also provide good growing conditions for weeds, fungi, viruses, and bacteria. Most of these do not affect garden production, and many improve it. Those that reduce production are called pests or pathogens.

Symbiosis is a general term for organisms living together, with their effects on each other either neutral, positive, or negative. Pests and pathogens harm or kill the crop plant they are living with as they grow and multiply. As we use the term, a *pest* is any organism that attacks plants externally, such as caterpillars eating tomato leaves. Usually pests eat leaves, fruits, seeds, roots or sap, but don't directly disrupt plants' physiological functioning. A plant *parasite* is an organism that lives on or in the plant that is its *host*, and derives nutrients from its host; fungal, bacterial, viral, nematode, and other microscopic parasites causing disease in their host are *pathogens*. In some contexts, all pathogens are considered pests, even those that do not live outside plant cells. *Disease* is abnormal physiological functioning detrimental to the plant's growth or productivity. Abnormal physiological functioning can also be caused by environmental factors like nutrient deficiencies, sunburn, extreme temperatures, and herbicides (Chapters 7 and 8).

Other organisms can form symbiotic relationships with crop plants that benefit the plant. A symbiotic relationship that benefits both the plant and the other organism is *mutualistic*, for example *mycorrhizae*—associations between mycorrhizal fungi that infect the roots of most garden crops (and other plants) and increase availability of phosphorus and other nutrients to their hosts, and the host plants that provide carbohydrates to those fungi (Box 7.1). An important variable in all of these relationships is the degree of specificity. While some pests, pathogens, and beneficial organisms are *specialists* that only form relationships with certain plant species or families, others are *generalists* that form relationships with a range of species or families.

In the first section of this chapter we describe the ecological approach to managing non-crop organisms in the garden. In the second section we briefly describe the common types of pests, pathogens, and other organisms found in gardens. In the five sections after that we cover five basic management strategies, emphasizing pests and pathogens: managing the garden environment, choosing crop plants, biological control, physical control, and chemical control. For each of these we include both concepts to use for optimizing net benefits and avoiding major problems, and for the last three, concepts to use when a major problem does develop. Section 9.8 is a guide to diagnosing and managing specific problems, and includes tables and accompanying figures summarizing some key concepts with examples.

Managing pests, pathogens, and beneficial organisms in the garden is a balance between losses due to pests and pathogens and the time and resources in managing them—the point is not completely eliminating pests and pathogens or the damage they do. While many broad principles for management are becoming clearer, like the benefits of biodiversity, the details are controversial and the subject of ongoing research (Isbell et al., 2017). The larger goal, like that of food gardens in general, involves going beyond the individual gardens and gardeners, and thinking in terms of the health of the environment and our communities, from local to global, and into the future.

9.1. The Basic Ecological Approach

Most pest and pathogen problems in the garden are influenced by environmental conditions such as too high or too low temperatures, too much or too little sunlight, too much or too little water, poor soil quality, planting inappropriate crops, or planting at the wrong time of year. Unstressed, healthy plants resist pests and pathogens much better than stressed, unhealthy ones, and some genotypes or varieties remain productive even when pests or pathogens are present, while others don't. An *ecological approach* to pest and pathogen management involves creating optimal environments for crops

and beneficial organisms by managing soil structure, pH, organic matter and water, as well as the effects of temperature, sunlight, wind, and relevant social variables. It also means meeting the gardeners' goals by choosing crop species and varieties gardeners want to eat, and that are adapted to the garden environment, and that maintain sufficient crop genetic diversity to encourage diversity in the animals and microorganisms in the garden (Section 10.2).

9.1.1. The long-term strategy

The long-term ecological approach we use has five main components: environmental management, crop choice, biological control, physical control, and chemical control (Fig. 9.1) (cf. Bridge, 1996). When major short-term problems develop in spite of this preventive long-term strategy, the last three components can also be used for short-term, problem-specific, curative management. This ecological approach works as a coping strategy to deal with familiar variation, like the growth of aphid populations in the spring, but can be adjusted to adapt to the unfamiliar challenges brought by trends, like the effect of higher average temperatures due to anthropogenic climate change (ACC) on the arthropod species in your garden (Chapters 2, 3).

However, when major, long-term pest or pathogen problems develop as the result of trends, and can't be solved by adaptation, then transformation will be needed, which can include changing garden locations, crops grown, or major modifications of control strategies. The need for transformation is likely to increase because of the trends discussed in Chapter 2, especially ACC.

While gardeners try keeping losses to a minimum, some pest or disease damage is a natural part of a healthy garden, so it's important not to overreact to a few chewed, spotted, or dead leaves. The goal is not having a perfect harvest from every crop in the garden, but harvesting the most food and other benefits for the smallest amount of work and resources invested, while working for prosocial and environmental goals. When problems develop, diagnosing the cause and responding appropriately can limit the damage, but identifying the specific causal agent (e.g., the insect or bacterium) is often unnecessary. The goal of diagnosis is to understand the situation well enough to choose the most appropriate management strategy, which sometimes may be doing nothing. The first step is careful

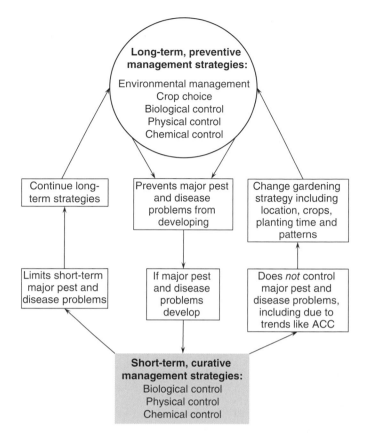

Figure 9.1. An ecological approach to managing organisms in the garden

observation (Section 3.2.2). Looking at the garden in terms of seasonal cycles, and comparing it to other gardens, crop fields, and natural plant communities is a good way to start. As in other aspects of gardening social organization matters, e.g., it can be more effective to coordinate some control measures with others (Box 9.1).

Most plants produce more leaves and flowers than required so partial defoliation or attacks on buds, blossoms, or young fruit by pests may not decrease yield significantly (Stacey, 1983). Young leaves are usually more important to plant yield than older leaves (Section 5.4), so damage to older leaves is generally not as critical as damage to younger leaves (Stacey, 1983; Van Dam et al., 1996). In a mixed garden with a number of different crops, if one or two crops do succumb, others can grow to take their place.

The majority of animals like insects and other arthropods, and microorganisms like bacteria and fungi either cause no damage, are beneficial to crop production, or are even essential for a good harvest. Generally, as the number of organisms increases, so does the rate at which they die from predators, disease, the weather, lack of food, or other causes (Alyokhin et al., 2005). In nature, most pests and pathogens do not have large enough populations to cause significant damage because they are kept in check by natural controls.

In addition to increased plant diversity in space, plant diversity through time, for example changing or rotating crops of different botanical families, can also be an effective response to some pest or disease problems (Section 9.3).

9.1.2. Short-term, curative responses

Even with an ecological approach, pest or pathogen populations sometimes get large enough to cause extensive damage, and may warrant targeted, short-term responses (Fig. 9.1), especially if they're

Our food gardens are part of larger environments, and many of the processes that are important for gardens occur across the landscape. For this reason, controlling pests or encouraging beneficial organisms sometimes works best if gardeners in an area cooperate to coordinate some of their practices. For example, in a community garden in San Luis Obispo, California, gardeners cooperated with a state-wide project to increase the number of native pollinating bees, with 16 of 18 gardeners adding plants that serve as bee resources (Pawelek et al., 2009). As a result, the number of bee species present in the garden increased from 5 to 31 from 2007–2009. After this experience the community garden adopted a mission statement supporting pro-pollinator practices.

Cooperation, including for pest control, can be encouraged by creating an institutional framework that ensures fairness and transparency, making clear the shared benefits and responsibilities (Box 3.3). This often means organizing on a local level. For example, a survey of 229 farmers in Missouri found that 91% would be willing to cooperate in coordinated, natural pest control efforts including pest scouting, targeted crop rotations, and other methods (Stallman and James, 2015). Even though some scientists assume regional efforts are best, organizing as small local institutions was the most acceptable strategy to the farmers surveyed, in part because they had experience with both organizing and environments at the local level, meaning minimal costs for pest control benefits. This highlights why it is important for institutions to be formed by those using them, whether across a state, a neighborhood, or a community garden, and not imposed from the top down.

---

affecting many crops, or one or two important ones, such as a stand of sweet corn that is almost ripe, a prize peach tree, or tomato plants loaded with fruit. For example, if *Phytophthora infestans*, which causes late blight disease of potato and tomato is already present, an explosion in the number of propagules of that pathogen can occur when there is a period of unusually cool, wet weather. Early detection and response is important because late blight is difficult to control without using synthetic chemicals (Stone, 2014). Some forecast systems, such as BLITECAST, have been developed to help manage late blight (e.g., Krause et al., 1975). Even though developed in 1975, BLITECAST has been used to help manage late blight organically in Wisconsin as recently as 2015 (Gevens, 2015). The key is to use organically approved fungicides prior to the onset of symptoms based in part on the BLITECAST model and then to continue with protective organic fungicides as long as blight conditions persist. Sanitation and planting resistant varieties are also important.

Gardeners also need to consider the tradeoff between the value of the plants needing protection (Are they for market or a special occasion? Are they rare seed you want to conserve?), with the environmental and economic cost of providing the protection (How much time do I have to spend? How would the methods impact nearby plants, animals, soil, and water?). While biological, physical, and chemical control are all used preventatively as part of the ecological approach, they are also used in a targeted way as part of a curative, short-term approach for responding to major problems.

9.1.3. Understanding causes vs. managing symptoms

In Section 9.2 we discuss different types of organisms you see in your garden, and their effects, which can help in figuring out how to manage for long-term productivity, or respond to major crises. However, understanding the basic principles of an ecological approach is more important than trying to remember all of the specific pathogens and pests that cause problems, and actions tailored to managing them. This is because many problems don't require identification of the specific pathogen or pest. It's also because different pests, pathogens, and environmental stresses can have similar symptoms, the same pest or pathogen can create different symptoms, and similar problems may require different responses depending on the crop, location, weather, time of year, gardeners' time and expectations, and other factors. In other words, there are lots of variables involved.

Farmers and gardeners have been developing successful strategies to deal with pests and pathogens for a long time without being able to understand the exact mechanisms involved. For example, in East

Africa, farmers successfully manage bean diseases without understanding that microorganisms they cannot see are the cause (Trutmann et al., 1993). Farmers do this by observing the association between moisture and disease, and controlling microclimates, e.g., by staking bean plants, by selecting bean seed from "rain-resistant" plants, and by sanitation, such as rotating the dominant crops in a field.

9.2. Common Types of Pests, Pathogens, and Beneficial Organisms

The following are brief descriptions of some common garden arthropods, nematodes, large animals, microorganisms, and non-crop plants, with an emphasis on pests and pathogens. We give some management suggestions, with many more in Sections 9.3–9.8.

9.2.1. Insects and other arthropods

Arthropods are small animals with segmented bodies and hard outer skins (*exoskeletons*). Nearly 85% of all animal species on Earth are arthropods (Giribet and Edgecombe, 2012). *Insects* are the arthropod group with the largest number of species, ~1 million species described and several more millions of species still unnamed (Chapman, 2009), and with individual animals alive at one time ~10 quintillion or 10^{18}. Insects are arthropods with a three-part body and six legs in their adult stage, while the immature stages of many insects are caterpillars, grubs, or maggots, some with no legs and all wingless. Most insect species are neutral or beneficial for gardens and people. For example, insects are extremely important in pollination, directly pollinating over $15 billion worth of crops each year in the US (Calderone, 2012). In the western US, bees, especially honeybees (including *Apis mellifera*, the domesticated and most common *Apis* species) and bumblebees (*Bombus* spp.), are probably the most important pollinators. Other minor pollinators include ants, beetles, butterflies, moths, and thrips. Spiders, centipedes, wasps, and many other arthropods are important pest predators or parasites, but other arthropods, such as sowbugs and mites, can be pests in the garden. Most acute arthropod pest problems can be controlled biologically, physically, or using non-synthetic chemicals with low impacts on other organisms. The benefits of having pollinators and pest predators and parasites are a strong argument against using broad-spectrum pesticides that indiscriminately kill all arthropods.

When arthropods do cause damage, it varies according to the way they eat. Many, such as ants, grasshoppers, caterpillars, and many beetles chew leaves, stems, flowers, fruit, and roots, but as mentioned earlier, in some cases light "damage" to a plant is inconsequential, or may actually stimulate production.

Sucking arthropods have piercing mouth parts that they use to penetrate plant cells and suck out the contents, and can spread disease pathogens in the process. Sucking arthropods include whiteflies, aphids, scale insects, thrips, plant hoppers, leaf or plant bugs, and mites. Common signs of damage caused by these animals are leaves that are curled, twisted, or have dry spots, and abnormal-looking shoots. Many other sucking insects and predatory mites are beneficial predators of garden pests.

Boring insects make holes in stems, roots, and fruits and do most of their damage from the inside. Often a small hole and perhaps some *frass* (feeding debris or feces) will be the only evidence on the outside. Some species of fly larvae (*maggots*), adult beetles and beetle grubs, and caterpillars (moth and butterfly larvae) are among the most common boring insects.

As mentioned, arthropods can spread plant pathogens, including viruses, fungi, and bacteria. They do this directly as pathogen vectors, by making feeding wounds where pathogens can easily enter, and by weakening plants by eating roots and leaves, making them more susceptible to disease (Agrios, 2005, 42–45). Some even spread pathogens as part of their life cycle. For example, the polyphagous shot hole borer (*Euwallacea* spp.) is a beetle from southeast Asia that recently arrived in southern California, probably on a cargo shipment (UCANR, 2018a). Females bore through the bark of some trees, including avocado, fig, and olive, to lay eggs and plant a species of *Fusarium* fungus that grows to provide food for larvae, and clogs the vascular system of the tree, which can eventually kill it.

9.2.2. Nematodes

There are several thousand species of nematodes, or roundworms. Most live freely in the soil and feed on microscopic animals including other nematodes, bacteria, and fungal spores. Some nematodes cause human disease, others parasitize insect pests and are an important biological control. Several hundred nematode species feed on plants and cause disease, with a number being widespread and serious pests of annual and perennial garden crops.

Most nematodes that attack plants are microscopic and live below ground where they feed on roots, although some feed on flowers, seeds, and leaves. Some nematodes feed on plant roots from the outside (*ectoparasites*), while others live inside of the roots (*endoparasites*), causing root galls, root lesions, excessive branching, injured root tips, or root rot. Nematodes also spread viral pathogens and interact with some disease-causing fungi and bacteria, resulting in greater damage to plants than each would cause separately. For example, root rot of green bean caused by the fungus *Rhizoctonia solani* is much more severe in plants already infected with root knot nematodes (*Meloidogyne incognita*) (Al-Hazmi and Al-Nadary, 2015).

Nematode problems in the garden can often be managed using low-cost techniques that help keep nematodes out of uninfected areas, and when present, keep their numbers in both soil and plants low enough that a decent harvest is still produced (Section 9.5). Because nematodes are such a significant and widespread problem, many researchers are studying holistic control methods that use multiple strategies (Collange et al., 2011), or use specific bacteria that are pathogens of nematodes (Kokalis-Burelle, 2015; Xiong et al., 2015; Meyer et al., 2016).

9.2.3. Larger animals

Because of their size, hungry rabbits, ground squirrels, raccoons, gophers, groundhogs, chickens, ducks, deer, or goats can wreak havoc on a garden. Wild birds can also be major garden pests, eating seeds, seedlings, leaves, and fruit. With increasing heat and aridity, and spreading towns and cities, these animals may cause more damage to your garden because they have fewer alternative food sources. In most cases the best approach is physical control, keeping larger animals out of the garden with fencing, and noise-makers, and locating gardens close to places with a lot of human activity. Snails and slugs can also cause significant seedling and leaf damage when the garden environment is cool and moist, vegetation is dense, and there is debris for them to hide under—picking them off by hand, clearing vegetation and debris, or using bait are common responses.

9.2.4. Microorganisms

While most microorganisms in the garden are neutral or beneficial, some cause important diseases. An *infectious* plant disease is caused by pathogens that can spread from plant to plant. *Noninfectious* plant diseases, like lettuce tip burn and blossom end rot of peppers and tomato (UCANR, 2018b), are caused by environmental or nutritional conditions (Section 7.3).

Fungi

Fungi are non-photosynthesizing organisms that obtain their food through the decomposition of organic matter, or from living organisms (Ellouze et al., 2014). Fungi play an important role in soil structure by secreting a sticky substance that helps form water-holding soil aggregates (Section 7.2.2), and they break down organic matter into forms that are easier for plants to use. Some fungi also attack and devour nematodes, including harmful ones, while others help to control other disease-causing fungi by competing with them. Mycorrhizae are associations of plants and types of fungi that are extremely important for most garden crops to thrive (Box 7.1).

However, fungi also cause most plant diseases, and can be spread by wind, water, insects, humans, and other vectors. The most severe fungal pathogens in dry environments usually occur below ground in the root system, but symptoms are expressed above ground as wilting or chlorosis. Fungi affecting leaves and other aboveground tissues typically grow well in the warm, moist conditions that occur near the soil surface and in the humid conditions in the middle of lush garden beds, even when the weather is hot and dry. When the environment is dry, fungal leaf spotting diseases are less common because spores require a layer of moisture on the surface of a leaf (boundary layer) to germinate and penetrate it. An exception are the fungal species in a number of different genera that cause powdery mildew, which can grow in moist conditions, but thrive in warm, dry conditions, and cause greater losses in crop yield globally than any other plant disease (Agrios, 2005, 448). Because most fungal pathogens thrive in warm conditions, it is thought that ACC will bring more fungal diseases to warm-season crops, and may cause wider distribution of fungal pathogens (Hartmann et al., 2002, 220).

Bacteria

Bacteria are simple, one-celled organisms that play an important role in gardens, and together with viruses constitute the most abundant organisms on Earth. Bacteria are very small, 0.001–0.003 mm (0.00004–0.00012 in) in diameter, about the size of clay particles. Most bacteria found in the garden are beneficial, helping decompose organic matter in the soil, and making nutrients like nitrogen more available to crops (Section 7.5.1). Some of these bacteria live only as parasites in plants, while others can also survive for a period in the soil, living on organic matter. In warm, moist environments bacteria can multiply very rapidly and are easily spread by insects, people, and water. The bacteria causing a number of important bacterial diseases of plants are transmitted by seeds or seedlings (Dutta et al., 2014) (Section 6.1).

Viruses

A *virus* is a parasite that can only reproduce by invading and taking over cells of other organisms, and cannot lie dormant in fallowed soil, because it cannot live outside of a host. Plant viruses are primarily spread by infected seeds, cuttings, grafts, or sucking insects. Viruses harm their hosts by diverting the resources and processes of cells they infect into the production of more virus. Often viral symptoms in plants are most obvious on new growth as deformities, die-back, and discoloration. Most importantly, plant diseases caused by viruses are almost always systemic (Section 9.8.1).

In the last three decades, virus diseases caused by begomoviruses (in the geminivirus family, *Geminiviridae*) have emerged as a major threat to crops globally, including vegetables grown from the tropics to the warm temperate regions of the world (Navas-Castillo et al., 2011; Saeed and Samad, 2017), such as okra, pepper, squash, and tomato, and also infect weeds that then act as reservoirs for the virus that can be spread to crops. Some think begomoviruses are the most damaging plant viruses in the world (Leke et al., 2015). Begomoviruses are transmitted by sweet potato whiteflies (sucking insects in the *Bemisia tabaci* species complex in the family *Aleyrodidae*), which feed on and spread the virus to a wide range of plant species. Management is complicated by the fact that there are at least 20 *biotypes* (different populations of a species having distinct genotypes) of the whitefly, and when a new biotype replaces an existing one, management strategies may need to shift. Although researchers are only beginning to understand alternative hosts for the begomoviruses, it is clear they are important (Leke et al., 2015), so environmental management strategies include regular weeding, because many weeds are alternative begomovirus hosts supporting virus population growth.

Early detection is helpful in managing the whitefly vector, and in areas where whiteflies are a problem, population levels can be monitored by checking for the insects on the underside of leaves, and using yellow sticky traps. Heavily infested leaves can be removed, or sprayed with water or insecticidal soap. Mechanical control measures for whitefly include reflective mulches, delayed sowing, and physical barriers and intercropping to impede their movement. Floating row covers can be used to protect young seedlings, but the benefit of virus reduction needs to be balanced with yield loss if the cover is left on too long (more than seven weeks in the case of tomatoes in one experiment) (Al-Shihi et al., 2016). Alternatively, UV reflective mulches (either plastic or straw) have reduced both whiteflies and aphid-transmitted virus diseases in cantaloupe (Summers et al., 2005).

Among insect vectors of plant pathogens—including viruses—and insect pests of plants, generalist vectors have a broad range of host plants, and specialists have a narrow range of host plants, that is, the number of plant species they damage. For example, Flint (2015) describes the host plant ranges of 10 species of whitefly that are potential garden pests in California. Some, such as the wooly whitefly (*Aleurothrixus floccosus*), have a narrow host range, whereas others, such as the sweet potato whitefly, have a very broad host range. To add further complexity to the picture, as noted above, the sweet potato whitefly has at least 20 biotypes, each with its own broad set of host plant preferences. Nevertheless, differences in host preferences in specific biotypes of the sweet potato whitefly have led to novel management strategies. For example, the host range of the silverleaf sweet potato whitefly includes both squash and tomato, but it prefers squash to tomato. In three controlled field trials, there was a reduction in the incidence of symptoms of the tomato yellow leaf curl virus (a begomovirus) in tomatoes surrounded by squash acting as trap plants compared with tomatoes surrounded by other tomatoes (Schuster, 2004). Because of this complexity the best strategy for gardeners against whiteflies and their associated viruses is having a diversity of plants and insects in the garden, including a variety of naturally occurring whitefly enemies, both generalists and specialists (Section 9.5.2) (Flint, 2015).

9.2.5. Other plants

When you look at the plants growing in your garden, you are likely to see some "volunteers" that you didn't plant. These can grow from seed of crops left in the soil or blown in from other gardens, and from seed of weed species (Box 9.2). A *weed* is any plant not intentionally sown, but especially one not wanted, and is therefore a "plant pest" (Ellstrand et al., 2010). A plant may be a weed when it competes directly with more valued crop plants for limited garden resources like light and water, and when it attracts or encourages crop pests or pathogens. But "weed" is a very subjective concept, and its definition differs between people, as well as in time and space. There are weed species that are familiar and welcomed by some gardeners who may even save and plant the seeds. For example, *verdolaga* (*Portulaca oleracea*), a succulent plant found throughout western North America, is classified as a weed by some, but is also prized by others as a delicious, nutrient-rich green vegetable, and is sold in some markets. Some plants are desirable leaf vegetables when they are young, but if left unharvested become weeds, including some *Amaranthus* and *Chenopodium* spp.

Box 9.2. Weeds and evolution

Like all living organisms, the plants that we call weeds are the products of the diversity shaping processes of evolution described in Chapter 10. For example, some weeds are invasive species that have moved into new regions like the US through intentional (Bermuda grass) or unintentional (field bindweed) gene flow. Others have evolved from crops, for example, the artichoke thistle from cultivated artichoke, or their close wild relatives, while some are the result of crop x wild plant hybridization (e.g., Johnson grass from sorghum x wild plant, and California wild radish from radish x wild plant, Ellstrand et al., 2010). The result is that many common weeds are similar to their crop relatives, and are hard to control because they look like the crop, so that until they have grown fairly large, or even flower, they are very hard to tell apart from the crop itself, for example weedy rice from rice, and wild radish from radish.

But it is not just how weeds look that can be important to their success, it is also how they grow. Weeds quickly take advantage of disturbed habitats like gardens, which in dry areas are oases of favorable growing conditions. Other common weed characteristics make them strong competitors with domesticated species, including shorter lifespans and greater hardiness, as well as shattering or dehiscent seed heads, and production of many seeds, with staggered seed dormancy periods and germination.

Like many other pests, weeds commonly have short lifespans with more than one generation per year possible, which means they can evolve relatively rapidly in response to selection by their environments, including by chemical weed control. In the US, the number of unique cases of evolved herbicide resistance has increased from 0 in 1950 to 161 in 2018, and in Australia during the same period the number went from 0 to 90. In California there have been eight cases of unique evolved resistance to glyphosate (the active ingredient in RoundUp®) between 2008 and 2016 (Heap, 2018). Glyphosate is the most widely used herbicide in the world, especially since commercialization of glyphosate-resistant transgenic crops starting in 1996 (Benbrook, 2016). The evolution of resistance in both weed and pest populations is one reason why transgenic varieties are now being produced with multiple ("stacked") herbicide-resistant and pest-resistant transgenes. Glyphosate was recently declared a probable carcinogen (Guyton et al., 2015).

ACC is also affecting weed evolution and movement, with weeds moving into new areas as temperatures rise, and increasing their biomass production with rising CO_2 concentrations. For example, with an increase in atmospheric CO_2 from 400 to 600–800 parts per million (ppm)—likely catastrophic levels for humans in multiple ways—the projected increases in biomass production for the weeds wild oat, *Datura*, and field bindweed are 84%, 76% and 36%, respectively (Ziska et al., 2011).

9.3. Environmental Management

The key to keeping pest and pathogen populations low and supporting beneficial organisms is managing the garden environment, including soil, water, and temperature, and growing crops when conditions are favorable for them, which will likely change with ACC. This strategy encourages healthy plants that are better able to withstand stress, and are less susceptible to pests and pathogens. As we discuss in the next section, it's also important to choose crops and varieties that are adapted to the environment in your garden.

Very young or old plants are most frequently affected by pests and pathogens. Plants that are stressed due to too much or too little water or shade, or growing in soil that is too low in organic matter or available nutrients, or with a pH too high or too low, are also more susceptible to pests and pathogens. However, there can be tradeoffs. Some conditions that would otherwise be good for the plant can also encourage problems. For example, mulch—which is so helpful under hot, dry conditions—may harbor insects, fungi, and bacteria that can harm plants, especially young plants (Section 9.8.2). This is why rigid management rules are not useful—every situation is unique, and strategies need adjusting based on analyzing your observations in terms of the key ideas in the Introduction, basic concepts, and evaluation of benefits and costs.

Many species of nematode and other soil borne pathogens are sensitive to high temperature, so their populations can be reduced by *soil solarization*. Heating moistened soil to 71°C (160°F) for about 30 minutes kills most bacteria and viruses and some fungi and nematodes (Ben-Yephet et al., 1987). When daytime temperatures are ≥32°C (90°F), soil can be solarized by putting moist soil in a covered metal pot or closed plastic bucket or bag and leaving it in the sun for several days. One of the most effective ways to solarize soil is to place it under two layers of clear plastic sheets with an airspace of about 6 cm (2.4 in) between layers that acts as insulation (Stapleton et al., 2002). Air space between the sheets can be created with plastic bottles, PVC pipe, or other spacers. This technique can raise temperatures 1.1° to 5.6°C (2° to 10°F) more than using a single layer of clear plastic, by retaining the heat from direct solar radiation, and reducing the loss of thermal energy that is re-radiated from the heated soil. Under these sheets the soil

should be in small (≤17.6–18.9 L; 2–5 gal), dark containers, or in small piles ≤30.5 cm (12 in) high, and watered at or close to field capacity (Section 8.2.1). Solarization like this can be used on your planting soil, and can even be extended to treat a garden bed, although beyond a maximum of about 30 cm (12 in) below the soil surface it's not effective (Stapleton et al., 2002). Similarly, solarization can be done with a cold frame built with used glass windows, either ones with double glass or windows themselves doubled up and sealed (Section 4.3.2). Solarization of potting soil is a garden task that makes good sense to do as a group, everyone making use of the solarization set up, or receiving solarized soil. Because heating the soil also kills beneficial microorganisms it should only be used when absolutely necessary.

Many disease pathogens thrive under moist, warm conditions, so staking or trellising plants and pruning lower leaves helps by increasing air circulation. Placing dry mulch, sticks, or a flat stone under fruits like melons keeps them from lying on the moist soil surface and rotting due to fungal or bacterial infections. Immediately removing and destroying infected fruit and leaves prevents spread of the pathogen by insects.

Pest and pathogen problems often increase with continuous planting of the same crop or closely related crops in the same location, so planting crops in rotation, especially those from different botanical families, is a good strategy (Section 9.5.1). Botanical family is the taxonomic grouping of genera that share morphological characteristics and evolutionary history (IARC, 2003). Examples of common garden crop families are Brassicaceae (cabbage, mustards), Leguminosae (beans, peas), and Solanaceae (peppers, tomatoes). One example where crop rotation can be successful is the golden or potato cyst nematode (*Globodera rostochiensis*), a major pest that attacks the roots of solanaceous crops (eggplant, pepper, potato, tomato), and solanaceous weeds (Ferris, 2013). In those plants, the immature female nematodes emerging from the roots appear as small, white bodies. As the infestation gets worse, plants have yellow, wilting leaves, similar to symptoms of water or mineral deficiency. Eggs of these nematodes can survive for over 30 years inside of cysts, but each year that non-host crops are planted some of the nematodes die, and after 5–9 years the population is low enough that susceptible solanaceous crops can be grown again (Ferris, 2013). Recent experiments in Mexico found that the potato cyst nematode population was reduced by 30% in just two seasons of growing non-host crops (López-Lima et al., 2013).

Still, in some cases, growing the same crop in the same location can also lead to reduction in disease, because it facilitates the growth of populations of antagonistic microorganisms, another example where observation is more important than following recipes. For example, continuous planting of cucumber led to reduction in *Rhizontonia* damping off, and continuous planting of 'Crimson Sweet' watermelon, a variety susceptible to *Fusarium* wilt, led to an increase in a competitive, non-pathogenic *Fusarium* species, resulting in a decrease in *Fusarium* wilt over time (Agrios, 2005, 304–305).

Another environmental management strategy is adjusting the planting schedule for the garden, or for specific crops, to minimize pest and pathogen problems. For example, cold temperatures reduce nematode reproduction and hatching, and delaying autumn carrot planting in California by only a few weeks in nematode infested soils eliminated yield losses, and in Spain, delaying lettuce transplanting into nematode infested soils from September to November prevented nematode root invasion (Collange et al., 2011).

Fungal root rots can increase when crops are grown out of season. The right soil temperatures favor seedling growth, and vigorous seedlings resist damage by pathogens, but when grown under unfavorable soil temperatures, seedling growth is very slow, and plants succumb to fungal disease. For example, cool soil temperatures favor some pathogenic fungi, like damping off in cucurbits (Section 9.8.2). To control seedling blight caused by the same strain of a fungal pathogen, maize should be planted in warm soils (20–25°C, 68–77°F) but wheat in cool soils (15–20°C, 59–68°F) (Dickson et al., 1923) (Section 5.8.2).

9.4. Crop Plant Choice for Management

As we discuss in Chapters 5 and 10, the evolution of our food crops through domestication of wild plants often involved a tradeoff between many of the characteristics favored by early gardeners and farmers and those that help plants survive in their natural environment. Humans originally selected crop plants for uniform germination, better taste, increased yield, and other desirable qualities, with the unintended result that, compared with their wild ancestors, crops have reduced resistance to pests, pathogens, and environmental factors like drought and extreme temperatures. Therefore, many crop species can't thrive, or even survive, on their own.

Crop plants are often adapted to many of the pests and pathogens they have evolved with. When we say that a crop is *locally adapted* to pests and pathogens, it means the crop has evolved adaptation to local pests and pathogens (Fig. 9.2). This evolution can happen over many years as farmers look for and select healthy undamaged seeds from healthy plants. For example, horizontal resistance (Section 10.3.3) to the fungal disease *Phytophthora infestans* has evolved in some potato landraces grown in Mexico over a period of 240 years (Grünwald et al., 2002). Crop adaptation in the form of disease resistance can

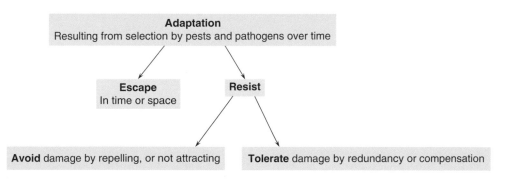

Figure 9.2. Plant adaptation to pests and pathogens

also happen more quickly as a result of gardeners and farmers collaborating with plant breeders, as happened, for example, in a project in the Andes that included barley, maize, and quinoa (Danial et al., 2007). As with adaptation to drought and heat (Section 5.6), plants can adapt to and survive pests and pathogens by escaping, avoiding, or tolerating them.

Gardeners can help plants *escape* damage by adjusting where and when crops are grown. For example, charcoal stem rot (*Macrophomina phaseolina*) can be a serious problem on common bean (UCANR, 2018b). If a gardener has planted beans in the same part of the garden many years in a row, the fungus inoculum can build up to high levels there, infecting any beans planted. By planting beans in a new location in the garden where the inoculum has not built up, the beans escape the problem.

Because many crops grown in temperate latitudes aren't photoperiod sensitive (Section 5.8.1), they can be planted and harvested across a range of dates, and adjusting these dates to optimize plant health can help them escape damage. Often, if a crop fails because the soil or air temperature was too hot or too cold, or because a pest or pathogen was especially abundant, it will prosper if planted on a different date, or in a different spot in the garden.

Plants can *resist* damage by *avoiding* it, even when pests or pathogens are present. Some plants have chemical odors that repel insects, while others lack the color or odor that attracts certain pests. Chemicals in the leaves may slow down eating by pests or even kill them. Plants can also make it physically difficult for animals to eat them. Thorns may repel rabbits; a thick cuticle prevents caterpillars from chewing; *trichomes*, small hairs on leaves and stems, inhibit pest movement and feeding; and silica (sand) particles that occur naturally in the leaves of many grasses like maize, rice, and sorghum, wear down the mouth parts of insects that chew the leaves, decreasing the damage any individual can do (Mitchell et al., 2016).

Crop plants can also resist damage by *tolerating* it, so that when they are eaten by pests or infected by pathogens the yield loss is minimized. The maximum proportion of defoliation a plant can experience without reduction in yield is known as the *damage threshold*. Root or tuber crops, for example,

may have many leaves removed or have extensive disease in their aboveground parts with little loss in harvest. One study found that potato had damage thresholds greater than 60% at emergence and after flowering, but less than 40% before and during flowering. However, if stems—not only leaves—were damaged, then damage thresholds dropped by about 10–50% (Stieha and Poveda, 2015). Sometimes damage to tubers can even result in increased tuber yield. For example, potato plants whose tubers had low levels of potato moth larvae damage in experimental plots in the Colombian Andes showed a 2.5 times increase in marketable potato (tuber) yield compared with undamaged plants (Poveda et al., 2010). Greenhouse experiments showed that this yield increase was caused by larval regurgitant stimulating tuber growth, resulting in larger tuber sizes.

When chewing removes the main growing tip in young plants of leaf crops like amaranth or basil, it can lead to fuller plants and increased leaf production, just as intentional pruning would. That's one example of why you can't always assume that you have a problem when you see your plants being eaten—you can also watch to see what the results are, which may mean waiting to see if the harvest is reduced compared with unaffected plants or previous years, keeping in mind the value and limitations of such informal experiments (Section 3.2.3). Assessing how to react to chewing damage depends on the plant, the pest and its population density, the growth stage of both plant and pest, the environment (temperature or season), and what has happened before in similar cases. As gardeners who have experience with them know, tomato hornworm is an example where a quick response is warranted because the larvae can grow into large caterpillars and totally defoliate a plant in just a few days. Aphid populations can also grow very rapidly (Section 9.6)

Some crop varieties may be able to tolerate systemic diseases, especially if the infection is not severe, or only becomes damaging near the end of the plant's life. We have seen both bean and squash plants infected with viral diseases survive to produce good harvests. In contrast, crop species and varieties that are not adapted to an area may be more vulnerable to the pests and pathogens there. Experienced local gardeners are often familiar with which varieties do best, and know how and when to plant them, how they grow, what pest and pathogen problems they have, and different ways to enjoy the harvest. You can also look for seed and other planting material of local varieties through personal contacts or community seed libraries, which may help minimize introduction of exotic pests or pathogens. Sometimes, if pathogens are well established, planting non-susceptible crops may be the only control method available short of abandoning the site.

An introduced pest or pathogen, for which local crops have no resistance, can be especially devastating, as is the case with huanglongbing (HLB, citrus greening) disease which has been spreading and causing large losses in citrus growing areas of the US since it was first discovered in Florida in 2005 (Section 3.1.3). HLB is caused by the bacterium *Candidatus* Liberibacter asiaticus, and is transmitted by an insect, the Asian citrus psyllid, *Diaphorina citri*. In addition to controlling the vector, the most promising response is identifying citrus relatives that have resistance, and introducing that resistance into cultivated varieties through breeding (Ramadugu et al., 2016). Responses using both transgenic and non-transgenic gene "editing" strategies (Section 10.4) are also in development and testing (APHIS, 2018).

The trends of ACC mean that some locally-adapted crops may become less adapted as the environment changes around them, inducing changes in pests and pathogens as well (Trumble and Butler, 2009). One study of the effect of climate warming on the number of generations of 13 important insect pests of crops in California found that all of them would experience an increase in the number of generations per year, especially in lower-latitude locations, and over time, through the 2090s (Ziter et al., 2012). For example, aphids will have about five more generations per year with a warming of 2°C (3.6°F) (Jones and Barbetti, 2012). This means that crops will be exposed to pest and pathogen pressure for longer periods each year, and pests will be able to evolve adaptations to new conditions and resistance to control methods more rapidly. For these reasons, crop varieties from other locations may be more adapted to future conditions than the current locally adapted ones. Experimenting with species and varieties new to your area, and growing promising ones for multiple generations, can help identify or develop adaptations appropriate to your area.

9.5. Biological Control

Biological control is the action of living organisms in controlling the population of other organisms. In a classic text, Van den Bosch et al. differentiated natural and applied biological control (1982). *Natural biological control* is always occurring everywhere, including in gardens, where it is supported by an ecological approach to management, the ongoing evolution of all organisms, and the dynamic interaction between gardens and the larger community and ecosystem they are part of (Section 9.1.1). In general, greater diversity of crop plants and other organisms in the garden provides biological control of pests and pathogens, although this is not always the case. While more research is needed on biodiversity and its effect on pests and pathogens, the evidence so far suggests that overall, increased predator and parasite diversity decreases pest populations. For example, one study in California showed that the level of damage to tomato crops was the same on more diverse organic farms as on conventional farms, even though the organic farms did not use pesticides (Letourneau and Bothwell, 2008, 433).

However, the effects of biodiversity can vary depending on how it is structured. For example, increased species diversity of predators and parasites can have a negative effect on pest regulation by increasing the probability of beneficial arthropods eating each other, or interfering with each other's behavior, and of a beneficial organism that is more efficient at controlling pests being outcompeted by a less-efficient beneficial (Letourneau and Bothwell, 2008, 433–434).

Applied biological control is human control of pest or pathogen populations to keep those populations at acceptable levels, but rarely eradicating them. In many cases, applied biological control is based on favoring certain trophic levels in the food web of the garden. *Trophic levels* are ways of identifying organisms in the garden based on what they eat and who eats them (Fig. 9.3). For example, biological controls often work by reducing the populations of pests or plant pathogens that feed on crops, or by encouraging the organisms that feed on them.

Next we describe four types of applied biological control using: plant diversity; natural enemies of pests; microorganism diversity; and genetically engineered, transgenic crops. We mention some of the mechanisms that make these types of control effective,

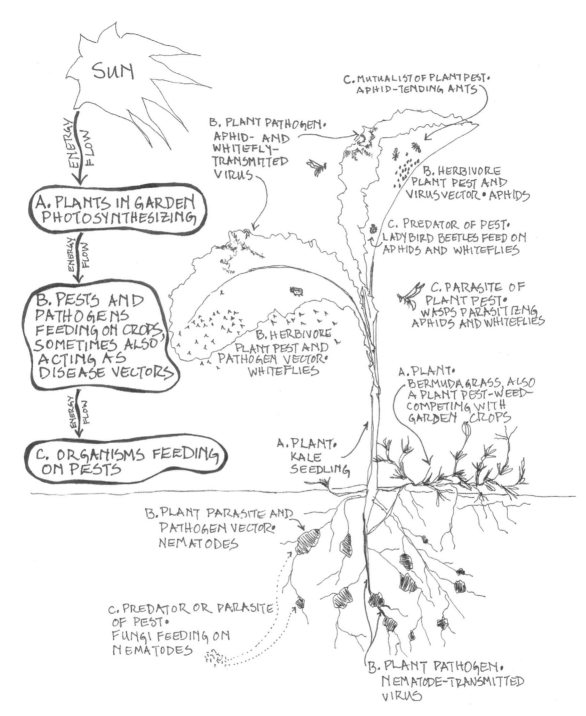

Figure 9.3. Trophic levels and functional groups in the garden important for pest and pathogen management

Based in part on Tooker and Frank (2012)

but also provide examples of when they don't work. The dynamics of biological control are complex, so outcomes can be quite uncertain, but if you apply it in your garden you can use observation and experimentation to assess its effects and adapt your management (Chapter 3).

9.5.1. Plant diversity to control garden organisms

Most gardeners practice biological control indirectly by maintaining a diverse garden environment that encourages a diversity of organisms, including beneficials. This includes plant diversity—different species, varieties within a species, and genotypes within varieties. A meta-analysis of the effect of organic farming and in-field plant diversification on arthropod diversity found the strongest effect was that this increased pollinator and predator populations, but not pest (herbivore) populations (Lichtenberg et al., 2017). An analysis of 552 experiments on the effect of plant diversification for reducing herbivores, increasing natural enemies of herbivores, and decreasing crop damage, found a significant positive effect (Letourneau et al., 2011). However, there was a significant negative effect of plant diversity on yield, due in part to the reduction in densities of crop plants when they are replaced by non-crop plants. But when greater diversity was achieved by adding crop species, for example as intercrops, overall crop yields increased significantly (see below).

Clearly the impact of plant diversity depends on how it is achieved, among other things. Because plant diversity can have mixed effects on control of garden pests and pathogens, observation and thinking about the garden's goals are important for figuring out the best strategy. Still, most research supports the conclusion of overall positive effects on yield of strategically increasing plant diversity in food gardens and surrounding landscape, especially when wider social and environmental benefits are also goals (Isbell et al., 2017). *Strategic management* of plant and other biodiversity, that is targeting specific components of biodiversity as opposed to increasing biodiversity in general, is key, because diversity can take many different forms (e.g., species *richness* [total number of species present] and relative abundance of species, which can have opposing effects).

A well-known series of experiments on intercropping in Yunnan, China, found that in comparison with monoculture fields, some combinations of short, modern hybrid rice varieties and tall, traditional rice varieties greatly reduced the damage from rice blast disease caused by the fungal pathogen *Magnaporthe grisea* (Zhu et al., 2000, 2003). The reduction in disease in successful combinations was likely due to changes in microclimate around individual plants, and increasing barriers to fungal spore dispersal created by the different plant heights. In addition, the greater plant diversity within rice fields led to an increased diversity of races of *M. grisea*, because different pathogen races preferentially attack either the modern or traditional rice varieties. The result was a higher probability that rice plants would be first attacked by a race of *M. grisea* that was virulent to the other rice variety, but not to it. This attack stimulated an immune response within the rice plant, which made it resistant to subsequent attack by a more virulent race of *M. grisea*. However, the researchers emphasized that these positive results occurred only with some, but not all of the combinations of modern and traditional rice varieties.

Plant diversity around gardens can also affect population dynamics within gardens. For example, research on the relationship between pest numbers and rates of parasitism on those pests in fields on the central coast of California found that species richness and abundance of tachinid flies, important parasites for biological control of vegetable pests, were positively associated with the presence of semi-wild vegetation near the fields (Letourneau et al., 2012), with different groups of parasites associated with unique types of natural vegetation (Letourneau et al., 2015). Research on four organic vegetable farms in this area also found that farmer-managed hedgerows with a diversity of native woody and herbaceous species had a higher ratio of natural enemies to crop pests compared with weedy borders (Pisani Gareau et al., 2013). Parks and vegetated vacant lots can serve as reservoirs of arthropod diversity, providing biological control for gardens, or potentially being sources of pests and pathogens (Gardiner et al., 2013). Again, observation is the key. If a serious plant virus disease problem occurs in the garden it is a good practice to check adjacent areas for weedy species with the disease and remove them if possible. This is another good reason for gardeners to work together and with their communities for more urban green space that supports beneficial garden organisms.

Sometimes wild plants, weeds, or other crops can serve as traps for pests and pathogens. *Trap crops* are plants that pests will preferentially feed on instead of the garden crops, and ideally will not also serve as pest breeding sites. For example, solanaceous weeds are important alternative hosts for pests of solanaceous crops, including eggplant, pepper, potato, and tomato, and all major sweet potato pests feed on the widespread field bindweed. Careful observation and experimentation is the only way to find out if a weed or other plant is attracting pests that damage crops, or is trapping those pests, thus reducing their harm to the rest of the garden. Once pests have settled on trap plants these plants can be discarded, for example by burying them, or putting them into a plastic bucket with a tight lid left in the sun, before the pests move to the crop plants.

In eastern Washington state, researchers have found that trap crops are an effective method for controlling the crucifer flea beetle (*Phyllotreta cruciferae*), a widespread major pest of brassica crops. A trap crop containing 'Pacific Gold' mustard (*B. juncea*), 'Dwarf Essex' rapeseed (*B. napus*), and pac choi (*B. campestris* var. *chinensis*) successfully protected broccoli (Parker et al., 2012). This method was equally effective whether the trap crop was planted 50 cm (1.6 ft) or 11 m (36 ft) away.

Plants that *repel* crop pests have been used by gardeners for a long time and function in different ways, including by masking the scent of crops with the aroma of the repellant plants' volatile oils (Parolin et al., 2012). Plants with documented effects as repellants include sacred or holy basil (*Ocimum tenuiflorum*), African marigolds (*Tagetes erecta*), and citronella grass (*Cymbopogon* spp). Other aromatic plants may also have some effect and are worth experimenting with. While research is just beginning, there is some evidence that certain combinations of oat genotypes stimulate production of volatile compounds that attract aphid predators and significantly reduce aphid numbers and damage, compared to plantings of those same genotypes alone (Glinwood et al., 2011; Tooker and Frank, 2012).

Push-pull is a strategy for insect pest control that combines a pest-repelling crop or other plants intercropped in the field or garden, with trap crops outside of the garden that attract pests. Results with small-scale maize farming in East Africa are very promising: a "climate adapted (i.e., drought adapted) push-pull system" combining a leguminous push (*Desmodium intortum*) and tropical grass pull crop (*Brachiaria* sp.) effectively control the major pests stemborer and fall armyworm (Midega et al., 2018). Control was observed to operate in multiple ways, including by attracting arthropods that are natural enemies of those pests. The application of push-pull systems to food gardens needs exploration.

9.5.2. Using natural enemies to control pests

Applied biological control (*biocontrol*) with natural enemies has a long history, including among gardeners. Medieval date growers on the Arabian Peninsula were among the first to practice biological control by seasonally transporting predatory ants from nearby mountains to their oases to control another species of ant that attacked the date palms (Van Den Bosch et al., 1982, 22–23). Citrus growers in parts of China 1,700 years ago intentionally introduced and managed predatory ants to control various citrus pests (Huang and Yang, 1987). The first major success of biocontrol in the US was of cottony-cushion scale (*Icerya purchasi*) on citrus crops in California in the 1880s. This pest was accidently introduced on acacia plants from Australia, and controlled by importing two natural enemies from Australia, the vedalia beetle (*Rodolia cardinalis*), whose larval and adult forms feed exclusively on *I. purchasi*, and the parasitic fly *Cryptochaetum iceryae* whose larvae feed on the scale from the inside when they hatch from eggs deposited by the adult. These two controlled the scale in less than two years (Bale et al., 2008). Most cases of successful modern biological control of insect pests involve using predators or parasites from the pest's area of origin (Van Den Bosch et al., 1982, 24–27).

As with pests and pathogens, control species are also described as *generalists* (attacking a broad range of organisms—both pest and non-pest species), or *specialists* (attacking only one or a few pest species). Many of the insects available commercially for biological control are generalists. Examples are lacewings, lady beetles, and *Trichogramma* wasps. However, generalists may not always be effective controls. For example, lady beetles need to be managed with great care to be effective or they will fly away if conditions are not right (e.g., if it is too hot) (Flint, 2014).

Specialists may be more effective than generalists in biocontrol. This seems to be the case for parasitic wasps evaluated for control of the olive fruit fly (*Bactrocera oleae*) in California. Two parasitic wasps currently being tested are specialists that probably evolved with the olive fruit fly in sub-Saharan Africa, and which oviposit into fruit fly larvae living in olive fruit (Daane et al., 2015).

In some years and in some locations, grasshopper populations can explode and devastate gardens. Biocontrol measures to manage grasshopper populations have been applied on a large scale (a rangeland, for example) and at the scale of a small garden (Schlau, 2010). Infectious spores of a biocontrol agent, the fungus *Nosema locustae*, can be spread in the garden area in a bran bait that the grasshoppers will eat. Some grasshoppers will eat the grasshoppers that have died from *N. locustae* infection and propagate the pathogen. As with many biocontrol agents, results are not immediate. Grasshopper populations need to be monitored to know when egg hatching is occurring, and the bait applied when most of the grasshoppers are early in their life cycle (the nymph stage), because only young grasshoppers are susceptible. If there are garden plant species that the young grasshoppers are strongly attracted to, those species can be grown as trap plants and the bait placed there. Together with protective plant covers, *N. locustae* can be a part of a grasshopper management approach that does not depend on toxic insecticides. *Nosema locustae* is a pathogen of many species of grasshoppers and some crickets, but for garden purposes, can be considered a specialist because it's host range does not go beyond these species. So far the results from using *N. locustae* at larger scales are mixed (Bomar et al., 1993; Lockwood et al., 1999), so observation will be critical, as will consideration of the impact of wide-scale applications because grasshoppers are an important part of many ecosystems (Lockwood, 1998).

If you want to increase the diversity of beneficial organisms, you could acquire small numbers from local gardeners and introduce them into your garden. Local sourcing can address at least two concerns when introducing insect predators, especially when they are collected in the wild and distributed widely, such as the convergent lady beetle. First, the introduced predators could be infected with pathogens that could move into and reduce local predator populations (Bjørnson, 2008). Second, the introduced predators can breed with the local population

of the same species, which could make that local population more, but also less, fit (Obrycki et al., 2001). In addition, while increasing the diversity of predators and parasites can reduce pest populations, it can also reduce their rate of attacking insect pests if a generalist predator, for example, preys on another species of predator or parasite. (Letourneau and Bothwell, 2008).

Interactions among plants and arthropods can also get complicated. For example, ants frequently protect or "tend" honeydew-excreting insects like aphids because the ants feed on the aphids' honeydew. This means that aphid populations can build up and reduce garden crop yields. However, if other herbivores, like caterpillars, are also eating the plants, the ants will attack them, which can result in the plants with more aphids producing a higher yield (Styrsky and Eubanks, 2007, 2010).

Another use of natural enemies is to control pathogens by introducing or encouraging predators of insects and other animals that are vectors of plant disease pathogens. For example, aphids, which are common pests of many different garden plants, and can infect plants with viral pathogens, can be kept at low numbers when they are parasitized by tiny parasitic wasps.

9.5.3. Microorganism diversity to control pests and pathogens

Encouraging a diversity of microorganisms, in addition to arthropod and plant diversity, is part of an ecological approach to controlling plant pathogens. A diversity of microorganisms generally decreases the chances of pathogen populations reaching damaging levels, and a diversity of plants often supports a variety of microorganisms, as in the example of rice diversification in Yunnan described earlier (Section 9.5.1).

Suppressive soils inhibit the growth of pathogens primarily because the microorganisms contained in them suppress pathogens through competition, parasitism, or producing antibiotics toxic to plant pathogens (Agrios, 2005, 304–307). Another way soil microbial diversity reduces disease is through induced resistance, an immune response in plants (Burketova et al., 2015), as also described for the Yunnan rice intercropping experiment. *General suppression* is when pathogen control is the result of several of these mechanisms, while *specific suppression* is when one factor is responsible (St. Martin, 2015, 35). A fungus that can cause damping off (*Rhizoctonia solani*) (Section 9.8.2) in young seedlings is parasitized by a soil-dwelling *Trichoderma* spp. fungus when the *Trichoderma* spp. can no longer easily access cellulose or glucose in the soil, present in new compost (St. Martin, 2015, 35). So, for some species of microorganisms that attack pathogens, fresh compost may inhibit their control activity (Chung et al., 1988), with inhibition decreasing as the compost decomposes. Other fungal species, e.g., *Penicillium*, have been found to be active when there are high glucose levels in the soil, another example of tradeoffs. Still, in general, supplying nutrient-rich organic matter to encourage the growth of a diverse mixture of soil microorganisms is the best way to take advantage of pathogen control by other microorganisms.

Probably the most widely known pathogen for control of agricultural pests is *Bacillus thuringiensis* (*Bt*), a soil bacterium producing crystalline proteins that kill certain kinds of caterpillars (butterfly and moth larvae) and beetle and fly larvae. *Bt* is not poisonous to other animals (Van Den Bosch et al., 1982, 63–64), although there are some reports of adverse effects in humans, including eye infections (Section 9.9.4). Starting in 1938 a spray containing *Bt* spores was sold commercially in France, and by 1958 in the US (Ahmad and Rasool, 2014, 127). *Bt* is cultured commercially in large-scale fermentation vats and sprayed onto crops, just like synthetic insecticides. While it is not an important control in nature, *Bt* sprays have been one of the most effective biocontrols available, and are approved for USDA certified organic production, but their effectiveness may be compromised by evolution of resistance in pests exposed to *Bt* genes in transgenic crops (Section 9.5.4).

Pest-specific viral diseases are also promising biocontrols. A solution containing the virus—basically a blend of water and diseased pests—is sprayed or washed onto the plant, and the pest feeding on the plant swallows the virus, sickens and dies. Research in East Africa found that diseased caterpillars can be used to make a very effective pesticide against the African armyworm (*Spodoptera exempta*), a major crop pest (Grzywacz et al., 2008). Some viruses that infect garden pests in the US, such as the cabbage looper and diamondback moth,

make excellent biocontrols because they are very specific and so do not harm beneficial arthropods. Gardeners can try making their own solution from diseased pest larvae. For example, caterpillars with viral infections commonly hang by their back "legs" from stems or branches (Van Den Bosch et al., 1982, 63), making it easier to identify them. Still, it can be difficult to make sufficient solution with a virus concentration high enough to be effective. There are some commercially available sprays, but because many of these are new, they should be investigated before using, including to see if they are approved for use in certified organic production (Section 9.9.5).

Fungal pathogens of aphids can be some of the most effective controls for those common garden pests (Barbercheck, 2014). For example *Verticillium lecanii* and a number of other fungi attack aphids. The fungi occur as airborne spores that infect and grow inside aphids, eventually killing them. Because the spores require moist conditions to start growing, dry hot conditions may inhibit them, and fungicides will kill the spores.

There is no convincing evidence that pathogens of insects infect humans, so it is generally felt that they can be safely used for pest biocontrol, though some research suggests that care should be taken. For example, there may be potential for bacterial pathogens of insects to evolve into human pathogens (Waterfield et al., 2004). Plant viruses are not considered dangerous to humans, however, certain plant fungal pathogens are quite dangerous to humans (Box 9.3).

9.5.4. Genetically engineered transgenic crop varieties

A *transgenic variety* is a crop variety with a transgene; a *transgene* is genetic material that is engineered into a species other than the one in which it originated. From our perspective transgenic varieties in their current form are not compatible with ecological and prosocial goals. The vertical resistance (Section 10.3.3) of transgenic varieties exerts high selection pressure on targeted pests, accelerating the evolution of resistance in them (Box 9.2),

Evolution of resistance has been documented for transgenic *Bt* crops, although this has occurred more slowly than expected. As already described, the

A number of common, universally occurring fungi that grow on grains, seeds, and fruits produce toxins (*mycotoxins*) that are extremely poisonous to animals, including humans, and cause chronic debilitating illness, cancer, or even death. Mycotoxin-producing fungi include *Aspergillus flavus*, that infects maize, peanut, pistachio, and many other species while still in the field or in storage; some *Fusarium* spp. that grow primarily on maize; and *Penicillium* spp. that grow on fruit and bread (Agrios, 2005, 39–41, 559–561). US Department of Agriculture researchers are

experimenting with using non-toxin producing strains of *A. flavus* to replace toxin producing strains in cultivated crops and thus reduce their mycotoxin contamination (Atehnkeng et al., 2014, 2016). *Aspergillus flavus*, along with other species of *Aspergillus*, are known to cause aspergillosis in humans, resulting in symptoms such as asthma or severe lung infections, especially in those with a compromised immune system. Never eat or feed to animals any moldy or suspect fruits, vegetables, grains, legumes, or other garden produce.

---

soil bacterium *Bt* has been widely used as a pesticide in organic agriculture for decades (Section 9.5.3). When *Bt* is used as an insecticidal spray, only the immature form of pests (e.g., moth caterpillars), that eat the bacteria, bacterial spores or toxin are affected, and the active ingredient degrades quickly with exposure to sunlight. But when *Bt* genes are engineered into crops to create transgenic, pesticidal crop varieties, all of the plant tissues above and below ground express the *Bt* insecticidal property, exerting a more sustained selection pressure. An analysis of 77 studies from five continents found that in five of 13 major pest species some populations had evolved resistance to the *Bt* transgene by 2011, compared with only one species with resistant populations in 2005 (Tabashnik et al., 2013). Pests resistant to *Bt* not only decrease the effectiveness of transgenic *Bt* crops, but also of *Bt* spray used by organic farmers and gardeners. *Bt* varieties also illustrate the potential for transgenic varieties to reduce the pest population to very low levels, opening up an ecological niche for minor pests, that are not susceptible to the *Bt* gene, to then become major pests (Catarino et al., 2015)

The novelty of *Bt* and other transgenic crops means we do not know their long-term effects on the environment, other organisms, and human health. Most transgenic crops are produced by a small number of very large corporations whose primary goal is the expansion of industrial agriculture and increasing profits. Alternative strategies exist, such as using genetic diversity present in crops and their close relatives, but are not being adequately researched because public and private funding is

overwhelmingly being invested in transgenic and other proprietary approaches (cf. NRC, 2000).

The most widely used transgenic trait globally is herbicide tolerance, which allows weed control by herbicide spraying without affecting the transgenic variety, and the most widely used combination is glyphosate-tolerant transgenic varieties and glyphosate herbicide, the most popular being Monsanto's Roundup®, a probable human carcinogen (Box 9.2). Since the introduction of glyphosate resistant crops in 1996 and the increased use of glyphosate, more than 40 weed species have evolved resistance (Gould et al., 2018).

While transgenic *Bt* varieties—mainly of canola, cotton, maize, and soybean—have been commercially grown since 1996, and are now grown on large areas in the US and other countries, transgenic vegetable and fruit varieties, including *Bt* and other transgenes, are still relatively rare. The USDA's official website shows the following deregulated (cleared for commercialization) since 1992, beginning with the FlavrSavr tomato, through potato and apple in 2016: apple, beet, chicory, flax, maize (including sweet corn), papaya, plum, potato, squash, tomato (USDA APHIS, 2018). Some transgenic varieties currently being grown are papaya in Hawai'i, developed and sold by a public-private consortium comprising the University of Hawai'i, Cornell University, Monsanto® (now owned by Bayer®), Asgrow®, Cambia®, and MIT (Gonsalves et al., 2007); and sweet corn and squash in the US and Canada, developed and sold by Seminis, owned by Monsanto® (Seminis, 2018). Transgenic vegetables are also being grown in other countries, for example, eggplant in Bangladesh, developed and sold by Mahyco, a subsidiary of Monsanto®, also with involvement of Cornell University (Lowery, 2018). Because major multinational agrochemical and biotech crop corporations are consolidating, for example, the Monsanto-Bayer merger in June 2018, and expanding rapidly into the vegetable seed market, transgenic vegetable and fruit seeds for gardeners are likely to increase. As discussed in Section 10.4, the assertions and motivations of the publicly traded, multinational, agrochemical companies that produce most transgenic seed need to be carefully evaluated in light of the five key ideas outlined in the Introduction, and the alternatives available.

9.6. Physical Control

Physical control of pests and pathogens is often part of the garden design, for example, fences around beds to keep rabbits and dogs out, mesh below garden beds to exclude gophers, or planting patterns that discourage disease. Physical controls can also be used to solve specific, short-term problems. Compared with fields, gardens are small, intensively managed production systems, and so physical control methods that might not be appropriate for large fields can be very effective.

Physical control of infectious plant diseases focuses on preventing the spread of pathogens by vectors like wind and insects, including removing and destroying diseased plants or plant parts. Barriers affecting air flow can also reduce airborne pathogen movement, as was found in the tall and short rice intercrop in Yunnan described in Section 9.5.1. Insects that commonly spread plant pathogens include aphids, leafhoppers, cucumber beetles, and whiteflies. While not all pathogens can be spread from infected plants or plant parts, many can (for seed-borne diseases, see Section 6.1). When there is a choice of planting material for vegetative propagation, take precautions to avoid using propagules that might be infested with pathogens and parasites. For example, sweet potatoes can be propagated by stem cuttings that do not become infested with nematodes, rather than root cuttings that do. See Table 6.2 for hot water treatment of vegetative propagules that have nematodes.

We recommend burying diseased plants away from the garden unless you are certain that your green waste is composted at a high temperature to prevent spreading the problem. With systemic diseases, the entire infected plant should always be buried well away from the garden or otherwise destroyed when it is removed. Placing infected plants in a plastic bucket with a tight lid and leaving it where it receives lots of sunshine, so that the inside temperature reaches at least 37.8°C (100°F) for a few hours a day over several weeks will kill most pathogens (Stapleton et al., 2008). When working around plants that may be diseased, cleaning your hands and tools with soapy water, 20% powdered nonfat milk solution (Section 6.1), or 10% bleach solution after contact can help prevent the spread of diseases like tobacco mosaic virus, which infects solanaceous crops.

Picking pests like beetles, caterpillars, and grasshoppers off garden crops by hand, and crushing or using water to wash off small arthropods like mites, aphids, thrips, or bagrada bugs, is practical and effective, especially if done before populations get large. Success depends on understanding the life cycle and habits of the pest. For example, watching for aphids and destroying them as soon as they are seen can prevent population growth. The earliest sign of an aphid infestation is the arrival of winged adults migrating into the garden. If you anticipate an aphid problem on a crop, such as broccoli, frequently checking the backs of leaves for the winged aphids' arrival can give you a head start on control. Yellow sticky traps can also be helpful in early detection. Usually, when an infestation is first noticeable, most aphids are wingless because the winged migrants have started reproducing wingless progeny. However, increasing stress from rising population density or disappearance of food sources will induce more and more aphids to be born with wings. Avoiding this winged stage keeps the aphids from spreading to new areas where their populations would grow rapidly once again. Removing heavily insect-infested plants and burying them can also help.

Other means of physical control include making barriers that prevent pests from getting to the plant, such as collars against cutworms (Fig. 9.4). Floating row covers can be an effective physical barrier to early infestations of aphids, whiteflies, and beetles on some crops, such as cucurbits. This can delay infection by aphid- and whitefly-borne plant viruses and beetle-vectored bacteria (causing vascular wilts) to the point where fruit yield is not severely reduced.

A STEM COLLAR PROTECTS A SEEDLING FROM CUTWORMS

SOIL LEVEL

CUTWORM

Figure 9.4. A collar to protect a seedling from cutworms

Some slow-moving pests can be excluded from gardens. For example, nematodes only travel a few meters per season on their own, although they can be transported much further if infested water, soil, or plants are moved from one area to another (Agrios, 2005, 830). This means preventing the movement of soil, transplants, and irrigation water from infested to noninfested areas, for example, and by cleaning soil from tools and gardeners' hands and feet. However, insects can also transport nematodes, and inactive dried stages can be blown by the wind.

To exclude larger animals like rodents and poultry, fences or other barriers can be built around gardens, plots or individual plants, and made from a variety of different materials. We've had success using cloth or paper bags to protect seed heads of sunflower, amaranth, and sorghum from birds in our garden. We also lay small branches of thorny plants like mesquite, palo verde, and bougainvillea over planting furrows to protect seeds and seedlings from birds and other animals. Many gardeners use traps to remove or kill burrowing animals like gophers and ground squirrels, but aviary or poultry wire is more effective at protecting plant roots; the wire should span from at least 60 cm

(2 ft) below ground to 30 cm (1 ft) above ground (Salmon and Baldwin, 2009). When planting fruit trees, baskets of this wire can be used to protect roots until the tree is large enough to survive some gopher damage, usually when its roots have grown through the slowly rusting cage. Scaring animals away may also be successful. For example, anything that will make noise in the wind or reflect sunlight, like strips of cloth or mylar, clusters of sticks, tin or aluminum plates, gourds, or old CDs can be fastened to poles or tree branches in the garden to help scare off birds.

Another kind of physical control is *fallowing*, simply not planting any crop for one or more seasons in an area. A clean fallow, one with no growing plants, also helps to break the reproductive cycles of pests like soil nematodes and some soil-borne pathogens by separating them in time from their host plants. However, for pests or pathogens that move long distances easily, including with wind, water, and soil such as viruses transmitted by aphids, fallowing does not work.

The most obvious and often the most effective way to control weeds is to pull them out when you notice them. But if you suspect that weeds will be a problem there are other practices that can help: keeping weed seeds or reproductive parts such as stolons and rhizomes of Bermuda grass out of the compost pile; solarizing the soil (Section 9.3); leaving a bed fallow for a season or more. *Pre-irrigation*, watering a garden bed to wet the top 7.5 cm (3 in) or so of soil, a week or more before planting seeds or transplants, germinates some weed seeds so they can easily be pulled or hoed before crop plants are sown. Because some weeds have asynchronous seed germination, doing this a couple of times can help, especially with particularly troublesome weeds. Once the garden is planted, focusing irrigation water on the planted areas helps keep weeds from getting out of hand, and this is an advantage of surface and subsurface drip, vertical mulch, and pot irrigation, that concentrate water delivery to crops.

Covering any bare soil deprives weed seedlings of light, and reduces evaporation. For well-established seedlings or transplants, heavy mulching works well, but be sure to watch for pests that may hide under the mulch. Dense crop mixtures create a living mulch, and combining ones with upright growth forms with those that sprawl or grow as vines on the ground works especially well. For example, the squash plants commonly grown together with maize and beans in some areas of southern Mexico and Central America send out rapidly growing vines with large leaves that quickly cover exposed soil and reduce early weed establishment (Fujiyoshi et al., 2007). Squash vines may also suppress weed growth through allelopathy (Section 9.7.2).

9.7. Chemical Control

Chemicals produced by many living plants are the basis of their ability to attract or repel organisms (Section 9.5.1), and are an aspect of ecological management that can be used when choosing crops and planting patterns. Plants inhibit or kill pathogens via chemicals exuded through their surfaces above and below ground, and within their cells (Agrios, 2005, 211–212). *Pesticides* are naturally occurring or synthetic chemicals applied by humans that can repel, sicken, or kill pests and pathogens, and are another way of addressing specific, short-term problems. Pesticides have different modes of action against pests: *contact* poisons are absorbed through the body surface, *stomach* poisons have to be eaten directly by the insect or other animal, and *systemic* poisons are first absorbed into the plant tissues that are then eaten by the pest or affect the pathogen. Pesticides also differ in their *specificity*, that is the range of organisms they affect, and their *persistence* as toxins in the environment, that is how long it takes for them to break down either chemically (e.g., by exposure to UV light), or biologically by being metabolized by microbes, into other chemicals, often with less toxicity than the pesticide itself. Synthetic chemical pesticides often exert powerful selection pressure on the targeted pests, resulting in the evolution of resistance in those pest populations (Section 10.3.3). One outcome can be what has been called the *pesticide treadmill* (Howard, 2009), the need to keep replacing pesticides that lose their effectiveness as pest populations become resistant.

9.7.1. Botanical and other naturally occurring chemicals

Botanical chemicals occur naturally in plants and plant products like oils, juice, and powders, and some of them, for example those including chile, citrus, garlic, neem, rosemary, and thyme, can be used as pesticides to protect crops. In most cases they are safer than synthetic pesticides because the active ingredients are less toxic, their concentration is lower, and they persist for shorter periods in the environment, reducing the likelihood that nontarget organisms will be harmed. Botanical chemicals work through contact and so require that you spray or paint them directly on plants where the pest or pathogen is, or where you want to repel them. Botanical pesticides are commonly used in organic agriculture (Isman, 2006), and many are acceptable for use in USDA certified organic agriculture, although synthesized versions of them may not be. However, because a pesticide is botanical doesn't mean it isn't harmful, or can be used casually.

Some botanical chemicals, like nicotine from tobacco, are very poisonous to people and animals (Isman, 2006, 53, 60), and in its common form (nicotine sulfate dust) it is not permitted by the USDA Organic Program (Section 9.9.5), and we do not recommend using it. Pyrethrum is a botanical pesticide widely used in agriculture. It is extracted from dried chrysanthemums and contains active ingredients (pyrethrins) used to control some garden and household pests. Most pyrethrum compounds used currently are synthetic versions (pyrethroids) that are more potent and stable in light (Isman, 2006), but are not acceptable according to current federal organic agriculture standards in the US (Caldwell et al., 2013). Though it causes few lasting problems in humans, pyrethrum is very toxic to fish and cats.

Common garden plants can produce useful pesticides too. Garlic juice is effective against aphids, some caterpillars, and beetles, and some plant pathogens including seed-borne *Alternaria* in carrot, and *Phytophthora* (leaf blight) in tomato (Slusarenko et al., 2008). It may also be helpful against soil-borne fungal pathogens in certain circumstances (Sealy et al., 2007). There is evidence that the bulbs of some flower species such as *Narcissus* spp. (e.g., daffodils) and *Colchicum* spp. (e.g., crocus) are avoided by rodents (Curtis et al., 2009), but how effective these are as a deterrent for damage to food garden plants has not been studied, and we haven't had good luck with them.

Some common household, non-botanical chemicals have also been used as pesticides. For example, soap sprays—1 to 2% dilutions of plain, mild, biodegradable, liquid dish soap in water can help control soft-bodied pests like aphids and whiteflies on contact, probably because the soap compromises their exterior cuticle, and they desiccate (Oneto, 2015). Another non-botanical permitted for USDA certified organic production is Bordeaux mixture, a mixture of the chemicals copper sulfate and hydrated lime (calcium hydroxide). It is used to control a number of diseases of fruit trees and vines, including fire blight on pears and apples caused by the bacterium *Erwinia amylovora*, leaf curl on peaches and nectarines caused by the fungus *Taphrina deformans*, and downy mildew on grapes, caused by the fungus *Plasmopara viticola* (UCANR, 2018b).

There are other issues to consider with botanical chemicals for pest and pathogen management in your garden. If you want to follow USDA organic guidelines, you cannot use synthetic (manufactured by humans) forms of many of these chemicals, for example, pyrethroids. Agrochemical corporations see organic agriculture as a growing market niche and are producing some natural control products, so purchasing those products supports those corporations. For example, the insecticide spinosad is made up of two products of fermentation of a soil bacterium from the Caribbean (Caldwell et al., 2013), and manufactured by a division of Dow AgroSciences, the second largest global chemical company, and the eighth largest seed and agrochemical corporation in the world (Howard, 2009), with a poor environmental and human rights record (Corporate Research Project, 2016). Dow spinosad products (including GF-120 Naturalyte® Fruit Fly Bait) are currently the primary method for controlling the olive fruit fly for USDA certified organic olive producers.

9.7.2. Allelopathy

Allelopathic plants produce chemicals that in some way affect the growth of other plants. (Some researchers use this term as originally coined to refer only to negative effects, others use it for positive effects as well.) For example, as mentioned, the leaves of a squash (*Cucurbita pepo*) grown in southern Mexico appear to suppress growth of many common local weeds, but not the maize and

bean crops they are traditionally interplanted with (Fujiyoshi et al., 2007). Similarly, sweet potato is used as a weed control both because it has allelopathic qualities, and because it rapidly covers bare soil (Anaya, 1999). A study in Greece found that allelopathic chemicals in some common aromatic herbs like anise, dill, and oregano can effectively control weeds in maize fields without harming the maize plants or their productivity (Dhima et al., 2009). The herbs were grown in fields and turned into the soil as green manure, before the maize was planted. This may be worth trying in the garden, but its effectiveness will depend on which weeds are present.

However, allelopathic interactions do not only affect weeds. Some varieties of sunflowers, asparagus, eggplant, and Jerusalem artichokes can produce allelopathic chemicals that affect other plants of the same species, and sometimes other crop species (Bhowmik, 2003). Leaf and stem residues of Jerusalem artichoke did not affect germination and early growth of maize, but did suppress that development in lettuce, tomato, and zucchini (Tesio et al., 2010).

Sometimes allelopathic plants are obvious because no other plants will grow around them, and they can be well-known among local gardeners and farmers. Placing a garden in the shade of a eucalyptus tree (*Eucalyptus* spp.) would be a waste of the gardener's time and energy because volatile oils in the leaves of many eucalyptus species inhibit germination and growth of plants, and hot, dry conditions stimulate the release of these oils. In Mexico for example, one species of eucalyptus was found to suppress the growth of maize, common bean, and watermelon while a different species affected the growth of squash (Espinosa-García et al., 2008).

Walnut wilt is a disease affecting garden plants, like tomato, maize, cucumber, and watermelon that are growing in the root zone of a walnut tree (*Juglans* spp.) (de Albuquerque et al., 2011; Appleton et al., 2015). The pith tissues in the garden plant's stems turn brown, the plant wilts, and eventually dies. The cause of walnut wilt is the allelopathic chemical juglone produced by all parts of the walnut tree. Juglone is exuded from the roots and remains in the soil even after the tree is dead, so putting a garden near where a walnut tree is or was growing is not a good idea.

Tamarisk (*Tamarix* spp.) is a halophyte (Section 5.7) from Asia that accumulates salts taken up from the soil in its leaf tissues, and is widespread in parts of the US and Australia. The leaves of these trees have salt crusts on them, and their leaf litter creates such salty soil conditions that most garden plants cannot survive. Pine (*Pinus* spp.) trees also have allelopathic qualities. The extent of the root system of known allelopathic trees and the area that would be affected by their leaf litter should be considered when establishing a garden; root pruning may be helpful (Table 6.4). Similarly, it's best to keep material from allelopathic plants out of the compost, but it can be used as a mulch to cover bare soil.

9.8. Diagnosing and Managing Pest and Pathogen Problems

While it is often difficult to diagnose the specific cause of a garden problem, this isn't always necessary in order to identify an effective solution, because identifying the symptoms of types of diseases or pests is often adequate when choosing how to control them. The key to diagnosing pest and pathogen problems is careful observation of how plants, animals, insects, microorganisms, and their environment interact. Signs of damage include the results of pests chewing, cutting, and sucking leaves, stems, and roots; disease symptoms include slow growing or sickly plants and discolored or misshapen plant parts. When the days are hot and dry, night is an active time for many arthropods and other animals, and a good time to check the garden for them.

Because pathogens cannot usually be seen with the naked eye, diagnosing disease relies even more on the plant's symptoms than is true with pests. While some crop pathogens are fairly easy to identify by the symptoms of the disease they cause, for many diseases only trained experts can identify the pathogen. In these cases, microscopic examination and laboratory tests by plant pathologists using specialized equipment and processes, including genetic screening, may be required. Even then, such testing is not always successful in identifying the cause of the problem (Agrios, 2005, 71–74).

When trying to diagnose problems, it is important to look at the garden as a whole. Is there a pattern among the plants that are not doing well? Are the affected plants young or old, scattered or

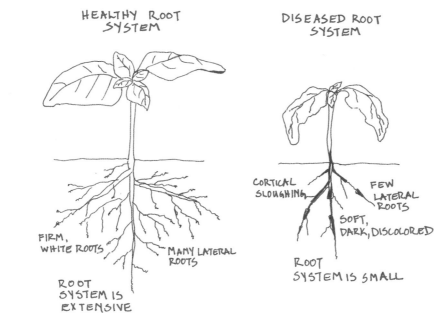

Figure 9.5. Comparing healthy and diseased root systems

concentrated, of the same variety, species or family, or different ones? Does the spatial pattern of plants with the problem correspond to the pattern of water drainage, soil, shade, wind? What was grown there before? What is the history of the garden site? Was anything buried there? Have chemicals been applied nearby? Has there been standing water? What is the source of compost or mulch brought into the garden?

Examining individual plants also helps. Are there pests or eggs on the undersides of leaves, in the soil and mulch around the plants? How do the color, pattern, and shape of leaves and stems of sick plants compare with those of healthy plants? Although not necessary, a small hand lens or magnifying glass can sometimes help, for example, in determining if leaf spots have small insect holes in the middle.

If there's room, it's a good idea to plant a few "extra" plants of each crop in the garden that can be dug up if needed to check on progress or to diagnose problems below the soil surface. Many problems—especially soil conditions, fungal and bacterial pathogens, and some insects—cause damage below ground in the root system, but are expressed above ground in the stems, leaves, and fruit.

Learn what a healthy root looks like by digging up a healthy plant and washing the roots carefully in water. A diseased root system may have fewer lateral roots, be off-color (tan or brown), and collapse when squeezed between your finger and thumb (Fig. 9.5). *Cortical sloughing* is the loss of cortex (cortical tissue) on the roots and occurs in several plant diseases such as fungal root rot. If a plant is suffering from a disease that causes cortical sloughing, but that tissue has not yet been shed, it can be easily tested by pinching the root lightly and pulling toward its tip. This will pull off a "sleeve" of cortical tissue, revealing a core of vascular tissue underneath (Fig. 9.6). It is not necessary to dig up a whole tree to examine the roots (Fig. 9.7). At a place on, or just inside the *drip line* (the line indicating the maximum spread of aboveground growth from the central stem or trunk), a 20–50 cm (8–20 in) deep hole can be dug and roots exposed. These can be checked for cortical sloughing, discoloration, deformities, and infestations.

Many synthetic agricultural and other chemicals can damage plants. Symptoms include abnormal stem and leaf growth, and leaf chlorosis, curl, and mottling. For example, poisoning by herbicides such as 2,4-D or clopyralid can cause symptoms that look like a viral infection. Some herbicides like clopyralid do not break down readily in the environment, including through the composting process, and can cause major

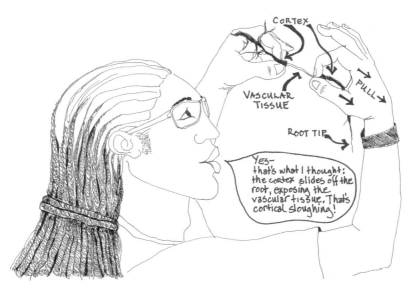

Figure 9.6. Testing for cortical sloughing

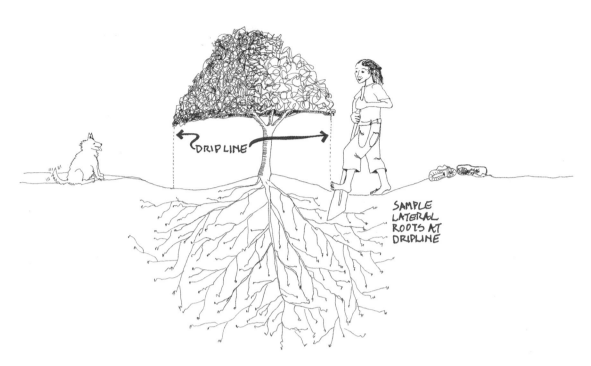

Figure 9.7. Collecting a root sample from a tree at the drip line

damage to plants in the garden when contaminated plant material, manure, or compost is used (Box 7.4).

Check for evidence of nearby spraying, or contamination by tools, containers, soil, shoes, or clothes brought into the garden. Be sure a poisonous chemical was not accidentally put in a bucket, sprayer, or on other tools used in the garden, and find out the chemical history of the garden plot (Section 4.2.1). This is especially important in urban areas where waste is often dumped in empty lots where gardens are later located.

In the rest of this section we provide detailed guides to diagnosing and managing pest and pathogen problems, and problems caused by the environment that can be mistaken for pest and pathogen problems. The guides are not comprehensive either in terms of problems covered or the possible management responses; their purpose is to illustrate diagnosing and managing pests and pathogens using *FGCW*'s emphasis on concepts and the key ideas in the Introduction.

9.8.1. Systemic vs. localized diseases

Distinguishing between systemic and localized (non-systemic) plant diseases can help gardeners plan their disease control strategies. *Systemic* diseases, including vascular wilts and almost all viral diseases, are spread throughout the plant through the vascular system (Section 5.1). *Localized* diseases are only active in certain parts of the plant such as the roots or leaves and fruits. Some localized diseases or pest problems, especially in the roots, may first become evident to the gardener in symptoms that look systemic, like wilting. So check the roots of garden plants before you assume a problem is systemic. Examples of pests affecting plant roots include soil nematodes, grubs, and larvae.

Localized fungal and bacterial diseases often appear first on lower stems, trunks, fruit, and older leaves near the soil surface. On leaves they show up as specks or spots that are water-soaked, dark green, or brown. Lesions may look like a target, having alternating dark and light concentric circles, or may have a furry, moldy appearance. The fruit may feel like a bag of water, look rotten and moldy, or have a scab-like wound, depending on the disease.

If caught early in their development, localized fungal and bacterial diseases can be controlled by removing the infected parts and using other management strategies. For example, if a fungal leaf spot is found on a few tomato leaves in the garden, these leaves can be removed and the plants pruned and staked to provide better air circulation. However, if leaves on the growing tip of a squash plant appear mottled and deformed, signs of a systemic viral disease, removing those leaves will not provide control because the entire plant is infected.

Management strategies for systemic disease will vary depending on the crop and severity of the disease. If other, healthy plants of the same crop are growing in the garden, immediately removing the diseased plants could help control the disease if it has not already spread. Gardeners need to decide if the plant is capable of surviving, if it is worth the extra work to save it, and if leaving it will lead to further spread of the disease. Some weeds may be alternate hosts for systemic or localized plant diseases or their vectors, and experienced local gardeners and scientists may know which these are for different plant diseases, so those weeds can be removed before planting.

9.8.2. Seedlings and recent transplants

The younger and more tender its tissues the more susceptible a plant is to pests, pathogens, and difficult growing conditions (Fig. 9.8). As described earlier, some garden crops contain chemicals that repel or are toxic to pests, but only start producing these chemicals as they grow and mature. In addition, young plants with small root and shoot systems are much more likely to be killed by pest or pathogen damage than more established plants. This makes careful observation and quick diagnosis and action especially important when caring for young plants.

The most common disease problem in seedlings is damping off, caused by several soil fungi and some bacteria that rot plant roots or stems at the soil line (Section 9.2.4, Fig. 9.5). Damping off diseases are localized, not systemic diseases. Infected seedlings wilt, the leaves dry out, plants fail to grow, or even fall over. You can check this diagnosis by digging out a seedling and washing it carefully in a container of water, to see if the roots and stem have soft brown areas and cortical sloughing. In advanced cases, the entire root system will be rotten and will disintegrate during washing. There are three common causes of damping off.

1. An unusually large population of the fungi or bacteria that cause the disease due to insufficient competition from other microorganisms in the soil. Adding compost can reduce damping off problems because compost contains many active soil organisms, some of which compete with and reduce the damping off fungi. Soaking seeds in a preparation of garlic in water at concentrations of between 4–20 crushed cloves in 220 ml (1 c) controlled damping off in carrot seeds caused by the fungal pathogen *Alternaria* (Slusarenko et al., 2008).
2. Waterlogged soil due to overwatering or poor drainage. This can be remedied by watering only as much as is necessary, and before planting by improving soil structure and drainage, for example, by adding organic matter to the soil and loosening any compacted layers in and below the root zone (Sections 6.2.2, 7.1.1).
3. Soil temperatures too hot or too cold for the crop variety result in lack of vigor and susceptibility to disease. This can be avoided by modifying soil temperatures (use a covering of surface mulch to cool soil, remove it to allow soil warming), or replanting at a better time of year.

Problems with transplants are often caused by transplant stress or exposure to pests or pathogens in the planting site. Transplant stress weakens the plant, making it vulnerable to pests or pathogens that were not a problem before transplanting. Bruised and broken tissues from handling during transplanting create openings where pests and pathogens can enter the plant. Hardening off (Section 6.6.2) and careful transplanting reduce these problems.

If a tree seedling or cutting grown in a container was root-bound at the time of transplanting, the roots can twist around each other and in three to four years choke each other off. Aboveground the leaves may become chlorotic, turn orange, or wilt. Cutting the sides of the root-bound root ball and loosening the

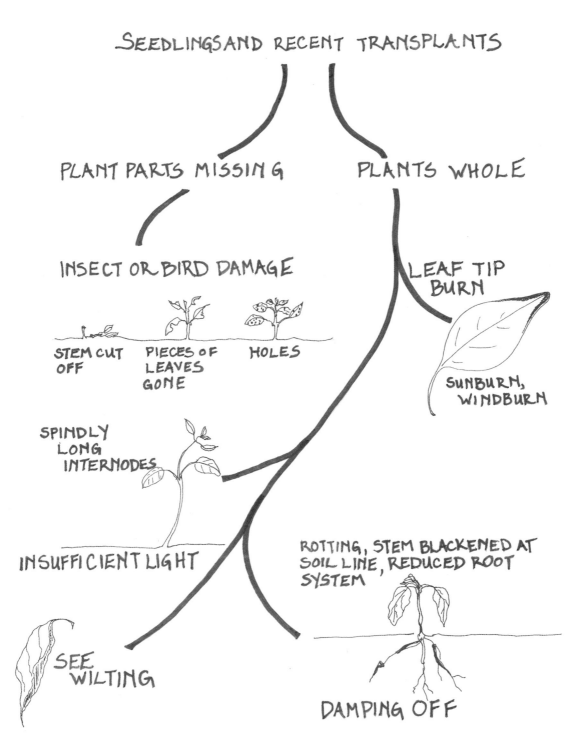

Figure 9.8. Diagnosing problems with seedlings and recent transplants

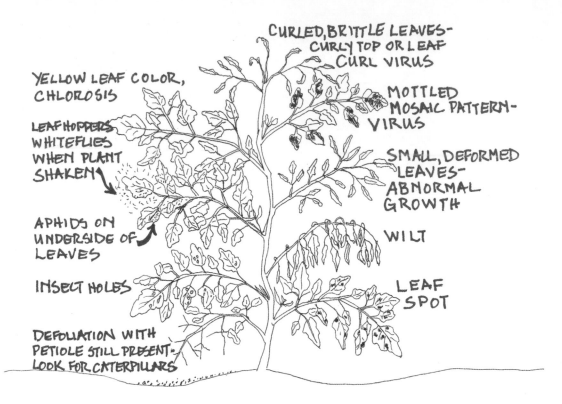

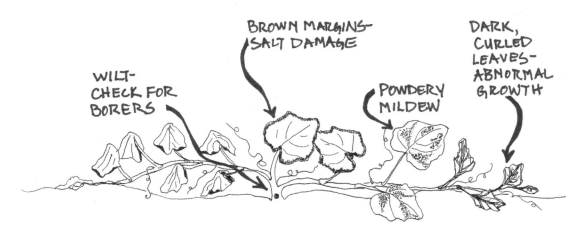

Figure 9.9. Problems with established garden plants observed aboveground

roots at the time of transplanting helps prevent this problem (Section 6.6.2).

One of the best ways to overcome problems with transplants and especially seedlings is to just keep trying. For example, try replanting or transplanting when the soil is a little warmer or cooler, wait until insect populations have died down, or plant where there is more or less sunlight. Often these informal experiments lead to an understanding of how to work with the garden environment instead of fighting it, with better long-term results using fewer resources, and more enjoyable gardening.

9.8.3. Established plants

Most of the symptoms of disease in established herbaceous annuals, such as wilting and chlorosis, are observed aboveground (Fig. 9.9), yet the majority of these symptoms, except those due to some insects and viruses, are caused by problems belowground. Many fungi, and a variety of nematodes can attack the roots of annuals and woody

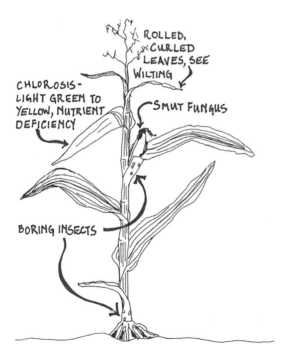

Figure 9.9. Continued

perennials, and checking the roots is important for diagnosing these symptoms. Impervious soil layers can slow or stop growth of perennials and can also contribute to waterlogging and salt buildup. Aboveground the most common problems are insects (shoot damage, leaf eating, or causing abnormal growth), powdery mildew, birds and other large animals, and wind damage. See Section 9.8 for discussion of diagnosing root problems, including in woody perennials (Figs 9.5–9.7). Wilts and chloroses caused by diseased roots are localized diseases as opposed to wilts and chloroses caused by viruses and fungal vascular wilts, which are systemic diseases.

All gardeners at one time or another have plants that just do not grow. The plants may look perfectly fine or show a variety of difficult-to-classify symptoms. Failure to thrive and grow may have one cause, but just as likely it is the result of a combination of interacting factors. Planting again at different times of the year, in different soil conditions, checking the roots of a few plants to look for clues, and talking with other gardeners to see what is happening in their gardens all help. If other gardens have similar problems, then a common climate (Section 5.8), water source, or seed source may be the cause. Sometimes it just takes time for a problem to resolve on its own. Some citrus trees we have planted have failed to thrive and grow for five or more years, for reasons we could not diagnose, and then grew into productive trees.

The following four sections include figures and tables to help in diagnosing garden problems according to types of symptoms. Figure 9.10 is a guide to these sections.

9.8.4. Wilting

When a plant wilts it loses its rigidity, its leaves and stems droop due to lack of water, often because there is inadequate water in the root zone (Table 9.1, Fig. 9.11). Check the plants in the early morning during the hot season to see if they have recovered. Shading plants and providing windbreaks also helps by reducing excess evapotranspiration, making it easier for the plant to maintain adequate water uptake. On very hot days, some plants like cucurbits, will wilt even with sufficient soil moisture, but will recover in the cool of the evening. Plants may

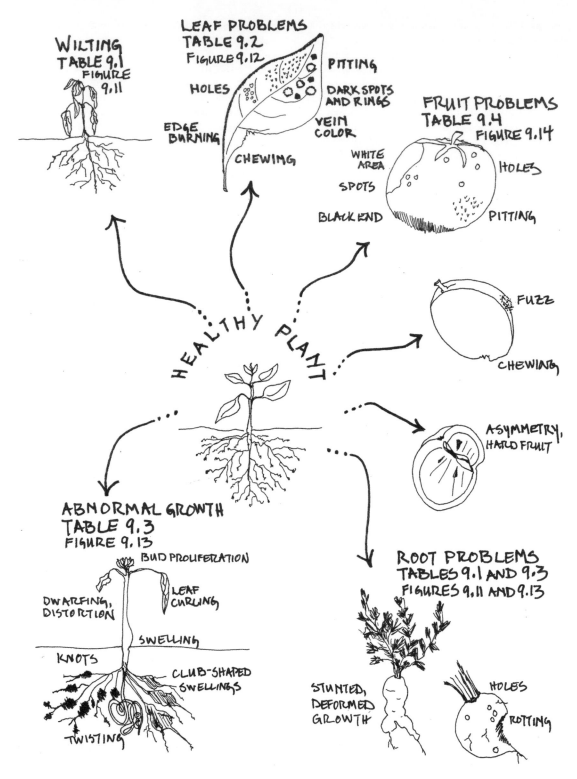

Figure 9.10. Diagnosing plant problems in the garden

also wilt while being hardened off for transplanting (Section 6.6.2). Other important causes of wilting include fungal and bacterial pathogens, insects, nematodes, and gophers.

Larvae or grubs in the soil that eat roots can cause plants to wilt. When they attack the storage roots of crops like carrot and Jerusalem artichokes, the plant may not wilt, but those wounds are sites for pathogens to become established that can destroy the root in the ground, or later in storage. Sometimes, however, damage can stimulate increased production (Section 9.4). Clusters of frass along a plant's stem are a common sign of stem-boring caterpillars, like the squash vine borer (*Melitta curcurbitae*).

Some nematodes infect roots and cause wilting. Identification of particular species requires microscopic examination by trained observers, but nematode damage can often be identified by gardeners, who can then take simple actions to control the problem. Aboveground symptoms are wilt, *chlorosis* (a light green or yellow color that contrasts with the normal, darker green leaf color) of the whole leaf, failure to thrive, and poor yield. One of the most common and destructive nematodes, especially in gardens in warm or hot areas with mild winters, is the root knot nematode (*Meloidogyne* spp.), which causes knots or galls on the root (a localized disease). Unlike nodules caused by nitrogen-fixing bacteria (Section 7.3.2), nematode galls will not rub off without breaking the root apart, and a layer of soil often sticks to them. Sweet potatoes and other tuberous roots have cracked skin when infested with nematodes.

In general, adding organic matter to the soil will help avoid or reduce diseases of the root system, because the microorganisms present in the organic matter will compete with the pathogenic microorganisms, or directly inhibit their growth (Section 9.5.3).

9.8.5. Leaf problems

Many soil nutrient deficiencies (Section 7.3) show up as discoloration of the leaves (Table 9.2, Fig. 9.12). A common sign of many problems is chlorosis. The pattern of the chlorosis is often a good clue to the cause of the problem. Leaf chlorosis or other discolorations due to nutritional deficiencies are often *symmetrical* (the same on both halves of the leaf), and follow leaf vein patterns, while leaf discoloration caused by diseases is often *asymmetrical*. Leaf mottling, mixed patches of light green or yellow with normal green leaf color, is often caused by viral diseases, which are systemic.

Leaves that are yellow or have brown edges, and poor plant growth may indicate damaging amounts of salt in the soil or water, which is common in arid areas (Section 8.8.5). Blowing sand, also common in some drylands, can tear and pit leaves, resulting in tiny scars. Localized fungi and bacteria affecting primarily the leaves are less common in dry weather because of the lower humidity, but when the leaves are wet from rain or overhead watering, fungi and bacteria can become a problem, especially on lower leaves. An important exception is powdery mildew, a disease caused by a number of different fungi that thrive in dry conditions (Section 9.2.4). Powdery mildew is a localized disease and when caught early in the garden, removing infected leaves can be helpful. However, just because a disease is localized does not mean it can be ignored—when conditions are right, powdery mildew can be a serious problem. Insect damage of leaves can be severe, but in gardens can often be controlled by hand picking, washing the leaves, or replanting.

9.8.6. Abnormal growth

Abnormal growth—resulting in misshapen roots, stems, leaves, or flowers—is a response to stress from pests, pathogens, the physical environment, management practices, chemical damage, or even genetic abnormalities (Table 9.3, Fig. 9.13).

9.8.7. Fruit problems

Fruit problems can be especially disheartening for the gardener because the harvest that took so long to develop seems so near. Fruit can show abnormal growth and the mottling symptoms of systemic virus infections (Table 9.4, Fig. 9.14), or the soft and fuzzy spots of localized fungal infections. Bacterial spots on fruit are also localized infections. Fruit symptoms are usually associated with leaf or shoot symptoms. Some of these symptoms can also develop in harvested, stored fruits.

Table 9.1. Wilting

Observations of conditions or additional symptoms	Possible causes	Suggested actions
	TEMPERATURE AND SOIL	
Heat stress	High temperatures and transpiration rates	Shade, mulch, check soil moisture, consider adjusting planting time
Soil in root zone dry	Underwatering	Irrigate, mulch, and shade
Water-saturated soil	Overwatering or high water table	Reduce watering, try raised beds if there is high water table, add compost for better structure and drainage
Chemical injury	History of herbicide use or chemical dumping in area	Move garden site or use containers and soil from other source
Poor soil tilth	Compaction, poor water quality, repeat cropping without fallow or rotation	Add compost, rotate with cover crop with deep roots
	LEAVES	
Many tiny insects on underside of leaves	Aphids, mites	Wash off, crush, spray, use a yellow sticky trap, trap crops
Mottled, ring spots, curled, misshapen, dwarfed	Viral disease (systemic)	Discard plant, clean hands and tools, plant resistant varieties
Young tomato leaves stunted, curled inward, bumpy on lower surface, brittle	Curly top virus spread by insects (leafhoppers) or one of the leaf curl viruses transmitted by whiteflies (systemic)	Shade tomato plants to discourage insect vectors. Try floating row covers to protect young plants from whiteflies
Wilting may be sudden, only in some parts, with other symptoms	Bacterial or fungal vascular wilts (systemic)	Discard plant, clean hands and tools, plant resistant varieties, add compost to soil
Dwarfed, curled, misshapen	Chemical injury	Move garden site or use containers and soil from other source
	STEMS	
Ring of vascular browning seen in stem cross section in eudicots	Fungal vascular wilt (systemic)	Plant resistant varieties, add compost, do not overwater, vary planting times; if problem severe try rotation of monocot, e.g., maize, or green manure like rye
White ooze from cut stem, discolored tissue	Bacterial wilt (systemic)	Plant local resistant varieties; rotate crop families for 5 years, or 2–3 years clean, un-irrigated fallow
Small holes on main stem	Stem- and vine-boring caterpillars	Remove borer, cover stem with soil; pile soil on squash stems encourage rooting elsewhere
Chew marks on base of stem	Caterpillars and sow bugs	Hand pick pest, especially at night, remove mulch immediately around stem, use stem collars
	ROOTS	
Roots brown and soft, cortical sloughing	Fungal root (localized)	Add compost, do not overwater
Club-shaped swellings (galls) on crucifers	Clubroot fungus (localized)	Rotate non-crucifers or resistant cruciferous varieties, raise pH above 7.2
Round knots, lesions or swellings	Nematodes (localized)	Rotate crop families, add compost, use non-infested planting material, solarize and then dry soil
Small root system, few lateral roots	Larvae or grubs eating roots	Dig around plants or turn soil to expose and kill pests, allow domesticated or wild birds to forage for larvae or grubs

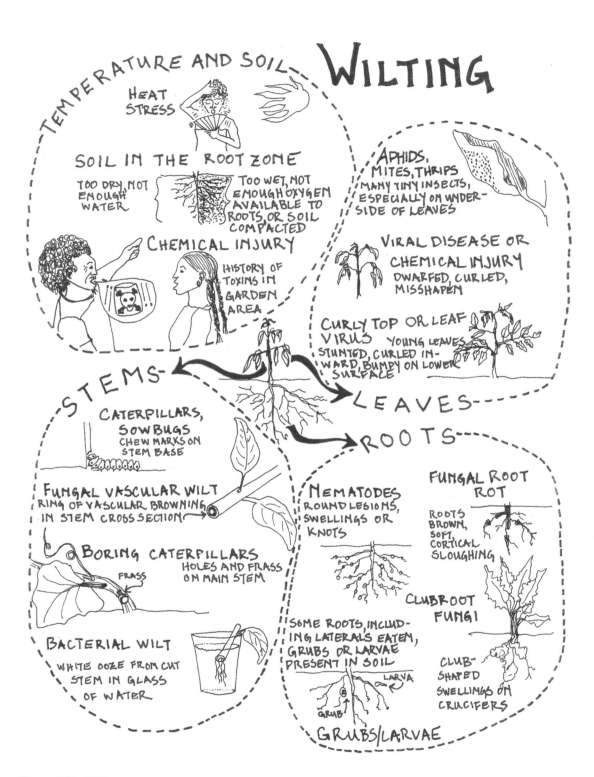

Figure 9.11. Wilting

Table 9.2. Leaf problems

Observation	Possible causes	Suggested actions
	CHLOROSIS (leaf yellowing)	
Whole leaf (older leaves)	Nitrogen deficiency	Add high N organic matter
Whole leaf, plus swellings or knots on roots that don't rub off	Nematodes (localized)	Rotate crops, add compost, fallow and/or solarize soil
Whole leaf, plus roots with browning and soft tissue	Fungal root rots (localized)	Add compost, do not overwater
Whole leaf, plus tangled, twisted root growth	Root-bound	Untangle roots, transplant to larger container or into the ground
Younger leaves pale yellow to light green	Sulfur deficiency	Add elemental powdered sulfur to the soil
Between veins of new leaves	Iron or zinc deficiency	Lower soil pH, add compost, especially in case of zinc deficiency
Veins in older leaves	Viral disease (systemic) or herbicide damage	Discard plant, clean tools and hands, control insect vectors, rotate to resistant variety or family
Asymmetric, e.g., half of leaf, plus vascular browning	Fungal vascular wilts (systemic)	Plant resistant varieties, rotate crop families, add compost
Mottled or mosaic pattern	Viral disease (systemic) or herbicide damage	Discard plant, clean tools and hands, control insect vectors, rotate to resistant variety or family
Chlorosis or necrosis between veins of mature leaves	Sulfur dioxide damage, e.g., from acid rai	Rinse leaves thoroughly with fresh water immediately following exposure to acid rain, raise pH of irrigation water, try resistant crops including asparagus, cabbage, celery, maize, onion, potato
	OTHER	
Leaves unusually purple	Phosphorus deficiency	Add phosphorus-rich organic matter such as a chicken manure compost, or add rock phosphate or fish bone meal
Black spots or rings	Fungal leaf spot (localized)	Remove and discard affected leaves, clean hands and tools, add compost to soil, open canopy for better air circulation, rotate in different crop family
Salt burn: edges brown, white, or yellow	High salt concentration in soil or water	Check water quality, water deeply, flush soil
Dry, brown patches, holes visible with lens	Sucking insects	Pick off or crush, wash or spray with water or a repellent, use traps
Chew marks on edges, small holes in center	Chewing insects	Pick off or crush, wash or spray with water or a repellent
Pitting and tearing	Windblown sand	Install windbreaks
White, powdery spots	Powdery mildew fungi (localized)	Remove and discard infected plants or parts, add compost to soil, plant resistant varieties
Small (5 mm [0.02 in] diameter) spots, dark with yellow margins	Bacterial spot (localized)	Remove and discard affected leaves and fruit, stake plants to increase air circulation, wash hands and tools, minimize leaf wetness, plant pathogen-free seed
Rusty yellow, brown, or white spots, also on stems, can form galls	Fungal rusts (localized)	Remove and discard affected leaves and fruit, stake plants to increase air circulation, wash hands and tools, add compost to soil
Young tomato leaves stunted, curled inward, bumpy on lower surface, brittle	Curly top virus spread by insects (leafhoppers) or one of the leaf curl viruses transmitted by whiteflies (systemic)	Shade tomato plants to discourage insect vectors, try floating row covers to protect young plants from whiteflies

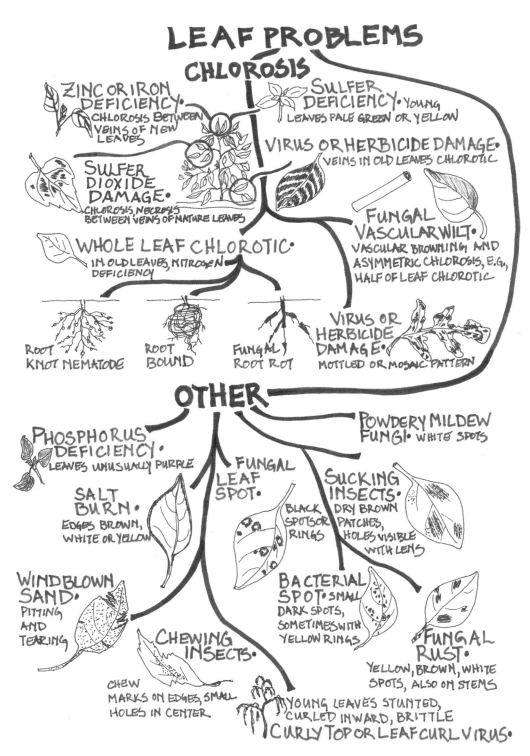

Figure 9.12. Leaf problems

Table 9.3. Abnormal growth

Observation	Possible causes	Suggested actions
ROOTS		
Round swellings or knots that don't rub off	Root knot nematodes (localized)	Rotate crop family, solarize soil, turn soil regularly during hot season, add compost, fallow
Spindle- or club-shaped swellings on crucifers	Clubroot fungi (localized)	Rotate in noncruciferous crops, raise soil pH above 7.2
Plant in container with roots growing in circle, crowded	Root-bound	Untangle or cut roots as necessary, transplant into larger container or into ground
Plant in ground with stunted, twisted, forked roots	Impermeable layer in soil, e.g., caliche	Dig deep planting hole, fill with good soil mixed with compost
	Transplanting too late or inappropriate crop e.g., carrot	Plant seeds directly instead of transplanting
STEMS		
Galls or swellings on crown or higher, especially stone fruits, grapes	Crown gall bacteria (localized)	Remove and discard infested young perennials, clean hands and any tools used near or on diseased plants
Large, dark, sunken spot at soil line	Collar rot fungi (localized)	Remove and discard affected plants, clean hands and tools, add compost to soil, do not overwater, plant resistant varieties
Swollen growths with black dust inside, at the joints and on flowers, in maize and teosinte	Maize smut fungus (localized)	Remove and discard affected stalks, ears, tassels before growths open, even better—when correctly identified on maize, this can be cooked and eaten, a Mexican delicacy—*huitlacoche*
Swellings or holes higher on stem, dieback above these	Insect eggs	Allow beneficials to hatch, remove others
Many buds	Feeding by thrips, mites, other insects; vectored viral diseases (systemic)	Remove and discard affected parts or entire plant if virus suspected, eliminate insect vectors, clean hands and tools
Pale, internodes long and spindly	Insufficient sunlight	Gradually reduce shade and/or remove mulch
LEAVES		
Misshapen, curled	Mites, thrips	Remove and destroy affected parts, crush, spray, trap pests, encourage predators
Misshapen, curled, sometimes sticky	Aphids, other sucking insects	Wash off, crush, spray, use yellow sticky trap
Dwarfed, misshapen, chlorosis	Chemical damage	Move garden or use containers with soil from another source
Misshapen, dwarfed mottled, especially the new growth	Viral disease (systemic)	Discard plant if symptoms severe, monitor carefully if symptoms mild and plant maturing, clean hands and tools, use disease-free seeds or other propagule
Curled and brittle, thickened midribs, leafhoppers active	Curly top virus (systemic)	Discard plant if symptoms severe, monitor carefully if symptoms mild and plant maturing, shade plants, plant resistant varieties, try floating row cover to protect young plants from leafhoppers
Misshapen, dwarfed with wilt or asymmetric chlorosis, vascular browning	Fungal vascular wilts (systemic)	Discard plant if symptoms severe, monitor carefully if symptoms mild and plant maturing, add compost, do not overwater
Grape or fruit tree leaves rolled, shredded	Leafroller caterpillars	Monitor plants to assess level of damage, and for parasitism of caterpillars, remove affected leaves and caterpillars, if damage is severe, use *Bacillus thurengiensis* (*Bt*) spray

ABNORMAL GROWTH

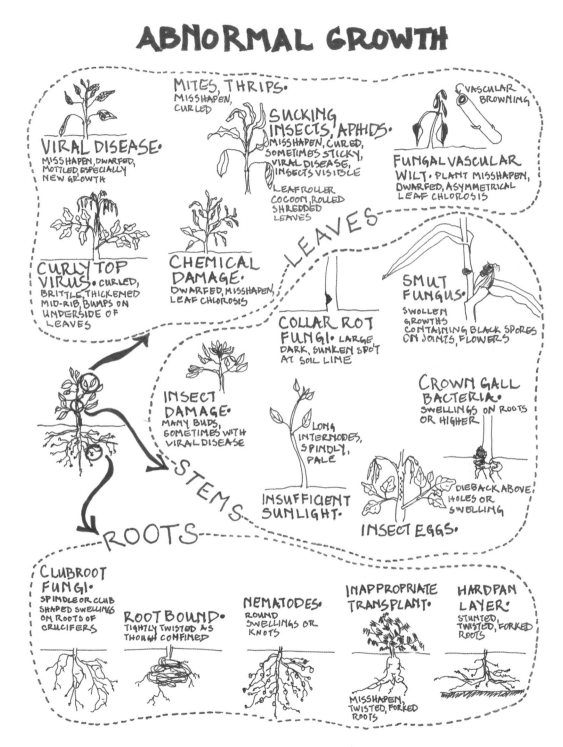

Figure 9.13. Abnormal growth

Table 9.4. Fruit problems

Observation	Possible causes	Suggested actions
	FRUIT HARD	
Citrus fruit that are small, green colored at blossom end, misshaped, asymmetric when cut open, Asian citrus psyllid (pathogen vector) present	Huanglongbing (HLB, citrus greening) caused by a phloem-limited bacterium (systemic)	Newly invasive disease organism in some areas. Contact government office for advice on proper tree removal and disposal
Mottled yellow/green, faint yellow rings	Viral disease (systemic)	If severe, remove and discard fruit and plant, control insect vectors, do not plant seeds produced by these plants
Small, scab-like dark spots with yellow margins	Bacterial spot (localized)	Remove and discard affected fruit and leaves, prune and stake plant, add compost to soil, plant resistant varieties
In tomatoes, citrus, a black/tan hard spot on end (blossom end rot)	Calcium deficiency	Water more regularly, mulch to maintain even soil moisture
Hard white lesion or scar on exposed part of fruit	Sunburn	Shade plant
Yellow spots with holes in middle	Sucking insects	Remove, trap or crush insects, wash or spray fruit with water
Brown pinpoint scarring on immature olive fruit	Olive fruit fly	Remove and discard infested fruit while green, do not allow fruit ripening and emergence of adults, discard all dropped fruit, kaolin clay sprays may reduce infestations
Pitting with no holes in middle	Windblown sand	Windbreaks, harvest early if danger of rotting, replant if possible
Bruised or scarred, chunks of skin missing	Hail	Harvest early if danger of rotting, replant if possible
Pomegranate fruit crack open	Uneven or inadequate water supply	Establish regular, deep watering schedule beginning just after flowering through fruit ripening
	FRUIT SOFT	
Soft spots where fruit touches the ground, e.g., melons	Fungal soft rots (localized)	Elevate fruit above moist ground on stones, sticks, mulch, or trellis
Fuzzy brown growth on stone fruits, also occurs in storage	Fungal brown rot (localized)	Remove and compost fruit before spores spread
Fruit feels like a bag of water	Bacterial soft rot (localized)	Pick and discard affected fruit to control fruit flies and other insect vectors, prune and stake plant to allow air circulation
Holes eaten in fruit, e.g., figs, dates, peaches	Fruit beetles or other boring insects	Harvest early, trap and remove beetles in morning, remove ripe, rotting fruit
Small holes on skin, inside eaten, rotten	Fruit fly larvae	Cover fruit to protect from egg laying, use homemade sprays, traps
Holes eaten in fruit	Birds	Frighten off, harvest early, cover fruit or whole plant

FRUIT PROBLEMS

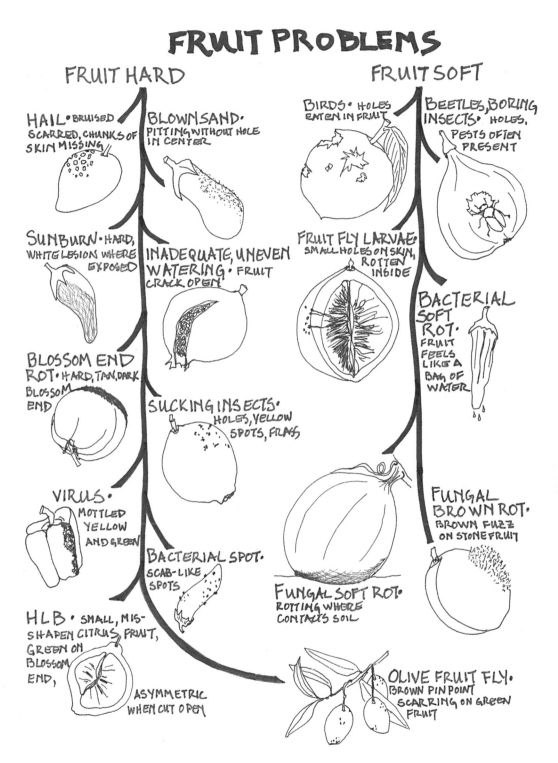

FRUIT HARD

HAIL • BRUISED SCARRED, CHUNKS OF SKIN MISSING

BLOWN SAND • PITTING WITHOUT HOLE IN CENTER

SUNBURN • HARD, WHITE LESION WHERE EXPOSED

INADEQUATE, UNEVEN WATERING • FRUIT CRACK OPEN

BLOSSOM END ROT • HARD, TAN, DARK BLOSSOM END

SUCKING INSECTS • HOLES, YELLOW SPOTS, FRASS

VIRUS • MOTTLED YELLOW AND GREEN

BACTERIAL SPOT • SCAB-LIKE SPOTS

HLB • SMALL, MIS-SHAPEN CITRUS FRUIT, GREEN ON BLOSSOM END, ASYMMETRIC WHEN CUT OPEN

FRUIT SOFT

BIRDS • HOLES EATEN IN FRUIT

BEETLES, BORING INSECTS • HOLES, PESTS OFTEN PRESENT

FRUIT FLY LARVAE • SMALL HOLES ON SKIN, ROTTEN INSIDE

BACTERIAL SOFT ROT • FRUIT FEELS LIKE A BAG OF WATER

FUNGAL BROWN ROT • BROWN FUZZ ON STONE FRUIT

FUNGAL SOFT ROT • ROTTING WHERE CONTACTS SOIL

OLIVE FRUIT FLY • BROWN PINPOINT SCARRING ON GREEN FRUIT

Figure 9.14. Fruit problems

9.9. Resources

All the websites listed below were verified on May 26, 2018.

9.9.1. Arthropods

College and high-school textbooks about entomology are good sources for learning more about insects, their life cycles and habits. Unfortunately, these texts often recommend using synthetic chemicals as solutions to pest and disease problems.

The website for a general entomology course created by Dr John Meyer at North Carolina State University has basic explanations and an identification guide, much of which is available to the public. https://www.cals.ncsu.edu/course/ent425/index.html

Linker et al. (2009) is a comprehensive extension publication on organic control of insect pests and includes information on microbial, botanical, and other organic pesticides.

An excellent guide for identifying insects in the garden is *Garden insects of North America* (Cranshaw, 2004), which is organized according to the part of the plant damaged, and contains many color photos and suggestions for control, including non-toxic options.

Tiny game hunting (Klein and Wenner, 2001) is full of practical, easy, non-toxic ways to control garden pests.

9.9.2. Nematodes

Nemaplex, a University of California, Davis, website, has extensive nematode information including identifications, plant host guides, and much more. http://nemaplex.ucdavis.edu/index.htm

9.9.3. Plant diseases

The non-profit American Phytopathological Society has excellent open access resources online, including a guide to management of many common garden plant diseases; however, the Society includes toxic synthetic chemicals in its recommendations. http://www.apsnet.org/EDCENTER/INTROPP/Pages/default.aspx

The most comprehensive and authoritative textbook, currently in its fifth edition is Agrios (2005). The American Phytopathological Society has published *Essential plant pathology*, a shorter, more practice-oriented text than Agrios (Schumann and D'Arcy, 2010). While these texts list some low toxicity methods for management, they also recommend toxic synthetic chemical treatments.

The University of Florida has a useful, online decision tree for diagnosing nutrient deficiencies. http://hort.ufl.edu/database/nutdef/index.shtml

9.9.4. Biological control

The University of Florida Extension has published guidelines for purchasing and using commercial natural enemies and biopesticides in North America. http://edis.ifas.ufl.edu/in849

Parker et al.'s "*Companion planting and insect pest control*" (2013) is an accessible overview of the scientific literature about companion planting that's freely available online.

For a fascinating history of the biological control of cottony cushion scale in California, see http://www.faculty.ucr.edu/~legneref/biotact/ch-35.htm

The US Centers for Disease Control and Prevention (CDC) has information on the health impacts of *Aspergillus* fungus. https://www.cdc.gov/fungal/diseases/aspergillosis/

The National Library of Medicine has information about *Bt* toxicology, including eye irritation. https://toxnet.nlm.nih.gov/cgi-bin/sis/search/a?dbs+hsdb:@term+@DOCNO+782

9.9.5. Chemical control

The US National Organic Program website has the list of allowed and prohibited substances for use in USDA certified organic agriculture. This list is kept current. http://www.ecfr.gov/cgi-bin/text-idx?rgn=div6&node=7:3.1.1.9.32.7

The University of California Cooperative Extension has an integrated pest management (IPM) website

with information for identifying pests and diseases including nematodes for home gardeners, substances allowed for certified organic production, and their application, such as Bordeaux mixture.
http://www.ipm.ucdavis.edu/index.html

Clemson University Extension has published an accessible short pamphlet about insecticidal soaps, although many of the pre-prepared ones suggested are brands of major agrochemical companies.
http://www.clemson.edu/extension/hgic/pests/pesticide/hgic2771.html

9.9.6. Synthetic chemicals

The Pesticide Action Network (PAN) is a well-established non-governmental organization (NGO), with a North American branch (PANNA) in Oakland, California. Unlike many government bodies and universities, PAN does not have ties to the synthetic pesticide industry. It has an excellent pesticide database on pesticides, including glossaries and much general information; especially valuable because its mission is education and public safety.
http://www.pesticideinfo.org/
PAN has also organized USDA data on pesticide residues on different foods.
http://whatsonmyfood.org/index.jsp

A number of non-profit organizations are developing tools and support services for communities concerned about contamination in their environment, including from pesticides, an especially important issue for low income communities near agricultural land. A popular method involves some form of simple, quick-catch air sampler; these two local examples also provide how-to-information and networks. PANNA's Drift Catcher program:
http://www.panna.org/resources/if-youve-been-drifted
and the Louisiana Bucket Brigade:
http://www.labucketbrigade.org/content/bucket

EXTOXNET (extension toxicology network) is an alphabetically organized list of agrochemicals with descriptions of chemistry, applications, precautions, and toxicology. It dates from the mid-1990s, but is still useful, and was compiled by USDA and major land grant university extension services.
http://pmep.cce.cornell.edu/profiles/extoxnet/index.html

9.10. References

Agrios, G. N. (2005) *Plant pathology*, 5th edn. Elsevier, Academic Press, Netherlands.

Ahmad, P. & Rasool, S. (2014) *Emerging technologies and management of crop stress tolerance: Volume 1, Biological techniques*. Academic Press, San Diego, CA.

Al-Hazmi, A. S. & Al-Nadary, S. N. (2015) Interaction between *Meloidogyne incognita* and *Rhizoctonia solani* on green beans. *Saudi Journal of Biological Sciences*, 22, 570–574, DOI: http://dx.doi.org/10.1016/j.sjbs.2015.04.008.

Al-Shihi, A., Al-Sadi, A., Al-Said, F., Ammara, U. & Deadman, M. (2016) Optimising the duration of floating row cover period to minimise the incidence of tomato yellow leaf curl disease and maximise yield of tomato. *Annals of Applied Biology*, 168, 328–336, DOI: 10.1111/aab.12266.

Alyokhin, A., Drummond, F. & Sewell, G. (2005) Density-dependent regulation in populations of potato-colonizing aphids. *Population Ecology*, 47, 257–266.

Anaya, A. L. (1999) Allelopathy as a tool in the management of biotic resources in agroecosystems. *Critical Reviews in Plant Sciences*, 18, 697–739.

APHIS (2018) APHIS-2017–0018. Animal and Plant Health Inspection Service, USDA, available at: https://www.regulations.gov/docket?D=APHIS-2017-0018 (accessed May 25, 2018).

Appleton, B., Berrier, R., Harris, R., Alleman, D. & Swanson, L. (2015) The walnut tree: Allelopathic effects and tolerant plants. Virginia Cooperative Extension, available at: https://pubs.ext.vt.edu/content/dam/pubs_ext_vt_edu/430/430-021/430-021_pdf.pdf (accessed Aug. 15, 2018).

Atehnkeng, J., Ojiambo, P. S., Cotty, P. J. & Bandyopadhyay, R. (2014) Field efficacy of a mixture of atoxigenic *Aspergillus flavus*. *Biological Control*, 72, 62–70, DOI: http://dx.doi.org/10.1016/j.biocontrol.2014.02.009.

Atehnkeng, J., Donner, M., Ojiambo, P. S., Ikotun, B., Augusto, J., Cotty, P. J. & Bandyopadhyay, R. (2016) Environmental distribution and genetic diversity of vegetative compatibility groups determine biocontrol strategies to mitigate aflatoxin contamination of maize by *Aspergillus flavus*. *Microbial Biotechnology*, 9, 75–88, DOI: 10.1111/1751-7915.12324.

Bale, J. S., van Lenteren, J. C. & Bigler, F. (2008) Biological control and sustainable food production. *Philosophical Transactions of the Royal Society B: Biological Sciences*, 363, 761–776, DOI: 10.1098/rstb.2007.2182.

Barbercheck, M. E. (2014) Biology and management of aphids in organic cucurbit production systems. *Organic Agriculture*, available at: http://articles.extension.org/pages/60000/biology-and-management-of-aphids-in-organic-cucurbit-production-systems (accessed Nov. 19, 2015).

Ben-Yephet, Y., Stapleton, J., Wakeman, R. & De Vay, J. (1987) Comparative effects of soil solarization with single and double layers of polyethylene film on survival of *Fusarium oxysporum* f. sp. *vasinfectum*. *Phytoparasitica*, 15, 181–185, DOI: 10.1007/bf02979581.

Benbrook, C. M. (2016) Trends in glyphosate herbicide use in the United States and globally. *Environmental Sciences Europe*, 28, 3, DOI: 10.1186/s12302-016-0070-0.

Bhowmik, P. C. (2003) Challenges and opportunities in implementing allelopathy for natural weed management. *Crop Protection*, 22, 661–671.

Bjørnson, S. (2008) Natural enemies of the convergent lady beetle, *Hippodamia convergens* Guérin-Méneville: Their inadvertent importation and potential significance for augmentative biological control. *Biological Control*, 44, 305–311.

Bomar, C. R., Lockwood, J. A., Pomerinke, M. A. & French, J. D. (1993) Multiyear evaluation of the effects of *Nosema locustae* (Microsporidia: Nosematidae) on rangeland grasshopper (Orthoptera: Acrididae) population density and natural biological controls. *Environmental Entomology*, 22, 489–497, DOI: 10.1093/ee/22.2.489.

Bridge, J. (1996) Nematode management in sustainable and subsistence agriculture. *Annual Review of Phytopathology*, 34, 201–225.

Burketova, L., Trda, L., Ott, P. G. & Valentova, O. (2015) Bio-based resistance inducers for sustainable plant protection against pathogens. *Biotechnology Advances*, 33, 994–1004, DOI: 10.1016/j.biotechadv.2015.01.004.

Calderone, N. W. (2012) Insect pollinated crops, insect pollinators and US agriculture: Trend analysis of aggregate data for the period 1992–2009. *PLoS ONE*, 7, e37235.

Caldwell, B., Sideman, E., Seaman, A., Shelton, A. M. & Smart, C. D. (2013) *Resource guide for organic insect and disease management*, 2nd edn. New York Agricultural Experiment Station, New York, NY.

Catarino, R., Ceddia, G., Areal, F. J. & Park, J. (2015) The impact of secondary pests on Bacillus thuringiensis (Bt) crops. *Plant Biotechnology Journal*, 13, 601–612. DOI: 10.1111/pbi.12363.

Chapman, A. D. (2009) *Numbers of living species in Australia and the world*. Australian Biological Resources Study (ABRS), Canberra.

Chung, Y., Hoitink, H., Dick, W. & Herr, L. (1988) Effects of organic matter decomposition level and cellulose amendment on the inoculum potential of *Rhizoctonia solani* in hardwood bark media. *Phytopathology*, 78, 836–840.

Collange, B., Navarrete, M., Peyre, G., Mateille, T. & Tchamitchian, M. (2011) Root-knot nematode (*Meloidogyne*) management in vegetable crop production: the challenge of an agronomic system analysis. *Crop Protection*, 30, 1251–1262.

Corporate Research Project (2016) Dow Chemical: corporate rap sheet, available at: http://www.corp-research.org/dowchemical (accessed May 25, 2018).

Cranshaw, W. (2004) *Garden insects of North America: The ultimate guide to backyard bugs*. Princeton University Press, Princeton, NJ.

Curtis, P. D., Curtis, G. B. & Miller, W. B. (2009) Relative resistance of ornamental flowering bulbs to feeding damage by voles. *HortTechnology*, 19, 499–503.

Daane, K., Wang, X., Nieto, D., Pickett, C., Hoelmer, K., Blanchet, A. & Johnson, M. (2015) Classic biological control of olive fruit fly in California, USA: Release and recovery of introduced parasitoids. *BioControl*, 60, 317–330, DOI: 10.1007/s10526-015-9652-9.

Danial, D., Parlevliet, J., Almekinders, C. & Thiele, G. (2007) Farmers' participation and breeding for durable disease resistance in the Andean region. *Euphytica*, 153, 385–396, DOI: 10.1007/s10681-006-9165-9.

de Albuquerque, M., dos Santos, R., Lima, L., Melo Filho, P., Nogueira, R., da Câmara, C. & de Rezende Ramos, A. (2011) Allelopathy, an alternative tool to improve cropping systems. A review. *Agronomy for Sustainable Development*, 31, 379–395, DOI: 10.1051/agro/2010031.

Dhima, K. V., Vasilakoglou, I. B., Gatsis, T. D., Panou-Philotheou, E. & Eleftherohorinos, I. G. (2009) Effects of aromatic plants incorporated as green manure on weed and maize development. *Field Crops Research*, 110, 235–241.

Dickson, J. G., Eckerson, S. H. & Link, K. P. (1923) The nature of resistance to seedling blight of cereals. *Proceedings of the National Academy of Sciences*, 9, 434–439.

Dutta, B., Gitaitis, R., Smith, S. & Langston, D., Jr. (2014) Interactions of seedborne bacterial pathogens with host and non-host plants in relation to seed infestation and seedling transmission. *PLoS ONE*, 9, e99215, DOI: 10.1371/journal.pone.0099215.

Ellouze, W., Esmaeili Taheri, A., Bainard, L. D., Yang, C., Bazghaleh, N., Navarro-Borrell, A., Hanson, K. & Hamel, C. (2014) Soil fungal resources in annual cropping systems and their potential for management. *BioMed Research International*, 2014, article ID 531824, DOI: 10.1155/2014/531824.

Ellstrand, N. C., Heredia, S. M., Leak-Garcia, J. A., Heraty, J. M., Burger, J. C., Yao, L., Nohzadeh-Malakshah, S. & Ridley, C. E. (2010) Crops gone wild: evolution of weeds and invasives from domesticated ancestors. *Evolutionary Applications*, 3, 494–504, DOI: 10.1111/j.1752-4571.2010.00140.x.

Espinosa-García, F. J., Martínez-Hernández, E. & Quiroz-Flores, E. A. (2008) Allelopathic potential of *Eucalyptus* spp. plantations on germination and early growth of annual crops. *Allelopathy Journal*, 21, 25–37.

Ferris, H. (2013) *Globodera rostochiensis*. Golden nematode, potato cyst nematode. University of California, Davis, CA, available at: http://nemaplex.ucdavis.edu/Taxadata/G053s2.aspx (accessed Aug. 15, 2018).

Flint, M. L. (2014) Lady bugs need special care to control aphids in the garden, available at: http://ucanr.edu/blogs/blogcore/postdetail.cfm?postnum=13933 (accessed May 25, 2018).

Flint, M. L. (2015) How to manage pests in gardens and landscapes: Whiteflies, available at: http://ipm.ucanr.edu/PMG/PESTNOTES/pn7401.html (accessed June 21, 2018).

Fujiyoshi, P. T., Gliessman, S. R. & Langenheim, J. H. (2007) Factors in the suppression of weeds by squash interplanted in corn. *Weed Biology and Management*, 7, 105–114, DOI: 10.1111/j.1445-6664.2007.00242.x.

Gardiner, M. M., Burkman, C. E. & Prajzner, S. P. (2013) The value of urban vacant land to support arthropod biodiversity and ecosystem services. *Environmental Entomology*, 42, 1123–1136, DOI: 10.1603/en12275.

Gevens, A. J. (2015) Managing late blight in organic tomato and potato crops, available at: https://fyi.uwex.edu/lincolncohort/files/2013/06/Organic-late-blight-control-publication.pdf (accessed June 10, 2018).

Giribet, G. & Edgecombe, G. D. (2012) Reevaluating the arthropod tree of life. *Annual Review of Entomology*, 57, 167–186, DOI: doi:10.1146/annurev-ento-120710-100659.

Glinwood, R., Ninkovic, V. & Pettersson, J. (2011) Chemical interaction between undamaged plants—Effects on herbivores and natural enemies. *Phytochemistry*, 72, 1683–1689, DOI: http://dx.doi.org/10.1016/j.phytochem.2011.02.010.

Gonsalves, C., Lee, D. R. & Gonsalves, D. (2007) The adoption of genetically modified papaya in Hawaii and its implications for developing countries. *Journal of Development Studies*, 43, 177–191.

Gould, F., Brown, Z. S. & Kuzma, J. (2018) Wicked evolution: Can we address the sociobiological dilemma of pesticide resistance? *Science*, 360, 728–732, DOI: 10.1126/science.aar3780.

Grünwald, N. J., Cadena Hinojosa, M. A., Covarrubias, O. R., Peña, A. R., Niederhauser, J. S. & Fry, W. E. (2002) Potato cultivars from the Mexican National Program: Sources and durability of resistance against late blight. *Phytopathology*, 92, 688–693, DOI: 10.1094/PHYTO.2002.92.7.688.

Grzywacz, D., Mushobozi, W. L., Parnell, M., Jolliffe, F. & Wilson, K. (2008) Evaluation of *Spodoptera exempta* nucleopolyhedrovirus (SpexNPV) for the field control of African armyworm (*Spodoptera exempta*) in Tanzania. *Crop Protection*, 27, 17–24, DOI: http://dx.doi.org/10.1016/j.cropro.2007.04.005.

Guyton, K. Z., Loomis, D., Grosse, Y., El Ghissassi, F., Benbrahim-Tallaa, L., Guha, N., Scoccianti, C., Mattock, H. & Straif, K. (2015) Carcinogenicity of tetrachlorvinphos, parathion, malathion, diazinon, and glyphosate. *The Lancet Oncology*, 16, 490–491, DOI: 10.1016/S1470-2045(15)70134-8.

Hartmann, H. T., Kester, D. E., Davies, F. T., Jr. & Geneve, R. L. (2002) *Plant propagation: Principles and practices*, 7th edn. Prentice Hall, Upper Saddle River, NJ.

Heap, I. (2018) The International Survey of Herbicide Resistant Weeds, available at: www.weedscience.org (accessed May 24, 2018).

Howard, P. (2009) Visualizing consolidation in the global seed industry: 1996–2008. *Sustainability*, 1, 1266–1287.

Huang, H. T. & Yang, P. (1987) The ancient cultured citrus ant. *BioScience*, 37, 665–671, DOI: 10.2307/1310713.

IARC (International Agency for Research on Cancer) (2003) *Fruits and vegetables*, Vol. 8. *IARC Handbooks of Cancer Prevention*, IARC, Lyon, France.

Isbell, F., Adler, P. R., Eisenhauer, N., Fornara, D., Kimmel, K., Kremen, C., Letourneau, D. K., Liebman, M., Polley, H. W., Quijas, S., et al. (2017) Benefits of increasing plant diversity in sustainable agroecosystems. *Journal of Ecology*, 105, 871–879, DOI: doi:10.1111/1365-2745.12789.

Isman, M. B. (2006) Botanical insecticides, deterrents, and repellents in modern agriculture and an increasingly regulated world. *Annual Review of Entomology*, 51, 45–66, DOI: doi:10.1146/annurev.ento.51.110104.151146.

Jones, R. A. & Barbetti, M. J. (2012) Influence of climate change on plant disease infections and epidemics caused by viruses and bacteria. *Plant Sciences Reviews*, 22, 1–31.

Klein, H. D. & Wenner, A. M. (2001) *Tiny game hunting: Environmentally healthy ways to trap and kill the pests in your house and garden*, New edn. University of California Press, Berkeley, CA.

Kokalis-Burelle, N. (2015) *Pasteuria penetrans* for control of *Meloidogyne incognita* on tomato and cucumber, and *M. arenaria* on snapdragon. *Journal of Nematology*, 47, 207.

Krause, R., Massie, L. & Hyre, R. (1975) Blitecast: A computerized forecast of potato late blight. *Plant Disease Reporter*, 59, 95–98.

Leke, W. N., Mignouna, D. B., Brown, J. K. & Kvarnheden, A. (2015) Begomovirus disease complex: Emerging threat to vegetable production systems of West and Central Africa. *Agriculture & Food Security*, 4, 1, DOI: 10.1186/s40066-014-0020-2.

Letourneau, D. K. & Bothwell, S. G. (2008) Comparison of organic and conventional farms: challenging ecologists to make biodiversity functional. *Frontiers in Ecology and the Environment*, 6, 430–438, DOI: 10.1890/070081.

Letourneau, D. K., Armbrecht, I., Rivera, B. S., Lerma, J. M., Carmona, E. J., Daza, M. C., Escobar, S., Galindo, V., Gutiérrez, C., López, S. D., et al. (2011) Does plant diversity benefit agroecosystems? A synthetic review. *Ecological Applications*, 21, 9–21, DOI: 10.1890/09-2026.1.

Letourneau, D. K., Allen, S. G. B. & Stireman, J. O. (2012) Perennial habitat fragments, parasitoid diversity and parasitism in ephemeral crops. *Journal of Applied Ecology*,49,1405–1416.DOI:10.1111/1365-2664.12001.

Letourneau, D. K., Allen, S. G. B., Kula, R. R., Sharkey, M. J. & Stireman, J. O. (2015) Habitat eradication and cropland intensification may reduce parasitoid diversity and natural pest control services in annual crop fields. *Elementa-Science of the Anthropocene*, 3, 000069, DOI: 10.12952/journal.elementa.000069.

Lichtenberg, E. M., Kennedy, C. M., Kremen, C., Batáry, P., Berendse, F., Bommarco, R., Bosque-Pérez, N. A., Carvalheiro, L. G., Snyder, W. E., Williams, N. M., et al. (2017) A global synthesis of the effects of diversified farming systems on arthropod diversity within fields and across agricultural landscapes. *Global Change Biology*, 23, 4946–4957, DOI: doi:10.1111/gcb.13714.

Linker, H. M., Orr, D. B. & Barbercheck, M. E. (2009) *Insect management on organic farms*. North Carolina Cooperative Extension Service, Raleigh, NC.

Lockwood, J. A. (1998) Management of orthopteran pests: A conservation perspective. *Journal of Insect Conservation*, 2, 253–261, DOI: 10.1023/a:1009699914515.

Lockwood, J. A., Bomar, C. R. & Ewen, A. B. (1999) The history of biological control with *Nosema locustae*: lessons for locust management. *Insect Science and Its Application*, 19, 333–350, DOI: 10.1017/S1742758400018968.

López-Lima, D., Sánchez-Nava, P., Carrión, G. & Núñez-Sánchez, A. E. (2013) 89% reduction of a potato cyst nematode population using biological control and rotation. *Agronomy for Sustainable Development*, 33, 425–431, DOI: 10.1007/s13593-012-0116-7.

Lowery, G. (2018) Genetically engineered eggplant improving lives in Bangladesh, available at: https://phys.org/news/2018-07-genetically-eggplant-bangladesh.html (accessed July 17, 2018).

Meyer, S. L. F., Everts, K. L., Gardener, B. M., Masler, E. P., Abdelnabby, H. M. E. & Skantar, A. M. (2016) Assessment of DAPG-producing *Pseudomonas fluorescens* for management of *Meloidogyne incognita* and *Fusarium oxysporum* on watermelon. *Journal of Nematology*, 48, 43.

Midega, C. A. O., Pittchar, J. O., Pickett, J. A., Hailu, G. W. & Khan, Z. R. (2018) A climate-adapted push-pull system effectively controls fall armyworm, *Spodoptera frugiperda* (J. E. Smith), in maize in East Africa. *Crop Protection*, 105, 10–15, DOI: https://doi.org/10.1016/j.cropro.2017.11.003.

Mitchell, C., Brennan, R. M., Graham, J. & Karley, A. J. (2016) Plant defense against herbivorous pests: exploiting resistance and tolerance traits for sustainable crop protection. *Frontiers in Plant Science*, 7, 1132, DOI: 10.3389/fpls.2016.01132.

Navas-Castillo, J., Fiallo-Olivé, E. & Sánchez-Campos, S. (2011) Emerging virus diseases transmitted by whiteflies. *Annual Review of Phytopathology*, 49, 219–248, DOI: doi:10.1146/annurev-phyto-072910-095235.

NRC (2000) *Genetically Modified Pest-Protected Plants: Science and regulation*, National Academy Press, Washington, DC.

Obrycki, J. J., Krafsur, E. S., Bogran, C. E., Gomez, L. E. & Cave, R. E. (2001) Comparative studies of three populations of the lady beetle predator *Hippodamia convergens* (Coleoptera: Coccinellidae). *Florida Entomologist*, 55–62.

Oneto, S. (2015) Soap sprays as insecticides. UCANR, available at: http://ucanr.edu/blogs/blogcore/postdetail.cfm?postnum=18009 (accessed May 26, 2018).

Parker, J., Miles, C. & Murray, T. (2012) *Organic management of flea beetles. A Pacific Northwest Extension publication, PNW640*.Washington State University Extension, Oregon State University Extension Service, University of Idaho Cooperative Extension system, available at: http://cru.cahe.wsu.edu/CEPublications/PNW640/PNW640.pdf (accessed Oct. 16, 2018).

Parker, J. E., Snyder, W. E., Hamilton, G. C. & Rodriguez-Saona, C. (2013) Companion planting and insect pest control. In: Soloneski, S. (ed.) *Weed and pest control: Conventional and new challenges*, 1–29. Intech Open, DOI: 10.5772/50276.

Parolin, P., Bresch, C., Desneux, N., Brun, R., Bout, A., Boll, R. & Poncet, C. (2012) Secondary plants used in biological control: A review. *International Journal of Pest Management*, 58, 91–100.

Pawelek, J., Frankie, G. W., Thorp, R. W. & Przybylski, M. (2009) Modification of a community garden to attract native bee pollinators in urban San Luis Obispo, California. *Cities and the Environment (CATE)*, 2, 7.

Pisani Gareau, T. L., Letourneau, D. K. & Shennan, C. (2013) Relative densities of natural enemy and pest insects within California hedgerows. *Environmental Entomology*, 42, 688–702, DOI: 10.1603/EN12317.

Poveda, K., Gómez Jiménez, M. I. & Kessler, A. (2010) The enemy as ally: Herbivore-induced increase in crop yield. *Ecological Applications*, 20, 1787–1793, DOI: 10.1890/09-1726.1.

Ramadugu, C., Keremane, M. L., Halbert, S. E., Duan, Y. P., Roose, M. L., Stover, E. & Lee, R. F. (2016) Long-term field evaluation reveals huanglongbing resistance in citrus relatives. *Plant Disease*, 100, 1858–1869, DOI: 10.1094/PDIS-03-16-0271-RE.

Saeed, S. T. & Samad, A. (2017) Emerging threats of begomoviruses to the cultivation of medicinal and aromatic crops and their management strategies. *VirusDisease*, 28, 1–17, DOI: 10.1007/s13337-016-0358-0.

Salmon, T. P. & Baldwin, R. A. (2009) Pocket gophers: Integrated pest management for home gardeners and landscape professionals. *Pest Notes, Publication 7433*. University of California, Statewide Integrated Pest Management Program, Agriculture and Natural Resources, CA, available at: http://www.ipm.ucdavis.edu/PMG/PESTNOTES/pn7433.html (accessed Oct. 11, 2018).

Schlau, J. (2010) Grasshopper management. University of Arizona Cooperative Extension, Yavapai County, available at: https://cals.arizona.edu/yavapai/anr/hort/byg/archive/grasshoppers2010.html (accessed Aug. 3, 2018).

Schumann, G. L. & D'Arcy, C. J. (2010) *Essential plant pathology*, 2nd edn. American Phytopathological Society, St Paul, MN.

Schuster, D. J. (2004) Squash as a trap crop to protect tomato from whitefly-vectored tomato yellow leaf curl. *International Journal of Pest Management*, 50, 281–284, DOI: 10.1080/09670870412331284591.

Sealy, R., Evans, M. R. & Rothrock, C. (2007) The effect of a garlic extract and root substrate on soilborne fungal pathogens. *HortTechnology*, 17, 169–173.

Seminis (2018) Seminis, fresh market seeds. Seminis, St Louis, MO, available at: http://www.seminis-us.com/products/results/categories/fresh-market/crops/all (accessed Aug. 14, 2018).

Slusarenko, A., Patel, A. & Portz, D. (10) (2008) Control of plant diseases by natural products: Allicin from garlic as a case study. In: Collinge, D., Munk, L. & Cooke, B. M. (eds) *Sustainable disease management in a European context*, 313–322. Springer, Netherlands, DOI: 10.1007/978-1-4020-8780-6_10.

St. Martin, C. C. G. (2015) Enhancing soil suppressiveness using compost and compost tea. In Meghvansi, M. K. & Varma, A. (eds) *Organic amendments and soil suppressiveness in plant disease management*, 25–49. Springer, Netherlands.

Stacey, D. L. (1983) The effect of artificial defoliation on the yield of tomato plants and its relevance to pest damage. *Journal of Horticultural Science*, 58, 117–120, DOI: 10.1080/00221589.1983.11515098.

Stallman, H. R. & James, H. S., Jr (2015) Determinants affecting farmers' willingness to cooperate to control pests. *Ecological Economics*, 117, 182–192, DOI: http://dx.doi.org/10.1016/j.ecolecon.2015.07.006.

Stapleton, J. J., Wilen, C. A. & Molinar, R. H. (2008) Soil solarization for gardens and landscapes. *Pest Notes*. Davis, CA: UC Statewide IPM Program, University of California, CA, available at: http://ipm.ucanr.edu/PMG/PESTNOTES/pn74145.html (accessed Oct. 16, 2018).

Stieha, C. & Poveda, K. (2015) Tolerance responses to herbivory: implications for future management strategies in potato. *Annals of Applied Biology*, 166, 208–217. DOI: 10.1111/aab.12174.

Stone, A. (2014) Organic management of late blight of potato and tomato (*Phytophthora infestans*), eXtension. Available at: http://articles.extension.org/pages/18361/organic-management-of-late-blight-of-potato-and-tomato-phytophthora-infestans (accessed June 21, 2018).

Styrsky, J. D. & Eubanks, M. D. (2007) Ecological consequences of interactions between ants and honeydew-producing insects. *Proceedings of the Royal Society B-Biological Sciences*, 274, 151–164, DOI: 10.1098/rspb.2006.3701.

Styrsky, J. D. & Eubanks, M. D. (2010) A facultative mutualism between aphids and an invasive ant increases plant reproduction. *Ecological Entomology*, 35, 190–199, DOI: 10.1111/j.1365-2311.2009.01172.x.

Summers, C., Mitchell, J. & Stapleton, J. (2005) Mulches reduce aphid-borne viruses and whiteflies in cantaloupe. *California Agriculture*, 59, 90–94.

Tabashnik, B. E., Brevault, T. & Carriere, Y. (2013) Insect resistance to *Bt* crops: Lessons from the first billion acres. *Nature Biotechnology*, 31, 510–521, DOI: 10.1038/nbt.2597.

Tesio, F., Weston, L. A., Vidotto, F. & Ferrero, A. (2010) Potential allelopathic effects of Jerusalem artichoke (*Helianthus tuberosus*) leaf tissues. *Weed Technology*, 24, 378–385, DOI: 10.1614/wt-d-09-00065.1.

Tooker, J. F. & Frank, S. D. (2012) Genotypically diverse cultivar mixtures for insect pest management and increased crop yields. *Journal of Applied Ecology*, 49, 974–985, DOI: 10.1111/j.1365-2664.2012.02173.x.

Trumble, J. T. & Butler, C. D. (2009) Climate change will exacerbate California's insect pest problems. *California Agriculture* 63, 73–78, DOI: 10.3733/ca.v063n02p73.

Trutmann, P., Voss, J. & Fairhead, J. (1993) Management of common bean diseases by farmers in the central African highlands. *International Journal of Pest Management*, 39, 334–342.

UCANR (2018a) *Polyphagous shot hole borer*. Division of Agriculture and Natural Resources, University of California Los Angeles, CA. Available at: https://ucanr.edu/sites/pshb/overview/About_PSHB/ (accessed Dec. 6, 2018).

UCANR (2018b) UC Statewide Integrated Pest Management Program, available at: http://ipm.ucanr.edu/index.html (accessed May 25, 2018).

USDA APHIS (2018) Biotechnology regulatory services: petitions for determination of nonregulated status. USDA, available at: https://www.aphis.usda.gov/aphis/ourfocus/biotechnology/permits-notifications-petitions/petitions/petition-status (accessed Aug. 3, 2018).

Van Dam, N. M., De Jong, T. J., Iwasa, Y. & Kubo, T. (1996) Optimal distribution of defences: Are plants smart investors? *Functional Ecology*, 10, 128–136, DOI: 10.2307/2390271.

Van Den Bosch, R., Messenger, P. S. & Gutierrez, A. P. (1982) *An introduction to biological control*. Plenum Press, New York, NY.

Waterfield, N. R., Wren, B. W. & Ffrench-Constant, R. H. (2004) Invertebrates as a source of emerging human pathogens. *Nature Reviews Microbiology*, 2, 833–841, DOI: 10.1038/nrmicro1008.

Xiong, J., Zhou, Q., Luo, H., Xia, L., Li, L., Sun, M. & Yu, Z. (2015) Systemic nematicidal activity and biocontrol efficacy of *Bacillus firmus* against the root-knot nematode *Meloidogyne incognita*. *World Journal of*

Microbiology and Biotechnology, 31, 661–667, DOI: 10.1007/s11274-015-1820-7.

Zhu, Y., Chen, H. R., Fan, J. H., Wang, Y. Y., Li, Y., Chen, J. B., Fan, J. X., Yang, S. S., Hu, L. P., Leung, H., et al. (2000) Genetic diversity and disease control in rice. *Nature*, 406, 718–722.

Zhu, Y., Wang, Y., Chen, H. & Lu, B.-R. (2003) Conserving traditional rice varieties through management for crop diversity. *BioScience*, 53, 158–162.

Ziska, L., Blumenthal, D., Runion, G. B., Hunt, E. R., Jr. & Diaz-Soltero, H. (2011) Invasive species and climate change: An agronomic perspective. *Climatic Change*, 105, 13–42, DOI: 10.1007/s10584-010-9879-5.

Ziter, C., Robinson, E. A. & Newman, J. A. (2012) Climate change and voltinism in Californian insect pest species: Sensitivity to location, scenario and climate model choice. *Global Change Biology*, 18, 2771–2780, DOI: 10.1111/j.1365-2486.2012.02748.x.

10 Saving Seeds for Planting and Sharing

D. Soleri, S.E. Smith, D.A. Cleveland

Chapter 10 in a nutshell.

- Seeds are the product of sexual reproduction, a process that includes pollination and fertilization, and can result in new genetic combinations and phenotypic diversity.
- Diversity at many levels in the garden can increase response capacity and reduce vulnerability.
- Maximum genetic diversity is not the goal, however, because too much diversity undermines adaptation and can be difficult to manage.
- Genetic diversity is shaped by four processes: mutation, gene flow, selection, and genetic drift; all except mutation are affected by gardeners.
- Different seed systems have different effects on diversity and access to seeds.
- Informal seed systems move genetic diversity to new environments and communities through personal social networks.
- Semi-formal seed systems distribute diversity through broader, non-commercial, less personal networks.
- Formal seed systems are commercially-based and the largest ones are dominated by multinational corporations that don't prioritize the environmental and social goals of gardeners.
- Saving and sharing seeds can save money, as well as select plant types that do best in local gardens. The effect on adaptation and diversity depends on how seed saving is done, on the crop, and on gardeners' goals.
- Knowing a crop's type of flower and pollination helps gardeners ensure seed production and control pollination, if that's desired.
- Saving seeds from many genetically different plants of a variety keeps the seed stock genetically diverse and healthy.
- Selecting mature, healthy seeds and storing them at low a temperature and humidity protects seed viability.

We have been planting *papaloquelite* (*Porophyllum ruderale*, in the sunflower family) seeds in our garden. The plants are beautiful and the fragrant leaves have become a favorite fresh herb. DS received the seeds from friends who are maize farmers in Oaxaca, Mexico. They had been growing *papaloquelite* for only a few years, from seeds their niece was given by her colleague from Michoacán when both of them were working as teachers in Vera Cruz. We planted these seeds in our garden, and harvested lots of seeds from the plants they produced. That same fall in Santa Barbara, DS first met Ángel, a gardener from Jalisco, Mexico, who has saved seeds of a *papaloquelite* variety originally from Jalisco, as well as a different variety—with a purple tinge to its leaves and flowers, and a distinct flavor—that he got from a friend he met in California, who is also from Mexico. Ángel grows a large patch of *papaloquelite* every year and enthusiastically describes having a heaping plate of the fresh greens on his table each evening during the growing season, how healthy it makes him feel to eat them, and how it reminds him of his childhood home. DS brought Ángel some of her seeds from Oaxaca, and he gave her some of his, so now they each grow three *papaloquelite* varieties, each with different histories and paths to their gardens in southern California (Fig. 10.1). Gardeners have been familiar with stories like this for a long time. Seeds are constantly being grown, saved and shared, and the plants that grow from them, and the flavors of foods made from them, are meaningful connections to other people, places, and times.

10.1. A Long Tradition of Seed Saving

We know that humans have been saving seeds of wild plants to eat since before the beginning of agriculture, because of evidence like the vast stores

---

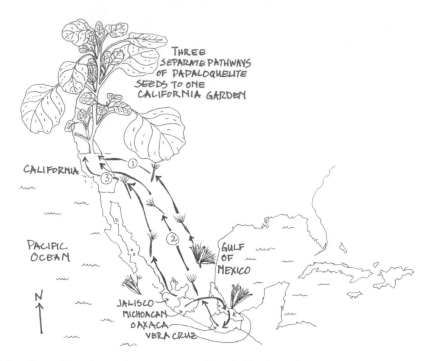

Figure 10.1. Map of *papaloquelite* seed routes to a southern California garden

Note: This seed exchange occurred many years ago; we recognize the purpose and significance of phytosanitary laws and standards (Section 10.7.2).

of wild barley and wild wheat grains found in archeological sites in the Middle East. Sometimes the saved seeds were later planted at a time or place more favorable for the plants, or for the people. For example, more than 7,000 years ago, when Olmec people migrated to the Gulf Coast from what is now central Tabasco, Mexico, they carried with them seeds of teosinte (*Zea* species), the wild ancestor of maize. In their new home they planted the seeds in favorable areas and harvested the grains (Pope et al., 2001). The same process likely occurred with many seeds, cuttings, and other *propagules* (plant parts capable of producing a new plant), such as olive seeds and cuttings brought from the eastern to the western regions of the Mediterranean Basin.

The domestication of food plants is an evolutionary process: genetic changes that adapt plants to human use through selection by people and by the environments they create in their gardens, fields, kitchens, and storerooms (Box 10.1). Evidence for food plant domestication is found almost everywhere there are people and plants, beginning over 12,000 years ago with the advent of agriculture (Fig. 10.2) (Larson et al., 2014). This is not surprising,

because working with plants, especially if you depend on them for food, leads to two observations: new plants grow from seeds or other propagules, and new plants resemble their parents, though they are rarely identical.

Virtually all of the food crops grown today are products of human selection acting on genetic diversity, and new varieties created by scientists, gardeners, and farmers are built on the work of previous gardeners and farmers. Cultivation, selection, and seed saving by gardeners and farmers continue to be key for conserving the genetic diversity we all rely on as the basis for adaptation to changing physical, social, and cultural environments, and for producing crop varieties suited to local needs. Selecting, saving, and sharing your own seeds is often the only way to maintain plant species or varieties that are only found locally, including ones with great personal or cultural significance, like Ángel's *papaloquelite*.

Saving and sharing seed is also a statement of independence from the centralized and corporatized global seed system that operates largely without regard for improving the health of people, communities, economies, or ecosystems, goals that many of us see as critical to building a positive future. In that sense, seed saving, or supporting open seed systems in other ways, is a political act (Kloppenburg, 2014; Soleri, 2017).

Most seed saving is relatively easy, and many gardeners already know how to do it. But the key to getting the most out of saving and sharing garden seeds is understanding how these affect diversity and adaptation, and what that means to us, and our communities. This chapter focuses on concepts that can make your seed saving and sharing more effective and consistent with your goals and values, and on practical suggestions for using those concepts.

10.2. Biological Diversity in the Garden

Diversity may be present at many levels in food gardens, including among the crops being grown (Fig. 10.3). Within the plants of a given variety there are thousands of individual genes for different traits. Genes are located on chromosomes and are coded for by the chemical sequence present within DNA (deoxyribonucleic acid) molecules. There is a copy of each gene, called an *allele*, on each of the (usually two) chromosomes in a set, and those alleles can be the same or different (Box 10.2). All the allelic combinations in an individual plant are its *genotype*, which together with the environment, produce a plant's *phenotype*, which is all of the characteristics of that plant, many of which are observable, like flower color, fruit yield, leaf aroma, or response to drought. There can be many genotypically distinct varieties within a crop species, and within a variety the genotypes in different populations and seed lots may also differ, all the result of the processes that shape diversity (Section 10.3).

Diversity in your food garden is part of the network of crop genetic diversity that extends through space to other gardens and green areas, both near and far, such as the many North American gardens

---

Box 10.1. Crop domestication

Crop domestication is human management of plant evolution. Through intentional and unintentional selection, plants are adapted to human needs, preferences, and practices. *Intentional selection* is the conscious identification of individual plants that we want contributing to the next generation of plants. *Unintentional selection* is the result of practices that are not meant to create selection pressures. For example, when early gatherers harvested seeds from wild bean plants, they unintentionally favored plants with pods that were less likely to burst open, or *dehisce* and scatter the seeds, which is essential for many wild plants to survive. These less-common plants were *indehiscent*, their mature pods did not burst, making their seeds easier to gather. People were unintentionally selecting for indehiscence.

Through selection in wild species, domesticated species evolved with genetic changes that gave them more characteristics that people like. It also made many domesticated plants more dependent on humans to plant their seeds, and to create the environments where they could grow, including removing competing plants, and providing fertile soils, sufficient water, and protection from pests like some insects and other animals. For example, compared with their wild ancestors, domesticated cereals like wheat, sorghum, and maize tend to have shorter seed dormancy periods (0–6 months) (Section 6.1.2)

and higher germination percentages (Section 6.1.1) that drop rapidly as their seeds age. This is because the cereals were selected for immediate germination and growth under relatively favorable conditions. Seeds of the wild plants they evolved from often have more stringent moisture and temperature requirements for germination, longer periods of seed dormancy, and staggered germination, all of which reduce the risk that all seeds germinate and then die in poor environmental conditions. Other characteristics often associated with the domestication of food plants include changes in biochemical composition that make crops less toxic and more flavorful, and changes in plant growth and form that make them produce more of the parts we like to eat, including larger fruits or seeds (Meyer et al., 2012). Again, often these changes also make them more attractive to insects and other animals.

Overall, crop domestication is the human innovation that has had the most far-reaching impact on human beings and the Earth. It made agriculture possible, which in turn required increasing management of the environment, and supported exponential growth in human populations and the flourishing of cultures and technologies. But agriculture has also had many negative consequences for our individual health, social equity, other living things, and the environment (see Chapter 2) (Cleveland, 2014).

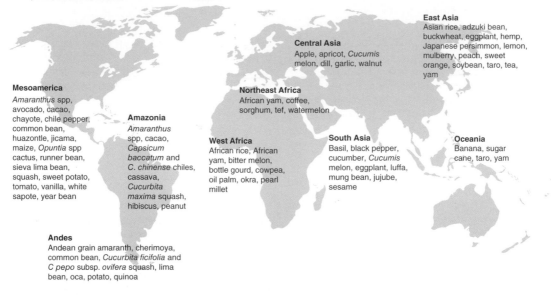

Eastern North America
American persimmon, cranberry, *Cucurbita pepo* squash, high bush blueberry, hops, Jerusalem artichoke, pecan, sunflower, wild rice

Mediterranean, Southwest Asia
Almond, apricot, artichoke, asparagus, barley, beet, cabbage group, caper, carob, carrot, celery, chard, cilantro, date, endive, fava bean, fig, garbanzo, grape, kale, lentil, lettuce, mint, oat, olive, parsley, pea, pistachio, pomegranate, rye, spinach, wheat

East Asia
Asian rice, adzuki bean, buckwheat, eggplant, hemp, Japanese persimmon, lemon, mulberry, peach, sweet orange, soybean, taro, tea, yam

Central Asia
Apple, apricot, *Cucumis* melon, dill, garlic, walnut

Mesoamerica
Amaranthus spp, avocado, cacao, chayote, chile pepper, common bean, huazontle, jicama, maize, *Opuntia* spp cactus, runner bean, sieva lima bean, squash, sweet potato, tomato, vanilla, white sapote, year bean

Amazonia
Amaranthus spp, cacao, *Capsicum baccatum* and *C. chinense* chiles, cassava, *Cucurbita maxima* squash, hibiscus, peanut

Northeast Africa
African yam, coffee, sorghum, tef, watermelon

West Africa
African rice, African yam, bitter melon, bottle gourd, cowpea, oil palm, okra, pearl millet

South Asia
Basil, black pepper, cucumber, *Cucumis* melon, eggplant, luffa, mung bean, jujube, sesame

Oceania
Banana, sugar cane, taro, yam

Andes
Andean grain amaranth, cherimoya, common bean, *Cucurbita ficifolia* and *C pepo* subsp. *ovifera* squash, lima bean, oca, potato, quinoa

Figure 10.2. Centers of domestication for some crops

Based in part on Gepts (2001); Doebley et al. (2006); Meyer et al. (2012); Larson et al. (2014). Base map, Wikimedia

contributing to current *papaloquelite* diversity, or gardens worldwide supporting diverse pollinators. Through choosing varieties, selecting, saving, and sharing seed, this diversity supports ongoing adaptation in response to changing conditions. Crop genetic diversity also extends through time from past to future generations of gardeners, and in formal seed or gene banks that conserve diversity for future use (Section 10.4.5), for example the "global" seed vault in Svalbarg, Norway.

Crop diversity in gardens provides a range of options for changing conditions such as anthropogenic climate change (ACC) or new food preferences (Foley et al., 2011). Still, the goal is not to have as much diversity as possible in your garden, because too much diversity at one time can decrease adaptation to garden conditions and gardener preferences (i.e., lower average performance, or less of a preferred flavor), and can be time consuming to maintain. Rice farmers in Nepal who grow local varieties in small terraced beds told DS they get rid of varieties with too much diversity, especially

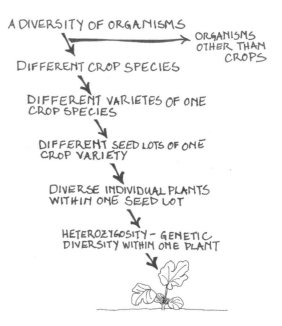

Figure 10.3. Levels of diversity in the garden

Box 10.2. Genetic diversity basics

A plant's genotype can be thought of as the genetic information in its alleles, half of which are inherited from the paternal (pollen) parent, and half from the maternal (ovule) parent (Section 10.5) (see also Fig. 5.4). We focus here on the genes located on chromosomes in the nuclei in a plant's cells, its *nuclear genotype*, but organelles outside of the nucleus—chloroplasts and mitochondria—also have genes and, therefore, genotypes. Together, all of the genes in the nucleus plus those in the organelles of an individual are its *genome*. The nucleus of each cell in a plant contains the same nuclear genotype (except the gametes and the special structures that produce them, as explained below, and in somatic mutations, Section 10.3.1). Genes contain the information—in the form of DNA sequences, or "codes"—that the plant needs to grow and reproduce. In each organism there is a base number of chromosomes, referred to here as a *set* and often represented as "*x*," for example in maize $x = 10$, and in common potato $x = 12$. In *somatic* cells, which are all plant cells except those associated with sexual reproduction, there are multiple sets, with diploid plants containing two sets ($2x$), for example maize with a total of 20 chromosomes, occurring in two sets of 10 chromosomes each. The copies of one chromosome in the sets are *sister chromosomes*. Common potato, a *tetraploid* ($4x$), has 4 sets of chromosomes in its somatic cells (indicated as $2n$), with a total of 48. However, only half of the total number of chromosomes is present in cells involved in sexual reproduction (indicated as *n*), including the gametes (Section 10.5). For example, two sets of 12 chromosomes each in potato pollen grains and the structures that will produce ovules (immature seeds), one set of 10 single chromosomes in these structures in maize. When the male and female

gametes are united at fertilization, they form an embryo that again has the same number of sets as other somatic cells, half from the male gamete, and half from the female gamete. This embryo will develop within the seed, and once germination occurs can grow into a plant.

The number of different alleles is often used as a gauge of diversity in individual plants and varieties. As gardeners we don't see alleles, but we do see phenotypes produced by different alleles and allelic combinations, and we see different populations of phenotypes resulting from the different allele frequencies. In diploids, each chromosome in a set has one allele for each gene, so individual plants have a maximum of two different alleles for each gene. However, within all the plants of a crop population, variety, or species, there can be many alleles for one gene, with individual plants in the population having different combinations of these alleles, which means that there can be a very large amount of genetic diversity. Genes that have only one allele are considered *homozygous*. Genes that are *heterozygous* have different alleles, two in diploids. A *homogeneous* population, variety, or species is one in which the plants have the same allelic combination for many genes, that is, they have the same genotype. A *heterogeneous* population, variety, or species is one in which there are diverse genotypes, with individual plants having different alleles for the genes we are interested in.

Estimating the number of possible different genotypes that can be produced in a new generation based on the number of genes with two or more alleles gives us a feeling for the incredible power sexual reproduction has to create genotypes with new allelic combinations (Table 10.1). Since so

Table 10.1. Estimating the number of possible different genotypes in a population of diploid plants produced by 1 to 5, and 10 heterozygous genes, each on a different chromosome (they are not linked)

Number of heterozygous unlinked genes in parent generation	Number of different genotypes possible in progeny generation when number of alleles per gene is:		
	2	4	8
n	3^n	10^n	36^n
1	3	10	36
2	9	100	1,296
3	27	1,000	46,656
4	81	10,000	1,679,616
5	243	100,000	60,466,176
10	59,049	10,000,000,000	3,656,158,440,062,980

Continued

Box 10.2. Continued.

much is known about maize genetics, we'll use that crop to illustrate. As already mentioned, maize is a diploid (two sets of chromosomes) with 10 base chromosomes and an average of 3,916 genes per chromosome, nearly 40,000 genes identified to date, and an average of eight alleles per gene. Many of the genes don't have a phenotypic effect we would notice or care about in our garden. Even if only *one* of the almost 4,000 genes on each of maize's 10 chromosomes was heterozygous, and each of those 10 genes had only two alleles, there would still be an incredible 59,049 different genotypes possible, and with the average eight alleles per gene it would be more than 3,000 trillion! Many crops have a large number of known genes: tomato, >34,000; potato, >32,000; Chinese cabbage, >41,000.

The many different possible genotypes provides the crop genetic diversity that selection acts on to create new populations, varieties or species, that can support response capacity (Chapter 3) in our gardens or fields.

As gardeners, we benefit from this enormous potential diversity when selection increases the frequency of genotypes that we like. However, many important traits like yield or drought adaptation are the product of many alleles for many genes, and the challenge of getting a phenotype you like is having the right combinations of alleles across many genes. This is why genotypes and populations with complexes of alleles and allelic frequencies that produce the phenotypes we want are so valuable.

those with diverse phenologies (timing of life cycle stages) because those varieties create more work, making tasks like irrigation, pest management, and harvesting go on much longer than with more uniform varieties. As gardeners we need to maintain, or be able to easily access, enough diversity to respond to familiar variation, or to larger trends over time. On the other hand, there are some phenotypes, varieties, and species we like a lot better than others, and some currently do much better in our gardens than others, for example they may be more heat adapted. So, it's a tradeoff between current biological adaptation and response capacity for the future (Box 10.3), but farmers and gardeners have found ways to organize gardens and seed systems that minimize the tradeoff (Section 10.4.5).

10.3. Processes Shaping Garden Crop Genetic Diversity

If you have ever grown squash from seeds you saved, you may have found that the fruit was a very different shape or color than the fruit you got the seed from. If you save seeds from the last eggplant to set fruit at the end of the summer season, you may find the plants grown the next year bear fruit later than you expect. Or maybe you've saved seed from a randomly chosen maize ear of an *open pollinated variety* (OPV, one that "breeds true to type," meaning progeny phenotypes are generally similar to parental ones, even without controlled pollination), and found the plants and ears

those seeds produced were different than the maize that you liked and saved seed from. These experiences are the result of the processes that shape genetic diversity (gene flow, selection, genetic drift in these examples). By "shape" we mean how much and which genetic diversity is present.

There are four processes that shape the genetic diversity in garden crops, and all living organisms. Being aware of these processes gives you a new way to see and manage what's going on in your garden and seed stocks, in order to keep varieties healthy and vigorous, and retain characteristics important to you. Two of these processes—mutation and gene flow into your crop population—usually increase diversity in the garden by adding new alleles; the other two—selection and genetic drift—typically decrease diversity by removing alleles. These processes result in *microevolutionary* changes in garden crops, that is, changes in frequencies of alleles and of genotypes over time. Over long periods these same processes can lead to *macroevolutionary* change, what is most often simply called "evolution," which can result in new species.

While genetic structures and processes are complex, you don't have to be a trained geneticist to be a good seed selector and saver. In fact, many of the most skilled seed selectors and savers are traditional gardeners and farmers around the world who are especially curious and observant, and have a basic feeling for these processes, even though most have little formal education. Studies with bean farmers in Rwanda and barley farmers in Syria, for example, have shown that they are better

Box 10.3. Adaptation

Biological adaptation is the result of selection acting on diversity, which over generations results in organisms with increased *fitness*, or reproductive success (Section 10.3.3). This is the core process of evolution. Because selection exerted by people and environments on crops is always changing, crop adaptation is an ongoing, dynamic process, resulting in crop populations, varieties, and eventually perhaps even species adapted to the climate, soils, insects, other plants, and the people where they have been grown and selected over time. When we intentionally and unintentionally select crop plants we are defining what adaptation is. So if you are a gardener interested in selection it is worth thinking about how you define adaptation.

What evolution by natural selection – genetic change as a result of more fit individuals leaving more offspring – does *not* do is adapt crops, or any organism, to novel, future conditions such as a changed climate, or new food preferences (Chapter 2). While a local crop variety may contain lots of diversity (see Table 10.1), it may not be the right combination or type of diversity needed to produce adaptations, especially given the rapid rates of change we are experiencing. That's why gene flow, the movement of alleles, populations, and species to new locations can be valuable (Section 10.3.2). A good example of this is the movement of sorghum varieties from north to south in West Africa. As early as the 1930s farmers in southern Mali noticed the climate was changing, getting drier with more variable rains during the sorghum growing season. In response, southern farmers went to northern Mali, closer to the Sahara Desert, where growing seasons have always been shorter and drier than in the south, and acquired seeds of sorghum varieties grown by farmers there, and took those home to plant. This north to south gene flow allows southern farmers to adapt to the new climate they are experiencing. Today they use both their original, longer-season sorghum varieties as well as the recently acquired shorter season ones, increasing diversity and reducing their vulnerability in the face of ACC (Lacy et al., 2006). Similarly, in the US, climate warming is resulting in a northern movement of commercial maize production, and in California some fruit varieties, and even certain species, need to be replaced by ones with shorter chilling requirements as the climate warms (Luedeling et al., 2009) (Section 5.8.2). More and more people and institutions are looking for varieties, crop species, and practices from areas that have typically had climates similar to what theirs is becoming (Jarvis et al., 2014).

Adaptation, including in garden crops, is often described as specific or wide. Selection in one homogeneous environment may result in genotypes or populations that are very well adapted only to that particular set of environmental variables (e.g., the climate, soil, organisms, and management practices in gardens in a specific location). But these *specifically adapted* genotypes may perform poorly in environments other than the one where they were selected. In contrast, genotypes selected for general or *wide adaptation*, that is the best average performance across a range of different environments, may do poorly in any specific environment compared with genotypes selected just for that environment (see Appendix 3A, Fig. 3A.4, Approach 2ii). Across different low-input gardens where conditions can be variable in space, and especially over time, generalist genotypes adapted to that variation can be valuable. While their performance may not be the best in any one environment and year, when there is lots of variation a generalist's overall performance is better than that of a more specifically adapted genotype. Genotypes with wide adaptation may become more important as ACC makes growing environments more challenging and variable, or if locally specific varieties are not being developed.

An important aspect of adaptation for generalist genotypes is the ability to respond to environmental variation via change in phenotype, a characteristic known as *phenotypic plasticity*. For example, in both domesticated cassava and its wild relatives, a single genotype will grow as a bushy shrub in an open growing environment, but as a climbing vine in a wooded environment (Ménard et al., 2013). Different lima bean genotypes show varying degrees of phenotypic plasticity for production of the toxin hydrogen cyanide, or its biochemical precursors, in their leaves, depending on the kind of pests damaging them (Ballhorn et al., 2006). Spider mites (sucking arthropods) cause greater production of the toxin in lima bean than bean beetles (chewing arthropods), and plants have multiple hydrogen cyanide phenotypes, but only one genotype. These phenotypes are a result of an individual plant's environment, so selection for one phenotype vs. another would not have any genetic effect beyond selecting for phenotypic plasticity itself. In these examples genotypes with phenotypic plasticity adapt to different environments in ways that help to maintain fitness.

A qualitative *genotype by environment interaction* (GEI) occurs when differences in phenotype between two or more genotypes (or varieties) in response to different environments results in a change in rank among them for a particular characteristic, such as yield. Barley breeder Salvatore Ceccarelli drew

Continued

Box 10.3. Continued.

attention to GEI in yields among barley varieties grown in the fertilized, high rainfall fields of experimental stations compared to unfertilized, low rainfall farmers' fields in Syria (Fig. 10.4a). Modern varieties had higher yield than farmers' *landraces* (traditional, local varieties) in better environments, but lower yields in stressful environments, showing that those landraces were more productive for farmers in their own fields.

The same qualitative GEI was observed among cowpea varieties grown with and without water stress in Maryland (Fig. 10.4b). The variety 'Quickpick Pinkeye' had higher yield under all levels of water stress except the most extreme, so would be the best choice for most gardeners if yield was the main criterion for choosing. However, if you wanted to grow either 'Whiteacre' or 'Two Crop Brown' because of other criteria, like phenology, color, or flavor, in addition to yield, then the choice would depend on whether water stress would not be a problem (Whiteacre), or if water stress was likely to occur (Two Crop Brown). There is also a general discussion of GEI in terms of water stress in Section 8.4.

These examples show the value of diversity for gardeners, and for adapting to new conditions. Using crop diversity for adaptation means working with the four processes that shape the diversity in our garden crops, and all living organisms. Whether you only want to save seeds for next year, or become an amateur plant breeder, a basic understanding of these processes will help you achieve your seed saving goals.

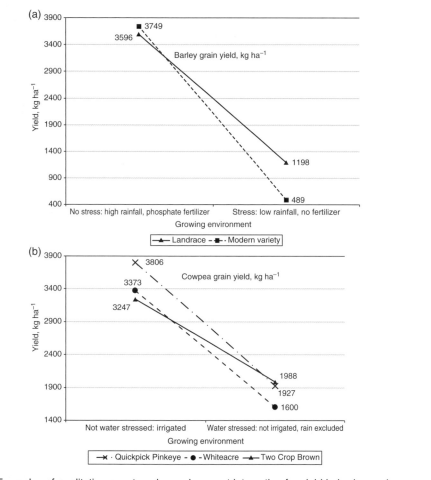

Figure 10.4. Examples of qualitative genotype by environment interaction for yield in barley and cowpea a) Ceccarelli (1996, Table 8); b) Dadson et al. (2005, Table 3)

than formally trained scientists at identifying plants that will perform well in their fields (Sperling et al., 1993; Ceccarelli and Grando, 2009). Through their knowledge gained from experience and reflection, these farmers are capable of consciously and unconsciously integrating many characteristics that they observe in the plants to identify the best ones for their needs, an example of the combination of informal science and art in agriculture. Because most gardeners in the US lack those farmers' years of experience, understanding some of the science makes for more effective seed savers, and seed activists. In the next four sections we describe these processes and their roles in food gardening.

10.3.1. Mutation

A *mutation* is a permanent change in DNA molecules or chromosomes, and is the ultimate source of all genetic diversity. There are many causes of mutations, and mutations can take many forms. Mutations can be caused by environmental factors like ultraviolet radiation, and if they occur in reproductive cells or tissues that produce them, those mutations can be inherited during sexual reproduction, producing new genotypes and perhaps new phenotypes. Inducing heritable mutations by irradiating seeds was a technique used by plant breeders for a time, but is no longer popular because of its unpredictable results, and since the vast majority of mutations have negative effects on the phenotype.

Mutations that occur in somatic cells may also have effects on phenotype, but are not heritable. In perennials, somatic cell mutations are the source of *bud sports*, vegetative structures with distinct genotypes and phenotypes. A familiar example is the 'Washington Navel' orange, propagated from a bud sport on the 'Selecta' orange variety by Portuguese colonialists in Goa, India, then carried to the Portuguese colony of Brazil, and was later among samples sent to California in the 1870s (Boulé, 2016). Desirable somatic mutations can be propagated vegetatively (Section 6.7) by grafting the mutant tissue onto a known root stock. Mutation is the only one of the four diversity-shaping processes that gardeners can't control.

10.3.2. Gene flow

The movement and integration of genetic material into new populations or locations is *gene flow*, or

migration. Gene flow is now recognized as a significant force in plant evolution (Ellstrand, 2014), including in domesticated plants, and gardeners can play a major role in this process. The effect of gene flow on diversity is positive when it adds useful alleles to the population, and negative when it adds unwanted alleles that replace desirable ones. The movement of pollen between different populations or varieties, followed by successful fertilization, is a common source of crop gene flow, as is moving seeds of a variety to a new location where they establish a new population (see Fig. 10.1), or their genes become established in an existing population through pollen flow. Those unfamiliar squash shapes and flavors you may have seen in your garden are the result of gene flow via pollen between individuals of different varieties the previous generation. For gardeners, gene flow that introduces new varieties or species occurs when people migrate and bring seeds with them to their new home, borrow seeds from their local seed library, or import soil, compost, or manure that contain seeds. The extent of gene flow via pollen is also affected by how closely in space and time you and your neighbors grow different varieties of cross-pollinated species.

Unintentional gene flow in crops can introduce useful new alleles, or it can be unwanted because it undermines local adaptation and the characteristics gardeners prefer. This possibility entered public discussion in the late 1990s with the unintended flow of transgenes (Section 9.5.4) via pollen of transgenic varieties. For example, escaped populations of transgenic canola (*Brassica napus*) have had different transgenes for herbicide resistance move into them by gene flow through pollen from other fields. The result in the next generation included volunteer plants resistant to multiple herbicides, and so more difficult to control when they become weeds (Knispel et al., 2008). Pollen from transgenic plants can also pollinate non-transgenic plants, resulting in transgenic seeds. This can result in "contamination" of crops, and the loss of sales when certified organic harvests are suspected to contain transgenes, or leave gardeners and farmers unable to save seed of local varieties free of transgenes. Transgene flow can also have negative cultural effects, for example, when transgenes move into important food crops, as they have into traditional maize varieties in Mexico (Dyer et al., 2009). Many farmers, scientists, and communities in Mexico view this as a violation of their cultural heritage and political sovereignty (Esteva and Marielle, 2003; Sin maíz no hay país, 2018).

When you want gene flow, you can introduce new varieties or populations into your garden and allow cross-pollination with existing varieties and populations. You can also manage gene flow more directly, for example by using pollen from one maize variety to pollinate a different variety, or mixing arugula seeds from the local seed library with your own saved arugula when you plant. To avoid gene flow, limit the plants, seeds, or pollen of unwanted varieties or populations. For example, you can hand-pollinate plants you want to save seed from, and then cover the flowers to prevent the pollen of other plants from landing on their stigmas. Staggering the planting dates of varieties that can cross-pollinate also reduces the chances of gene flow. To control unintentional gene flow between neighboring gardens, gardeners could work together to plan crop selection and planting dates.

For gardeners growing small seed lots, managed gene flow can be a valuable tool for keeping varieties genetically "healthy"—diverse and capable of adaptation (Box 10.7). While gene flow usually increases total diversity present at a place and time, there are notable exceptions (see vertifolia effect below), related to the effect of the third diversity-shaping process, selection.

10.3.3. Selection

Selection is the identification of the parent plants whose genes will contribute (via pollen, seeds, or vegetative propagules) to the next generation, and the elimination of contributions from all others.

For selection to have an effect over generations, that is, to change the crop population genetically, there must be genetic diversity present in the first place, and selection that acts on that diversity. Selection is based on *phenotype*—discernable characteristic(s)—of a plant, fruit, or seed. If there is a genetic basis to differences in phenotype between plants that are selected and those that are not, then the selection criteria are *heritable*. When selection for a heritable phenotype is maintained over generations, allele frequencies in the population change and the frequency of the selected phenotype increases, it has greater fitness in that selection environment. Many, but not all, biologists (e.g., Endler, 1992) define selection as only including cases where there is a genetic basis, that is, they think of selection and the heritability of the characteristics being selected for together, but gardeners and farmers may not (Box 10.4).

When we save seed or other propagules for planting we can select intentionally for qualities we prefer, like taste or color, or time of ripening. Or we may save seed with no goal of changing the population genetically, but this can also change varieties through unintentional selection, as when early farmers favored seed from plants with indehiscent fruit because they stayed on the plant and so were easier to collect (Box 10.1).

Growing and seed storage environments can also be sources of unintentional selection. For example, when an unusually early first frost leaves only the shortest cycle winter squash plant with fruit mature

---

Box 10.4. Selection: it's not just about genes!

Seedlings growing from larger seeds or grains are generally more vigorous than those growing from smaller seeds with the same genotype. This is because larger seeds have more food to support growing embryos and seedlings before they establish their roots and start photosynthesizing. Farmers and gardeners around the world have known this for a very long time (Cleveland and Soleri, 2007). For example, barley farmers in northern Syria dedicate their best field, or part of a field, to growing plants they will select their seeds from. The seeds farmers sow in these chosen areas are exactly the same as those they plant in other fields, but the favorable environment with better soils and moisture content produces larger seeds. In Oaxaca, Mexico, maize farmers select the largest, healthiest kernels for

planting, because they know from their experience that large kernels have more reliable germination and more vigorous early season growth than smaller ones, and so have a greater chance of surviving. Experiments showed that the differences between selected and unselected kernels were not heritable (Soleri et al., 2000). Farmers already knew this, because in their fields they had observed that those differences in kernel size were primarily the result of the parent plant's growing environment.

It's an advantage to have plants with vigorous germination and early growth whether you are a barley farmer in Syria, maize farmer in Mexico, or home gardener in Australia. So it is always good to select larger, healthy seeds, whether that is your only selection criterion, or one among many.

---

enough to save seeds from. Or when a seed storage container gets hot and only the most heat-tolerant genotypes survive. Sometimes unintentional selection has no lasting effect. For example, if bagrada bugs attack the plants in a bed of kale early in the season, killing all but a few of the plants, selection occurred. If the survivors are genetically distinct, let's say they have alleles that make them produce more compounds in their tissues that the bugs avoided, compared with those that didn't survive, then the selection will change the allele frequency of that population. If there was no genetic difference for resistance to the bugs among the plants that died and those that survived, but survivors grew in a corner of the bed that was especially fertile and well watered, then their survival was neither adaptive nor heritable—it had no genetic basis, those were just lucky plants!

Intentional selection reduces genetic diversity in a useful way by eliminating alleles associated with phenotypes we don't want. But selection can also reduce diversity so much that a crop population cannot adapt to familiar variation or trends, or could even experience inbreeding depression (Section 10.3.4). An example of intentional selection reducing the ability of a crop to adapt was observed by potato breeder J. E. Van Der Plank in the 1960s (Van Der Plank, 1966). Van Der Plank identified resistance to fungal disease in potato that was the product of multiple genes with small effects on the phenotype, called *horizontal resistance* (HR). This contrasts with *vertical resistance* (VR), which is based on only a few genes that each have a major phenotypic effect. VR can be very effective, either killing or repelling most of a pest or pathogen population. But because it is so effective it creates strong pressure for the evolution of resistance in that pest or pathogen population. That is, by eliminating susceptible individual pests, VR selects *for* those individuals that are resistant to it, over time resulting in a pest population no longer controlled by VR. On the other hand, HR with multiple sources of minor control does not create strong selection pressure for the evolution of resistance in pests or pathogens. Van Der Plank noticed that when VR was introduced and selected for in a population, HR was lost (Robinson, 2009, 374). He called this the *vertifolia effect* because he observed it in a potato variety named 'Vertifolia' that was selected for VR to the fungal disease late blight, caused by *Phytophthora infestans*. As a result of selection for VR, the Vertifolia variety lost

HR for the disease, even though that HR had been shown in field evaluations to be effective and durable (Grünwald et al., 2002). When HR is lost in a plant population, that population is vulnerable to the evolution of resistance in the pest or pathogen controlled by VR. HR can also be lost when pesticides are used during plant breeding so alleles providing HR to pests are no longer selected for.

When the strong selective pressure exerted by genotypes with VR results in the rapid evolution of resistance in pests or pathogens, those genotypes must be replaced by new genotypes with different VR, a phenomenon known as the *breeding* or *varietal treadmill*. A similar phenomenon with selection exerted by powerful pesticides has resulted in the *pesticide treadmill* (Section 9.7). In contrast, plants with HR suffer some damage when attacked, but can often still produce good yields, and their polygenic defenses are more durable compared to VR (Van Der Plank, 1984; Robinson, 2009; Keane, 2012).

Maintaining HR is one reason some people favor using "*heirloom*" varieties, older OPVs that are not the product of modern plant breeding, because modern varieties may include VR for important pests or pathogens. However, heirlooms are not necessarily the best ones for all needs, or they may require improvement. Sometimes heirlooms have such low yields that they are not a viable option for small-scale farmers, a situation that's been called the "heirloom dilemma" (for examples, see Brouwer et al., 2016).

There is a tradeoff between reducing diversity to tailor a variety to meet your needs and maintaining enough diversity to adapt to ongoing changes. However, there are a number of simple strategies gardeners can use to address this tradeoff, including how selection is done (see Box 10.7).

10.3.4. Genetic drift

Imagine that every year at the end of the cool season you thresh all the seed heads from a large patch of cilantro plants to grind into coriander powder. If you only save a few seeds to plant, your seed lot will probably experience genetic drift. *Genetic drift* is the change in genetic diversity—most often a loss—that results from random reductions in population size. It is the result of selection that is random, that is, has nothing to do with phenotypes. Genetic drift can result in the loss of adaptation, genetic diversity, and the phenotypes you like, and of response capacity for the future, with rarer alleles especially likely to be lost. Populations of cross-pollinated species are most affected because they are more genetically diverse, and heterozygous, and rely on that diversity for creating a healthy population. Genetic drift can cause large decreases in that diversity, leading to populations with less heterozygosity and more inbreeding depression in subsequent generations. *Inbreeding depression* is the phenotypic expression of undesirable, recessive alleles normally hidden by more desirable dominant alleles, occurring especially in cross-pollinated species. Interestingly, some cross-pollinated garden crops like watermelon, cucumber, and melons generally display little inbreeding depression, and scientists speculate that because these are large, sprawling plants, early domestication may have been based on very small population sizes and so undesirable alleles were exposed and eliminated long ago (Wehner, 2008).

To avoid small population size whether from genetic drift or selection, especially in cross-pollinated species, you can save seeds from as many plants as possible, being sure to save about the same number of seeds from each plant; avoid seeds from populations with a history of multiple generations of severely reduced seed stock size; and encourage gene flow via pollen or seeds between populations with characteristics you like.

The diversity-shaping processes can add or eliminate alleles from a population or location. In contrast, recombination during sexual reproduction or cell replication does not add or eliminate alleles, but can produce new combinations of existing alleles at one or more genes. These new combinations of alleles can change the genotypic and phenotypic diversity of populations.

The four diversity-shaping processes are interconnected. For example, mutations can create new alleles and allelic combinations that could flow through pollen to other populations, and then be selected by gardeners and become more frequent, or could be lost through genetic drift when only a small number of random seeds are kept for planting. But it's not just our individual actions as gardeners that affect these four processes, the seed system also has a big impact.

10.4. Seed Systems: Seeds, Plant Genes, and People

When we were working with Zuni gardeners and farmers on a sustainable agriculture project for the Pueblo of Zuni in New Mexico, discussion of establishing a community seed bank was already underway. However, running a seed bank and maintaining accessions was not feasible at the time, so we talked about the idea of a seed network where everyone who wanted to participate could add their name to a list in the project office, and the project would facilitate the seed exchange. One Zuni colleague made an interesting observation during the discussion. He explained that the list would include families he would never share his seeds with directly, and others who would never share seeds with him because of long-standing disputes. But he said he wouldn't mind giving some seeds to a project seed network, and wouldn't mind if another family—even one he had a disagreement with—got them through the network. What he said pointed out how small changes in scale, institutions, and organization of a seed system can make a difference in who has access to which crop diversity, even within a small community.

Seed systems include the seeds we plant in our gardens, the sources of the diversity they contain, how that diversity is conserved (Section 10.4.5), controlled (see Box 10.5), and used to create and improve varieties, and how people access, grow, and use them. This means that the seed system determines the species and varieties available, and how genetically diverse they are. Seed systems tell us a lot about the people involved (e.g., the crops and varietal characteristics they value, and the relationships and institutions they engage in). For example, the increasing concentration, scale, and profit expectations of the largest corporate seed companies have meant that they offer fewer numbers of crop species and varieties for gardeners compared to smaller seed companies with other priorities. For example, a focus on high yields in commercial US vegetable and fruit breeding since about 1930, with selection for increased size, weight, and fruit durability, has unintentionally selected newer varieties to have lower concentrations of nutrients such as copper, calcium, protein, vitamins C and A, and riboflavin compared to older varieties (Davis, 2009).

Starting in the late 1970s, concern about the loss of publicly available crop diversity (Fowler and Mooney, 1990) stimulated the rise of small, regional seed companies and non-profit organizations worldwide, like Seed Savers Exchange in the US, and may have encouraged some existing seed companies, like Burpee®, to continue offering OPVs (Table 10.2). One study claims that the response has been so effective that there is now more varietal

Table 10.2. Garden vegetable variety diversity from three seed companies in 2016

Crop	Seminis® Home garden			Burpee®			Seed Savers Exchange[b]
	OPV	hybrid	total	OPV	hybrid	total	OPV
Broccoli	0	3	3	5	3	8	3
Carrots	0	4	4	13	5	18	8
Cauliflower	0	1	1	3	2	5	1
Cucumber	2	5	7	16	20	36	24
Eggplant	0	0	0	6	6	12	14
Green bean[c]	7	0	7	39	0	39	22
Hot pepper	5	7	12	9	8	17	31
Melon	0	1	1	10	19	29	29
Onion	1	4	5	19	3	22	5
Squash	9	3	12	30	25	55	37
Sweet corn/maize	0	4	4	2	17	19	3
Sweet pepper	7	9	16	23	22	45	24
Tomato	14	6	20	52	42	94	47
Watermelon	0	0	0	4	2	6	18
Total	45	47	92	231	174	405	266
OPVs as % of seller's total	49%			57%			100%

Number of varieties offered by seller[a]

[a]When seed of same variety is available as both organic and nonorganic, it is counted as one variety.
[b]Seed Savers Exchange does not sell hybrid seeds (see Box 10.6).
[c]"Green bean" refers to *Phaseolus vulgaris* bean grown primarily to cook and eat as fresh pods, including "snap," "filet," and "haricot vert."
All accessed August 9, 2016: http://www.seminis-us.com/products/results/crops/all/categories/home-garden;
http://www.burpee.com/vegetables/; http://www.seedsavers.org/department/vegetable-seeds

diversity, measured as the number of named varieties available to gardeners and small farmers, than was true in the early 1980s (Heald and Chapman, 2009). But the actual diversity in available garden seeds is a complicated variable to define and measure, especially over time, and this hasn't been done yet. However, a significant decline in *effective species diversity*, an estimate of the number of crop species dominating production, in US farming from an average of over five per county in 1987 to about four in 2012 has been documented (Aguilar et al., 2015).

Access to seed within a seed system is determined by its social and economic structure, including who is in control and who benefits. The multinational corporations that dominate the commodity crop seed systems of industrial agriculture are moving quickly into the vegetable seed business. Seminis® (Table 10.2), "the largest developer, grower and marketer of fruit and vegetable seeds in the world," is the most extreme example. A key tool of large seed corporations is intellectual property rights (IPRs), laws, and regulations they have used to control and limit access to genetic materials for their financial benefit (Box 10.5). In 2016, Seminis® offered 92 "Home Garden" varieties, and 226 "Fresh Market" horticultural varieties for field production, including nine transgenic squash and seven transgenic sweet corn (maize), and all 226 are patented or protected under the Plant Variety Protection Act. The IPR restrictions for the "Home Garden" varieties are unclear, but these made up only 4% of Seminis'® gross profits, and 0.2% of gross profits of the Seeds and Genomics division of its parent company Monsanto® (Monsanto, 2015). That is, garden seeds were a very small part of Seminis® and Monsanto's business, and the vast majority of these seeds have IPR restrictions. With two recent mergers (Dow® and Dupont® in August 2017, Monsanto® and Bayer® in June 2018), the "Big Six Life Science" corporations that dominate the global seed system (Howard, 2009, 2015) are now reduced to four, and a relaxation of IPR restrictions seems unlikely. These stockholder (investor) owned seed and agrochemical corporations present carefully crafted publicity about their benevolent interest in society and the environment, and indeed there are

Box 10.5. IPRs and seeds

The free or reciprocal exchange of diversity common among some gardeners and farmers, a form of gene flow in informal seed systems, was also true of formal plant breeding when it first began. This open access to diversity let farmers, gardeners, and eventually breeders find and select characteristics that met their goals. Still, this free exchange did not result in compensation for farmers and gardeners when genetic diversity they created was used to develop commercial varieties, a story repeated worldwide and over centuries. For example, early Native Americans domesticated and diversified maize throughout the Americas, and European American small farmers later developed widely used maize varieties, but neither groups were compensated when their crop and varieties were used extensively for commercial agriculture.

In the US, free exchange of diversity started declining with the advent of large-scale commercial sales of plant propagules early in the twentieth century. This was promoted by the 1930 Plant Patent Act, which gave protection to vegetatively propagated plant varieties for a limited time, after which anyone could propagate and sell them (Pardey et al., 2013). In 1970 the Plant Variety Protection Act (PVP) extended similar protection to sexually propagated varieties—including seeds—although breeders, farmers, and gardeners were given exemptions for breeding or saving their own seed for the next season. In 1980 the US Supreme Court ruled that all newly identified or modified living microorganisms, and all of their components, including genes, qualified for the same type of utility patent applied to other human inventions. This was followed in 1985 by the ruling *Ex parte* Hibberd that extended the same patenting to plants (Van Brunt, 1985) so that new varieties, described gene sequences, and transgenes all became patentable (Table 10.3). For example, a popular regional US seed company is now selling a patented lettuce mix (Salanova®) developed by a Dutch company.

Table 10.3. IPRs and fruit and vegetable varieties in the US

IPR regime	Basic conditions	Total number of crop varieties covered	Number of fruit and vegetable varieties covered[a]			Fruit and vegetable varieties as % of total crops under IPR regime
			Fruit	Vege-table	Total	
Plant patent, 1930	Patent protection for vegetatively propagated plants, but not tubers; breeders, farmers, gardeners exempt	20,982	2,976	79	3,055	15%
Plant Variety Protection, 1970	Protection is based on distinctiveness; prevents sale, and propagation that might precede sale, of seeds without consent of developer; for sexually propagated plants, with some exclusions in original Act (e.g., tomato, F_1 hybrids) added in later amendments; breeders, farmers, gardeners exempt; duration 20 years (25 years for vines and trees)	9,639	10	1,898	1,908	20%
Utility patent, 1980	Patent protection for living organisms and their components (genes, gene sequences); no exemptions; duration 20 years; made possible by Supreme Court decision in Diamond vs. Chakrabarty. Specific application to plants granted in 1985 under *Ex parte* Hibberd	3,719	2	173	175	5%
Total		34,340	2,988	2,150	5,138	15%

[a]"Vegetable" includes vegetable fruit, e.g., melon, tomato, etc.; "fruit" refers to tree fruit.
Data for number of varieties from Pardey et al. (2013)

Continued

Box 10.5. Continued.

More recently, large seed companies have added contractual controls on seed for commodity crops, making buying seed equivalent to renting it for a season—farmers are referred to as "technology users," and prohibited from saving seed for future planting. Contracts sometimes include stipulations requiring use of the company's agrochemicals as part of the technology package. Under these conditions a small number of corporations have enormous control over the genetic diversity of crops and what current and future conditions they are adapted to, with the goal of maximizing profit. These controls on genetic diversity and seeds create restrictions for gardeners and farmers, but also for plant breeders, including those with public, non-profit, or some private institutions.

Alternatives are being developed in response to the increasing proprietary control of seeds and genes. For example, a new non-profit, the Open Seed Source Initiative (OSSI) is promoting an approach based on the one used for open source software agreements (Section 10.7.1), which allows free use of seeds, including for breeding and sale, as long as the same open source agreement is passed on with any seeds or varieties derived from that material. However, the OSSI protection is an honor system pledge, not a legal contract.

Labeling produce sold in stores and farmers' markets with its seed type and seed system has also been suggested as a way to harness consumer support for alternative seed systems (Howard, 2015).

scientists with prosocial goals working for them. However, the reality is that it is illegal for such corporations not to have profit maximization as their primary goal, and if they stray from this their stockholders can take legal action.

DAC met with a group of Monsanto® executives at their headquarters in 2007 as the debate about transgenic maize in Mexico—maize's center of origin and diversity—was heating up. He suggested that one way for Monsanto® to demonstrate its stated goodwill toward Mexican farmers would be to issue a legally binding declaration that it would never seek to prosecute farmers for inadvertently having Monsanto's patented transgenes in their fields, as a result of gene flow via seed or pollen. He was told that Monsanto's lawyers would never allow this, because it might compromise the company's legal mandate to maximize profits. Interestingly, a new category of for-profit corporations has been recognized in some US states known as "benefit" or "B corporations," that are legally permitted to include social and environmental impacts as part of corporate decision-making, along with profit. We are not aware of any B corporation seed companies, but there are small, regional, non-profit ones (Section 10.4.3).

Recent development of gene "editing" technologies may have significant consequences for the users of formally improved crop varieties (Jaganathan et al., 2018). Notable among these is a rapidly evolving technology commonly known as CRISPR that permits targeted alterations in existing genes, including those associated with important traits. This is different than producing transgenic varieties by incorporation of genetic material from other organisms into plants. Many proof-of-concept examples for CRISPR exist in crop plants, and some varieties produced using these methods may soon be in use. In the US, these varieties are not regulated using the more stringent rules that transgenic crops are subjected to. Given the cost of CRISPR technology, its application is likely to be dominated by the major agrochemical and seed corporations, who are likely to seek strong intellectual property protection for the resulting varieties, making their use by gardeners less likely.

Seed systems range from individual gardeners who save, grow, and share their seeds to global, commercial seed systems that include institutions like multinational agrochemical corporations. Yet, across this entire range the same six basic functions are accomplished: conservation, improvement, multiplication, distribution, production, and consumption (Fig. 10.5) (Soleri, 2017). Also across this range, crop diversity is shaped through the same four evolutionary processes we described above. Below is a brief outline of four contrasting seed system types and how they affect the processes and crop diversity.

10.4.1. Informal seed systems

Informal seed systems of one or a few gardeners are made up of social networks based on geography, familiarity, and reputation, typically through face-to-face interactions. Gene flow via seeds may provide access to greater genetic diversity than that available to one gardener saving and using her own seeds. If only a few individuals in the network are seed savers,

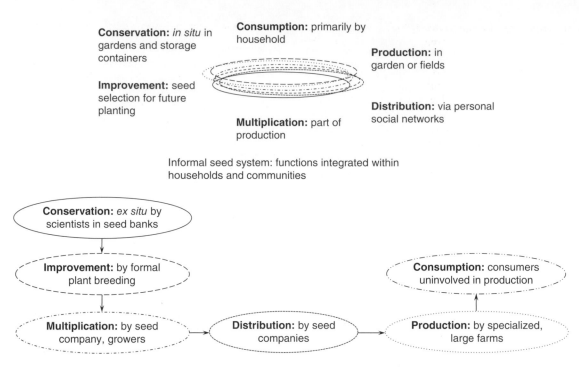

Conservation: *in situ* in gardens and storage containers

Improvement: seed selection for future planting

Consumption: primarily by household

Production: in garden or fields

Distribution: via personal social networks

Multiplication: part of production

Informal seed system: functions integrated within households and communities

Conservation: *ex situ* by scientists in seed banks

Improvement: by formal plant breeding

Multiplication: by seed company, growers

Distribution: by seed companies

Production: by specialized, large farms

Consumption: consumers uninvolved in production

Formal seed system: functions separated, specialized

Figure 10.5. Organization of basic functions in small-scale, informal and large-scale, formal seed systems

Copyright Daniela Soleri, used with permission

they can dominate gene flow in these systems, which further increases the risk of very small population size due to selection or genetic drift because of the small size of garden plots and seed lots (Section 10.3). Seeds are often given freely, or available through trade, although as our Zuni colleague explained, social disputes can create barriers in informal seed systems. Gardens are the selection plots and living seed banks, so that selection, multiplication, and conservation are not separated from growing food.

10.4.2. Semi-formal seed systems

Like informal seed systems, semi-formal ones do not use commercial seed distribution, but they include institutions that make their seeds available beyond personal social networks (Soleri, 2017). Community seed banks and networks, seed libraries, and organized seed swaps are examples of such institutions. Gene flow is enhanced when these institutions integrate varieties from sources in addition to gardeners, such as regional seed companies,

and when they conserve and multiply unique selections such as rare local varieties. Through their slightly impersonalized structure, social barriers can be overcome, so that diversity can move more easily across a region or community, as in the Zuni example. Quality and consistency are based on the credibility of the institutions, the gardeners and farmers contributing to them, and the quality of their seeds. Gene flow, selection, and conservation of diversity occur mostly as part of growing food and saving seed. There can also be intentional efforts to grow enough to have reserve seed stock on hand for more than one season in case of new demands, or poor harvests. Typically, labor is by seed users or others supporting the institution; seeds are available free, through reciprocal exchange, or for nominal costs that support the institution.

10.4.3. Regional, formal seed systems

Formal seed systems are commercial, and typically have specialized practices for each seed system function, but there are many different types of formal systems, and different institutions that support them. Independent or regional seed companies in the US are institutions supporting small-scale formal seed systems, and many focus on OPVs and local varieties, including heirlooms. In the US, commercial seed sales require adherence to standards for consistency and quality (Box 6.1) that protect seed buyers in transactions that are more impersonal than ones in semi-formal seed systems. Meeting the standards is the responsibility of the company selling the seeds.

Some regional seed companies are non-profit with educational or research objectives (e.g., Seed Savers Exchange; Native Seeds/SEARCH), others are private companies that are able to make decisions based on criteria other than just profit because they do not have publicly traded stock. Regional crop breeding and seed institutions are being established in many locations, specifically to serve the needs of organic farmers and gardeners (Section 10.7.4). Most of these institutions are driven by a mission, such as diversity conservation, gardener education, or support of alternative seed and food systems. Some of the mission-driven institutions have signed the "Safe Seed Pledge" indicating they do not use or sell transgenic varieties.

For simplicity we also include in this category the private seed companies that supply much of the US garden seed market, and are intermediate between the regional companies just described and larger corporations, and are best described as business- rather than mission-driven. Seed suppliers such as Burpee® (now owned by George Ball, Inc., a family business, see Table 10.2) and Park Seed® (now owned by Jackson & Perkins Park Acquisitions, a division of publicly traded Western Capital Resources, according to Wikipedia), offer heirloom and OPV as well as hybrid seeds (see Box 10.6), and some sell seed from multinational corporate firms such as Seminis®.

Overall, seed institutions in this category strive to avoid selection and gene flow that would alter a variety. They conserve diversity in their own seed or gene banks as well as making use of some public ones, and some sell their seeds with IPR restrictions. Institutions in this group doing crop improvement may incorporate new alleles and genotypes (gene flow), and select to achieve or maintain desired characteristics.

10.4.4. Formal, multinational corporate seed systems

Formal, multinational corporate seed systems are at the opposite end of the range from informal seed systems, and are centered around large institutions that often consolidate many seed system functions. The largest of these are publicly traded, multinational, for-profit corporations that also produce agrochemicals, and often promote or require these as part of technology packages to farmers growing commodity crops, and increasingly those growing horticultural crops. The most common examples are transgenic crop varieties with tolerance of a corporation's herbicide, for example Bayer's® Roundup Ready® varieties and their glyphosate herbicide, marketed as Roundup®. Powerful intellectual property tools like patents and contracts are used to control seeds, varieties, and the genetic materials used to develop them (Box 10.5), by prohibiting seed saving and replanting. Hybrid varieties, available for many garden crops (Box 10.6), are a form of biological intellectual property protection, since seed saved from these are different from, and often inferior to, the parent plants.

In formal, corporate seed systems selection is by professional plant breeders and other scientists, with high yield response to inputs a major goal; sources of diversity are farmer-created, from public, national and international gene banks, as well as a company's proprietary materials.

10.4.5. Conservation of crop diversity

In the informal seed systems of gardeners, and the semi-formal seed systems supported by institutions like seed libraries and community seed banks, conservation, selection, multiplication, and production may all occur together. Conservation in gardens and fields is on site, or in situ, and is dynamic, in the sense that plants are constantly exposed to the processes that shape diversity, so populations and varieties are adapting to current selection pressures. This diversity may be distributed among different gardeners who make it accessible to each other and their institutions. Sometimes traditional farmers in informal systems grow a main crop plus small quantities of other varieties of the same crop to keep seed of those varieties readily available should it be needed, what has been called "insurance diversity" (Jarvis et al., 2008).

Box 10.6. Hybrid seeds

Hybrid varieties can be very productive, they can also be a form of biological control of intellectual property in seeds because they do not reproduce true to type, which discourages seed saving. Many fruit and vegetable seeds available from regional and multinational seed companies are hybrids, but gardeners are often confused about what a hybrid is. In biology a *hybrid* is the product of the cross between any two genetically different individuals. In plant breeding, a hybrid is a particular kind of crop variety produced by crossing inbred lines that are homogeneous and homozygous (definitions in Box 10.2). However, in popular usage the word is often used to mean any modern commercial variety, which are not all hybrids. To help understand the process we'll briefly outline how single-cross hybrid maize seed is produced (Fig. 10.6). Nearly all of the maize seed planted in the US is hybrid; the exceptions are OPVs grown and maintained by Native Americans through informal seed systems, and some OPVs grown by gardeners and organic farmers, which are mostly available through small, regional seed companies.

In the first stage of creating a hybrid, heterozygous plants from the same species (*Zea mays* in this case) are inbred, using self-pollination, for about seven generations, while selecting among the offspring for those with desirable traits. The results are plants homozygous for many genes whose offspring have the same characteristics as the parents—they reproduce true to type. These breeding materials are *inbred lines*, for example, the two different lines A and B in Fig. 10.6. As an example, for the two alleles of the same gene, line A may have only r alleles, while line B has only R alleles for that same gene.

While inbreeding can *fix* alleles for desirable traits, that is eliminate all but one allele for a gene, it also fixes alleles for some undesirable traits. However, in creating inbred lines this is acceptable, as long as the second step produces desirable progeny.

In the second step, two inbred lines with different desirable alleles and traits are cross-pollinated, to produce more vigorous and productive progeny than either inbred parent. This is because a desirable dominant allele in one inbred can override the effect of an undesirable allele for the same gene in the

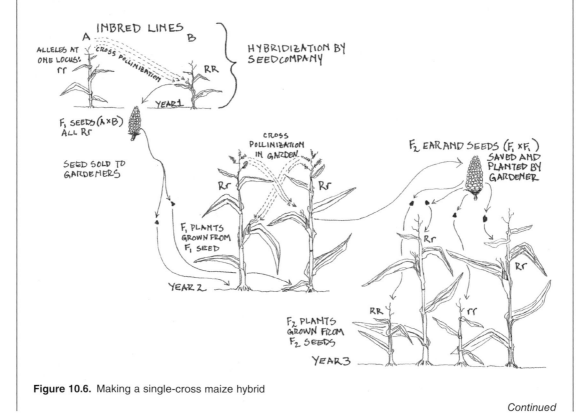

Figure 10.6. Making a single-cross maize hybrid

Continued

Box 10.6. Continued.

other inbred, and all of their progeny will have that same combination of alleles. To eliminate any self pollination and ensure that all of the seed produced is the result of cross-pollination, one line is used only as the female (seed) parent and the other only as the male (pollen) parent. Thus, all of the seeds borne by the plants of the female parent line are the identical cross of lines A and B (A x B); they are a homogeneous population made up of heterozygous individuals. In our example all would have alleles Rr for that gene. The seeds produced by this cross are F_1 (*filial one*, first generation progeny or offspring) hybrids, which are sold to farmers and gardeners. These seeds grow into strong, healthy F_1 plants because of their desirable traits and a high level of heterozygosity, a phenomenon called *heterosis* or hybrid vigor, the basis of which is not entirely understood.

When these identical F_1 maize plants cross-pollinate (F_1 x F_1), they produce ears with the F_2 kernels we eat. If these kernels are saved and planted they grow into a heterogeneous population of heterozygous F_2 plants, different both from their parents and from each other because those F_2 plants include some of the many genotypes possible when crossing two heterozygous individuals (Table 10.1). They are not true to type, having different combinations of alleles across many genes including the gene used in the example above. There will be weak plants showing some undesirable traits of the original inbred parent lines, A and B, and plants with some desirable traits. Because of this unpredictable variation, gardeners and farmers using hybrid seed usually buy new F_1 seed every year to ensure plants, kernels, and fruit with consistent characteristics.

Some gardeners and farmers do save seeds from hybrids, selecting their own OPVs from the diversity of plants these seeds produce, but this takes time. Recently a farmer in California started selecting among plants grown from the seeds of 'Early Girl' tomatoes, a popular hybrid variety with parent lines now owned by Seminis® (then Monsanto®, but now a part of Bayer®). That farmer is working to develop a locally-adapted variety, and avoid purchasing seeds owned by Monsanto® which his customers did not want (Duggan, 2014). Because Early Girl is a hybrid from the 1970s, the PVP protection on it has expired, but even if it hadn't, under PVP the farmer could have developed and sold his own variety from his selection, as long as it was distinct from Early Girl. Today, hybrid seeds are sold for many varieties of commercially produced and home garden vegetable crops, including cabbage, lettuce, maize, melon, onion, pepper, squash, and tomato.

Hybrid seeds are not the same as transgenic ("genetically modified") ones, although all of the few transgenic varieties in common garden crops (e.g., sweet corn, summer squash) are also hybrids. These are transgenic versions of modern hybrid varieties that plant breeders are already familiar with. Transgenes could be introduced into traditional, open-pollinated and heirloom varieties, but there is little profit motive for the seed companies since the market is relatively small, and stable introduction of an effective transgene takes time and money. In addition, many of the gardeners and farmers growing these varieties do not want transgenics, and these types of varieties would increase the difficulty of controlling unwanted transgene flow. It is also clear that crop species and their relatives contain vast amounts of diversity that can be used for crop improvement without the need for transgenics (McCouch et al., 2013).

In contrast, in the larger regional and corporate formal seed systems, crop diversity is primarily conserved in gene or seed banks, separate from the crops growing in people's gardens and fields. Material in gene banks can be in the form of seeds, vegetative propagules, tissue cultures, and genetic sequences. This is *ex situ* or off-site conservation, away from where the materials originated and are used, and every effort is made to prevent exposing the material to any of the four evolutionary processes that shape diversity, to keep it genetically the same as it was when collected. Such *ex situ* collections are maintained as resources for current and future research and crop improvement, for example as a source of material used to keep up with the breeding treadmill (Section 10.3.3). *In* and *ex situ* conservation are obviously different, but both have valuable roles to play in conserving and maintaining open access to crop genetic diversity.

The challenge for public seed banks is maintaining good seed or other propagule quality, avoiding selection or genetic drift during storage and regeneration, and ensuring equitable access to seeds. A widely recognized problem with formal, public seed and gene banks is that their collections serve researchers and plant breeders, but not necessarily gardeners and farmers. Even if gardeners and farmers did have access to those seeds, the way they are stored, and the environment and practices used when they are *regenerated* (periodically grown out to maintain viable seeds), can create unintentional selection pressures that change varietal genotypes. Even if seed could be stored and regenerated without exerting selection pressures that changed its genetic makeup, the originally collected seed genotypes may be poorly adapted to changing local environmental conditions in gardens and fields (Soleri and Smith, 1995; Tin et al., 2001). It is also impossible for seed banks to maintain collections of all locally adapted populations of crop species, especially because these are also always changing.

One response to this and the lack of easy public access has been the establishment around the world of community seed banks over the past 20 years, including in Oaxaca, Mexico, by farmers working with our colleague Flavio Aragón-Cuevas (Aragón-Cuevas, 2015). In community seed banks, collections are maintained by and for community members, with the institution providing the framework for overseeing seed stocks, multiplying rare varieties, ensuring members maintain backup stocks for emergencies, enabling gene flow within and between communities, and education for varietal improvement. Although called seed banks, the seed collections are frequently rotated with fresher, *in situ* regenerated seed replacing older seed, reflecting selection pressures occurring in field and garden environments.

10.5. Getting Ready to Save Seeds: Sexual Reproduction in Flowering Plants

Knowing how plants produce seeds helps you get the most out of seed saving. For example, you may want to control pollination to create certain genetic combinations, or increase production of seed or fruit. Because there are so many detailed resources about how to do seed saving (Section 10.7.4), we focus on the main concepts involved.

Most seeds and the plants that grow from them are the products of sexual reproduction, the combination of genetic material from the gametes of both parents through pollination and fertilization, usually necessary for fruit development. In plants, male gametes are produced within tubes that grow from pollen grains that develop in the anthers, and female gametes contained in ovules are produced in the ovaries. *Pollination* is the movement of pollen grains from the anthers where they develop, to the stigmas of the female flower; these plant parts are identified in Fig. 5.4. Plants are either predominantly *self-pollinated*, meaning the pollen goes mostly from anthers to stigmas on flowers on the same plant, or *cross-pollinated* (or "outcrossing"), where pollen moves mostly from the anthers of flowers on one plant to stigmas of flowers of another plant. We say "mostly" because frequently one form of pollination predominates, but the other is also present. For example, common bean (*Phaseolus vulgaris*) is typically classified as self-pollinated, and this is the case with most varieties, but some have been observed to have up to 70% outcrossing (Ibarra-Perez et al., 1997).

After a pollen grain lands on the stigma, and grows a pollen tube down the style, it will produce sperm cells, the male gametes. These enter the ovule and fertilization can occur, but pollination does not guarantee fertilization, for example if the pollen is not viable. Another reason that fertilization can fail is if the individuals or varieties are *self-incompatible*. For example, even though some members of the cabbage group (*Brassica oleracea*) have perfect flowers they cannot self-fertilize. However, genetically different individual plants of the same OPV can be fertilized following cross-pollination by insects. Growing these varieties in blocks or clusters encourages cross-pollination. Varieties of some species, like apple and pear, are self-incompatible and need to be grown with compatible varieties for pollination and fertilization to produce fruit.

Fertilization is the successful joining of a sperm cell nucleus with the egg cell nucleus, which is part of the female reproductive structure or female *gametophyte*. In most crops each pollen tube can deliver two sperm cells to the ovule: often one fertilizes the egg cell, which develops into the embryo with 50% paternal and 50% maternal genes. Another sperm cell nucleus fertilizes two other nuclei inside a single cell in the ovule to form the endosperm, with 33% paternal and 67% maternal genes. The fertilized ovule will develop into a seed, with an outer layer or *seed coat*, and the ovary around the seed(s) developing into a *fruit*. The embryo within the seed will grow into a seedling when the seed germinates. In many garden crops the endosperm supports the development of the cotyledons, and the mature seed has virtually no remaining endosperm, as is true of legumes such as common bean and pea. In others, including cereals such as maize, a large proportion of a mature kernel is made up of endosperm tissue, which is important for culinary and nutritional quality.

Lots of garden plants are grown for their fruit, including chile, squash, tomato, and tree fruits. Unlike the seed embryo, which is a product of male and female gametes and so a combination of male and female parent alleles and characteristics, as described above, a fruit is a mature ovary, and made up solely of female cells (Fig. 10.7). This means that if there are plants whose fruit you prefer but are not saving their seeds, then the source of pollen is not important. However, if you are saving seeds and want to maintain or improve the plants or fruits that will grow from them, then it makes sense to select both male and female parent plants. For example, if you grow more than one squash variety and you want to maintain seed for a kabocha variety with sweet, deep orange flesh, you will need to select both male and female parents, control pollination, and only save seeds from plants you intentionally cross.

A few garden crops produce seed or fruit without fertilization. *Parthenocarpy* is fruit development without fertilization, sometimes stimulated by pollination that does not lead to successful fertilization, as when there is an imbalance in the number of chromosomes from each parent. The oldest domesticated fig remains found in an archeological site in Jordan, at least 11,200 years old, may be of a parthenocarpic fig (Kislev et al., 2006). Completely parthenocarpic figs do not have viable seeds, but their fruit can be delicious. Some, but not all, fig varieties today are parthenocarpic, including one of the most widely used, the 'Mission' fig (Janick, 2005, 283). Other garden crops with some varieties that bear parthenocarpic fruit include banana, cucumber, eggplant, pear, sweet and chile peppers, and tomato. Parthenocarpic perennials like fig and banana are propagated vegetatively; varieties of parthenocarpic annuals can be the result of special crosses that gardeners do not make, as described below for watermelon.

Apomixis is another form of seed production without fertilization found in some citrus species and in mangos. In these types of citrus, fruit production involves the generation of one seed from fertilization, plus numerous others solely from the cells of the ovule or surrounding tissues with the same genotype as the tree bearing the fruit. Typically all the seeds will be viable.

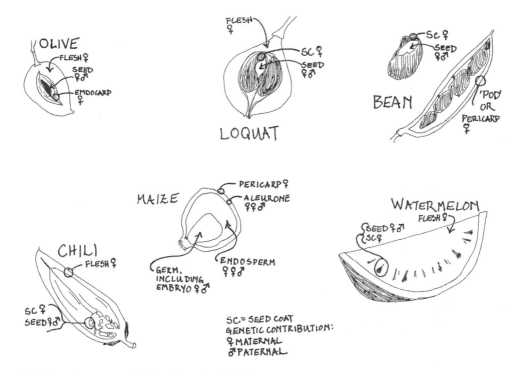

Figure 10.7. Genetic sources of different seed and fruit tissues

Seedless watermelon is produced in yet a different way. By crossing two parent plants with different numbers of chromosomes (one diploid, the other tetraploid), fertilization and fruit development occurs, but the resulting seeds fail to develop because the different number of chromosomes from each parent means that not all of them can form pairs. Some new varieties of seedless garden crops that are hybrids (Box 10.6) and/or transgenic are being developed with a much higher proportion of seedless fruit than occurs naturally.

10.6. Seed Saving Basics

Seed saving is a quick, easy, inexpensive way to obtain seeds for your garden. When we choose varieties to grow, select individual plants within those varieties, and save those seeds, over time we can create and maintain garden crops suited to local conditions, and our own preferences.

Observation is the first step in seed saving. You want seed from plants with *environmental fitness*, characteristics such as drought and heat adaptation, and pest and disease resistance so they are able to survive, reproduce, and produce food under current garden conditions with minimal inputs. The many characteristics contributing to fitness are embodied in a plant's ability to produce seed and a good harvest in your garden. When you save seed from these plants a lot of the selection is being done for you by the growing environment. Aesthetic and culinary traits don't always contribute to environmental fitness, but they do contribute to *cultural fitness* when you save seed from plants that have the color, aroma, flavor, texture, or cooking quality you prefer. Often cultural fitness is just as important as environmental fitness.

Farmers we know in Oaxaca, Mexico, grow small garden patches of runner beans for their seeds selected for eating quality, but also selected for the bright scarlet flowers that are used fresh to add color to soup. Selection for large, intensely colored blossoms is as important as seed characteristics for some farmers, and this flower color may also support environmental fitness by attracting pollinators. We also know farmers and gardeners who have selected maize for the beauty of its purple tassels or purple husks that color tamales, arugula for its leaf shape and size, or fenugreek for the intensity of the caramel flavor of its fresh leaves. One of the pleasures of gardening and seed saving is pursuing those qualities that you value and enjoy.

Still other characteristics are sought in garden plants used for medicine, crafts, and other purposes. By saving seeds of devil's claw (*Proboscidea parviflora*) fruits with dark, long fibers, the most desirable characteristics for their craft, Tohono O'Odham Native American basket weavers in the southwestern US and northern Mexico have created several new varieties (Nabhan and Rea, 1987).

When the characteristic you want to select for such as flavorful leaves, or early onset of flowering, will not be evident at harvest, you need to mark the plant to save seeds from as soon as you notice that characteristic. This is especially important if you want to control pollination in cross-pollinated crops. No matter what characteristics you prefer, the plants you identify will depend on the kind of selection you want to do (Box 10.7).

For most food garden species that produce dry fruits, the best way to determine seed maturity is simply looking for seeds that have reached full size and started to dry and harden on the plant. In most species that produce fleshy fruit, seeds are mature and can be cleaned for saving when the fruit is overripe. Examples of harvesting and processing seeds from different types of plants are summarized in Table 10.5.

10.6.1. Seed drying and storage

The goal of seed storage is to keep seeds alive and healthy for future planting. Seeds respire (Section 5.4), using oxygen and the carbohydrates in the endosperm or cotyledons to obtain the energy they need to stay alive in storage, and in the process they produce heat, CO_2, and H_2O. The more respiration before planting, the less energy the seed will have for healthy germination and emergence, so minimizing respiration is a seed storage objective (Chidananda et al., 2014). The longer seeds are stored, the longer they will respire and the more energy they will lose. Because respiration is greater at higher seed moisture levels, and pathogens and pests need water to live, it's also important to minimize the moisture content of stored seeds.

Seed conservationists have worked to find the combination of temperature and humidity that optimizes seed storage benefit:cost. For example, the guidelines developed in the early 1960s by Harrington, which he called "thumb rules," are

Box 10.7. Selecting garden seeds with diversity in mind

In seed selection there is a tradeoff between maintaining diversity that supports our response capacity for future adaptations and preferences (Box 10.3), and better adaptation to immediate conditions (selection pressures), which reduces diversity. Two ways of balancing this tradeoff are not selecting so strongly that diversity is greatly reduced, and/or having access

to additional diversity (gene flow) in case it is needed for adaptation as conditions change.

The number of plants of a variety grown in a garden is often very small—one butternut squash plant, five 'Serrano' chile plants, or 30 'Golden Bantam' maize plants—so selection can have a big impact, and a major loss of diversity is a real possibility. To minimize

Table 10.4. Basic differences in organization of genetic diversity in self- and cross-pollinated and vegetatively propagated garden crops

Primary form of propagation in the garden[a]	General characteristics of the organization of genetic diversity	Minimum number of individual plants recommended for seed bank collection[b]	Number of individual plants suggested for maintaining diversity in garden seed stock[c]	Examples of garden crops
Self-pollinated seeds	Male and female gametes contributing to an embryo are genetically similar	50	5–25	Amaranth, barley, common bean, lettuce, oat, okra, pea, quinoa, rye, tomato, wheat
	Mostly homozygous individuals			
	Inbreeding depression not significant problem			
	Population can be homogeneous, but traditional varieties may contain multiple different homozygous genotypes (also called lines in varieties), requiring larger numbers of individuals	>50[d]	>20[d]	
	Some outcrossing always present; resulting heterozygous individuals can be valuable source of diversity			
Cross-pollinated seeds	Male and female gametes contributing to an embryo often have different alleles for many genes	5,000	50–120	Brassicas, chile and sweet pepper, gourd, maize, runner bean, tomatillo
	Heterozygous individuals			
	Inbreeding depression with small populations is significant problem, especially for some species like maize			
	Population is heterogeneous			
Asexual, vegetative propagules[e]	Usually same genotype as parent plant, unless somatic mutation occurs (Section 10.3.1)	>50	5[d]	Bunching onion, garlic, potato, sweet potato; stem cuttings of many herbs; cuttings and root suckers, e.g., fig, jujube, olive
	Individual may be heterozygous or homozygous, depending on propagation history			
	Heterogeneity of resulting population depends on how many different genotypes provide propagules			

[a]Characteristics described here are for open pollinated varieties of either self- or cross-pollinated species.
[b]Based on Ellstrand and Elam (1993); Frankel et al. (1995); Rao (1998); Brown (1999).
[c]Based on SSE (2017). Number could be smaller if individual gardeners pool seeds.
[d]These are our estimates.
[e]Refers to using the vegetative propagules, not seed from these plants. Often for gardeners even smaller numbers are adequate.

Continued

Box 10.7. Continued.

the loss of diversity, it is important to save about the same number of seeds from each selected plant. When deciding how many plants to save seeds from the first consideration is the different way genetic diversity is organized in self- and cross-pollinated crops. Table 10.4 outlines the basic differences. As you can see, the ideal minimum number of individual plants is far more than most gardeners can or want to plant, and why semi-formal seed systems can help overcome small crop population sizes in gardens.

Figure 10.8 illustrates approaches to selection for larger tomato fruit size, or other characteristics like yield, plant height, or leaf size. These *quantitative characteristics* are expressed across a range of values, which often follow a normal distribution, with the mean and median being roughly equivalent as we show in the simplified figure. For *qualitative characteristics*, ones that occur in discrete classes such as seed coat color in beans, or tassel color in maize, plants with the desired characteristic can be easily chosen in many cases.

In the three approaches in Fig. 10.8, seeds from fewer and fewer individual plants are being selected

for growing as you move from a) to c). This means gene flow is increasingly important to maintain diversity, so a local seed system larger than an individual household is a very good idea. For example, a seed library could make seed available for a variety that has been locally grown and selected by multiple households and pooled, reducing the risk that very small population size from strong selection or genetic drift will be a problem. Plant breeders have developed techniques for effective selection that can be adapted for gardeners, but the small size of plant populations in the garden remains an issue. Resources for adapting some plant breeding techniques for selection by gardeners are given in Section 10.7.4.

Finally, with a small number of plants, and without intentionally controlling for environmental variation (see Appendix 3A, Section 3A.2), it is especially important to be aware of what a large effect the growing environment can have on quantitative characteristics, which can mean that expression of traits selected for may not have a genetic basis, and therefore not be heritable (Section 10.3.3).

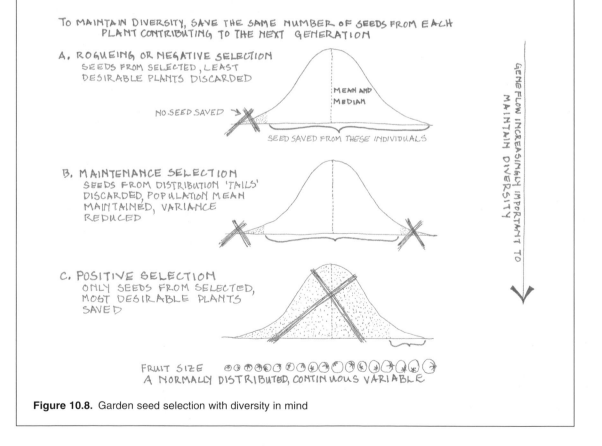

Figure 10.8. Garden seed selection with diversity in mind

Table 10.5. Examples of harvesting and processing garden seeds

Crop	When to harvest seeds	Processing
Fruit as pods or capsules: e.g., brassicas, okra, green/snap bean, pea, pulses (dry bean, lentil, garbanzo), sesame	For fruit that dehisce, harvest just before it dries, while turning brown but still pliable. Pea is indehiscent—pods don't burst open, so can be harvested even after dry	Dry on cloth or in paper bags, or on mats, so that seeds are collected when fruit open
Cucurbits: e.g., cucumber, gourd, luffa, melon, squash, watermelon	When fruit overripe, i.e., 2–6 weeks after-ripening depending on genus	Squash, gourd: embryos continue to mature after fruit ripe, store fruit in cool, dark place at least 6 weeks after fruit is considered ripe. Cucumber, Cucumis melons, watermelon: after-ripen about two weeks after ready to eat, makes separating seeds from pulp easier, improves germination. Even in slightly immature fruit, e.g., picked after early frost, viable seeds can sometimes be saved if fruit left to after-ripen. Rinse and separate seeds from pulp; discard small, flat, floating seeds, dry in well-ventilated place, not direct sunlight
Chile and sweet pepper	When fruit ripened on plant to red, orange, or black, not green	Remove seeds as fruit is eaten, either fresh or dried, dry at temperatures below 38°C (100°F)
Soft, small-seeded fruit: e.g., eggplant, tomatillo, tomato	When fruit ripe to overripe	Put mashed pulp of fruits in uncovered container. Leave to ferment 3–7 days, stirring occasionally; the warmer the air temperature the less time required. Fermentation is the result of microorganisms digesting the carbohydrate pulp, producing simpler compounds like acids. A sour smell and bubbles on the pulp's surface indicate fermentation is occurring. Fermentation destroys disease-causing pathogens on seeds, thins out pulp, allowing heavier, viable seeds to separate and sink to the bottom. Fermentation also removes gelatinous coating on seeds, changing their texture from slippery to rough when rubbing the seeds between your fingers. When seeds and pulp separate, discard floating seeds. Rinse remaining seeds, lay on a cloth, screen, or paper, dry in place protected from wind and direct sun. Eggplant seeds don't need fermentation. Mash pulp of soft, very overripe fruits. Separate seeds by rinsing, lay out to dry
Seed-bearing flower heads e.g., amaranth, basil, cilantro, *papaloquelite*, quinoa, sunflower	Just before seed head completely dry and brittle, or dehisces	Cut seed heads, lay on cloth or in bag, when dry remove seeds by rubbing gently
Maize	When kernel color developed; can be left on plant until dry if no pest or mold problems. Husks may slow drying. If high daytime temperatures may ferment, especially sweet varieties. If past milk stage ears can be harvested, allowed to dry in shade in husks, with husks open, or husks removed	Dry kernels on cob, with or without husk. Can store on cob, reducing pest damage to the seeds' pedicels, where attached to the cob. If space limited seed can be removed from cob. There are many traditional ways to store maize seed, with or without cob, husks, in containers, hung in bunches, or strung in hanging racks
Fruit trees: e.g., cashew, citrus, date, jujube, mango, olive, papaya, stone fruit	When fruit completely ripe	Remove fruit flesh and clean seed, plant fresh, dry, and/or stratify, depending on crop

approximations, but still widely used (Harrington, 1960): a) the sum of the air temperature in °F and the *relative humidity* (the water present as a percentage of the amount of water that would saturate the air at a given temperature) of a storage area should be <100; b) the time seed can be stored doubles for both of the following: every 1% decrease of seed moisture content (between ~5–14% moisture), and every 5.6°C (10°F) reduction in the storage temperature between 4.4–37.8°C (40–100°F). Moisture content of seed is directly related to relative humidity of the air (Table 10.6), but is not a 1:1 relationship, and varies according to crop species and air temperature. The important point for saving garden seeds is that the higher the relative humidity the higher the seed moisture content. High temperatures (>26.5°C, 80°F) and seed moisture content (>7%) encourage seed-damaging pathogens and increase embryo respiration, shortening the seed's life (Hartmann et al., 2002, 190). Extremely high temperatures kill garden seeds.

Refrigerating, or better, freezing, can be a good way to store small quantities of seeds, and freezing can also kill pests and some pathogens. Freezing bursts the water-containing cells in pests, but for the same reason seeds should be well dried (~<15% moisture content) before freezing or their cells will also burst.

The length of time seed can be stored depends on the type of seed, its quality, and the storage conditions. After complete development and any required dormancy period (Section 6.1.2), seeds lose viability as they age because the embryos weaken and die. So the longer they are stored, the lower their germination percentage becomes (Section 6.1.1). If stored for a long time, seeds like beans and peas can become so hard and dry that water cannot penetrate them to swell and break the seed coat, and you will need to scarify them. Growing out, or regenerating, saved seed (Section 10.4.5) helps avoid this problem and ensures fresh, viable seed stock, but must be done with care to avoid unintentional selection. To allow for losses during storage, germination, and early growth, we suggest saving at least 50% more seed than needed for planting.

People develop methods of storage that make sense for them. DS has seen gardeners in northern Nepal smear melon pulp and seeds on a shaded wall where they will dry slowly and stay pasted in place, can be checked for pests or mold, and are kept away from hungry chickens until needed for planting. In the global north there are lots of simple options, such as free or inexpensive used jars and other containers.

10.6.2. Controlling pests in stored seeds

There are two ways to manage pest and pathogen problems in stored seeds: avoid them by making the seeds or the storage environment inhospitable, or if present, destroy them.

Clean, dry storage areas and containers limit moisture, food, and breeding or hiding places for pests and pathogens. Some insects like weevils (*Apionidae* family) and bruchids (*Bruchidae* family) can lay their eggs in seeds such as bean while they are still in the garden. If conditions are right, the eggs will hatch later and the larvae will eat the seeds. Holes in the seeds and powder at the bottom of the container from the damaged seeds are signs of these pests, and their small white larvae can usually be found inside or around the seeds. These and other pests in stored seeds will die if seeds are frozen, or stored in sealed, airtight containers with as little extra air space as possible to minimize available oxygen.

Certain additives will also kill pests. Fine, sharp sand can be added to seeds in a 1:1 ratio by volume. Sharp sand is quarried sand with sharp edges, unlike beach sand which has been smoothed. Because it fills the spaces between seeds, reducing available oxygen, and because of its weight, the sand prevents insects from moving around easily.

Table 10.6. Examples of seed moisture content when seeds are stored at different levels of relative humidity

Crop	Seed moisture content, %, when relative humidity is:		
	10%	30%	75%
Green bean[a]	3.0	6.8	15.0
Carrot	4.5	6.8	11.6
Sweet corn/maize	3.8	7.0	12.8
Leaf mustard	1.8	4.6	9.4
Okra	3.8	8.3	13.1
Onion	4.6	8.0	13.4
Pepper	2.8	6.0	11.0
Winter squash	3.0	5.6	10.8
Tomato	3.2	6.3	11.1
Watermelon	3.0	6.1	10.4

[a]"Green bean" refers to *Phaseolus vulgaris* beans grown primarily to cook and eat as fresh pods, including "snap," "filet," and "haricot vert."
Justice and Bass (1978, 40), based on data from (Harrington, 1960)

This does not limit the amount of oxygen enough to harm the seeds themselves. Adult bruchid beetles, common pests in stored beans, are unable to move around enough to mate and reproduce in beans mixed with sand, and so the population dies out. The sand also scratches the thin wax coating of the insect's outer cuticle and its delicate limb joints, causing it to dry out and eventually die. Fine dusts such as limestone can also be used—some scratch while others absorb the insect's protective waxy layer, causing it to dehydrate. Care should be taken not to inhale these dusts.

Another way to control pests is by using a repelling substance. For example, the compound azadirachtin contained in the leaves and seeds of the rapidly growing neem tree (*Azadirachta indica*), which is native to India, has been found to be an effective insect repellent (NRC, 1992), and its seed is commercially available. Picante, strong-smelling dried chile peppers and onion leaves, are used in southern Nigeria in stored cowpea seed, and may also repel weevils. Some Native Americans traditionally used wild tobacco leaves (*Nicotiana rustica*) to repel insects from stored seeds.

10.7. Resources

All the websites listed below were verified on June 19, 2018.

10.7.1. Intellectual property

Jack Kloppenburg's *First the seed* is a good history of the development of the seed industry and crop IPRs, with a focus on the US (Kloppenburg, 2005).

The OSSI developed the OSSI pledge based on open source software licensing: "You have the freedom to use these OSSI-Pledged seeds in any way you choose. In return, you pledge not to restrict others' use of these seeds or their derivatives by patents, licenses or other means, and to include this pledge with any transfer of these seeds or their derivatives." More information about OSSI and the plant breeders working with it and selling their seeds through it can be found on the OSSI homepage. https://osseeds.org/

Pardey et al. (2013) review IPR laws and practices in the US.

10.7.2. Seed law and regulations

Transporting or importing seeds or other propagules across national borders often involves national and local regulations that are in place to safeguard agriculture, human and animal health, and natural environments. Following these regulations is generally not difficult, but does usually require advance planning. In the US, the Department of Agriculture's Animal and Plant Health Inspection Service (APHIS) is responsible for enforcing regulations related to importation of plants and their products into the US. https://www.aphis.usda.gov/aphis/ourfocus/planthealth/import-information

The Federal Seed Act (FSA) is available on the US Government Publishing Office's Electronic Code of Federal Regulations website. https://tinyurl.com/US-FSA

See the website of the Association of American Seed Control Officials (AASCO), the US and Canadian organization for discussion of issues related to seed regulation in those countries, including the development of RUSSL (Box 6.1), and other interfaces between state, federal, and national policies. AASCO membership voted in 2016 in support of an amendment to RUSSL that would provide a model for recognizing and exempting non-commercial seed sharing from many of the regulations applying to commercial seed. http://www.seedcontrol.org/

The Association of Official Seed Certifying Agencies (AOSCA—yes, it is different than AASCO!) operating in the US, Canada, and a few other countries, is a non-profit agency that establishes seed quality standards together with national and regional public seed regulatory agencies (e.g., state crop improvement associations). AOSCA has an organic seed-finding site with listings of sources of certified organic seed, sorted by type of crop, although it may not be updated frequently. http://organicseedfinder.com

10.7.3. Biology, genetics, and reproduction

Detailed basic explanations of seed biology can be found on the website of Gergard Leubner's lab, University of London, along with current research being done there. http://www.seedbiology.eu/index.html

---

A clear, accessible handbook of basic genetic terms and concepts originally intended for natural resource managers working with native plant revegetation projects (Smith and Halbrook, 2004) is available from Steven Smith's University of Arizona website.
https://arizona.box.com/s/bogxolc5bvcge98gvlkm4wb93epevxsi

A valuable compendium of research about flower structure and insect pollination in crops (McGregor, 1976) is available free online, including some updates.
http://www.ars.usda.gov/sp2userfiles/place/20220500/onlinepollinationhandbook.pdf

10.7.4. Seed saving and storage

Some seed libraries, banks and companies sign the "Safe Seed Pledge" to inform the public that they do not carry transgenic crop varieties. The pledge was developed in 1999 by the Council for Responsible Genetics in Cambridge, Massachusetts.
http://www.councilforresponsiblegenetics.org/Projects/CurrentProject.aspx?projectId=17

The most comprehensive information about starting and running seed libraries, with links to relevant networks and newsletters and other information is the Richmond Grows Seed Lending Library's website.
http://www.richmondgrowsseeds.org/

Legal work on seed libraries in the US by the Sustainable Economies Law Center (SELC) in Oakland, California.
http://www.theselc.org/save_seed_sharing

There are a number of high quality publications produced by non-profit organic farming resource groups that are available free online, and these are a good place to look for more detailed information about seed saving and processing.

The Organic Seed Alliance (OSA) in Port Townsend, Washington is a research and education organization with active plant breeding programs for organic crops, and many publications including a 2010 seed saving guide for gardeners and farmers (Colley et al., 2010).
https://seedalliance.org/

A seed saving chart for gardeners and farmers assembled by Seed Savers Exchange and based on one originally created by the Organic Seed Alliance includes pollination type, pollinator, isolation distance, seed maturity relative to harvest, and "population size," meaning number of different plants seeds should be saved from. Presumably this is to maintain 95% of genetic diversity, but the numbers are not possible for most gardeners, a reason why gene flow will remain important, for example through a seed library. This is currently housed by the Community Seed Network.
https://www.communityseednetwork.org/page/seed-chart-guide

Saving Our Seeds focuses on mid-Atlantic and southeastern regions, but has publications useful for gardeners anywhere, including a seed saving and processing publication (McCormack, 2010).

An international leader in participatory plant breeding (PPB), barley breeder Salvatore Ceccarelli and colleagues have conducted extensive research on crop improvement with and for communities farming in challenging, variable environments. This manual is geared toward agricultural projects, and has valuable insights and methods from years of PPB (Ceccarelli, 2012).
https://tinyurl.com/Ceccarelli2012-PB-with-farmers

Carol Deppe's *Breed your own vegetable varieties* (2000) is a classic, now in its second edition. With a PhD in biology and years of hands-on independent plant breeding, she provides lots of good information and plant breeding stories. Deppe is an active contributor of varieties to the OSSI (Section 10.7.1).

Raoul Robinson was a Canadian plant breeder and champion of horizontal resistance. His book *Return to resistance* (Robinson, 2007) (also in Spanish) is a good overview of crop breeding for horizontal resistance, and is freely available online, along with other books of his, including about amateur plant breeding clubs.
https://tinyurl.com/R-Robinson-books

Breeding organic vegetables (White and Connolly, 2011) is an excellent resource with clear graphics and synopses of basic plant breeding methods addressed

to farmers. It may be hard to locate in print, but look in libraries.
https://eorganic.info/sites/eorganic.info/files/u2/BreedingOrganicVegetables-2011.pdf

An USDA report with details of many studies regarding seed storage (Justice and Bass, 1978) is also available free online.
http://naldc.nal.usda.gov/download/CAT87208646/PDF

10.8. References

Aguilar, J., Gramig, G. G., Hendrickson, J. R., Archer, D. W., Forcella, F. & Liebig, M. A. (2015) Crop species diversity changes in the United States: 1978–2012. *PLOS ONE*, 10, e0136580, DOI: 10.1371/journal.pone.0136580.

Aragón-Cuevas, F. (2015) Mexico: Community seed banks in Oaxaca. In Vernooy, R., Shrestha, P. & Sthapit, B. (eds) *Community seed banks: Origins, evolution and prospects*, 136–139. Routledge, New York, NY.

Ballhorn, D., Heil, M. & Lieberei, R. (2006) Phenotypic plasticity of cyanogenesis in lima bean *Phaseolus lunatus*—Activity and activation of β-glucosidase. *Journal of Chemical Ecology*, 32, 261–275, DOI: 10.1007/s10886-005-9001-z.

Boulé, D. (California Forum) (2016) Navel orange produced a big bang in the Golden State. *Sacramento Bee*, January 2.

Brouwer, B. O., Murphy, K. M. & Jones, S. S. (2016) Plant breeding for local food systems: A contextual review of end-use selection for small grains and dry beans in Western Washington. *Renewable Agriculture and Food Systems*, 31, 172–184.

Brown, A. H. D. (1999) The genetic structure of crop landraces and the challenge to conserve them *in situ* on farms. In Brush, S. B. (ed.) *Genes in the field: On-farm conservation of crop diversity*, 29–48. Lewis Publishers; IPGRI; IDRC, Boca Raton, FL, Rome, Ottawa.

Ceccarelli, S. (1996) Adaptation to low/high input cultivation. *Euphytica*, 92, 203–214.

Ceccarelli, S. (2012) *Plant breeding with farmers: A technical manual*. International Center for Agricultural Research in the Dry Areas, Aleppo, Syria.

Ceccarelli, S. & Grando, S. (2009) Participatory plant breeding. In Carena, M. J. (ed.) *Cereals 3. Handbook of plant breeding*, 395–414, DOI: 10.1007/978-0-387-72297-9_13.

Chidananda, K. P., Chelladurai, V., Jayas, D. S., Alagusundaram, K., White, N. D. G. & Fields, P. G. (2014) Respiration of pulses stored under different storage conditions. *Journal of Stored Products Research*, 59, 42–47, DOI: http://dx.doi.org/10.1016/j.jspr.2014.04.006.

Cleveland, D. A. (2014) *Balancing on a planet: The future of food and agriculture*. University of California Press, Berkeley, CA.

Cleveland, D. A. & Soleri, D. (2007) Extending Darwin's analogy: Bridging differences in concepts of selection between farmers, biologists, and plant breeders. *Economic Botany*, 61, 121–136, DOI: 10.1663/0013-0001(2007)61[121:EDABDI]2.0.CO;2.

Colley, M., Navazio, J. & DiPietro, L. (2010) *A seed saving guide for gardeners and farmers*. Organic Seed Alliance, Port Townsend, WA, available at: https://seedalliance.org/publications/seed-saving-guide-gardeners-farmers/ (accessed Jan. 18, 2019).

Dadson, R., Hashem, F., Javaid, I., Joshi, J., Allen, A. & Devine, T. (2005) Effect of water stress on the yield of cowpea (*Vigna unguiculata* L. Walp.) genotypes in the Delmarva region of the United States. *Journal of Agronomy and Crop Science*, 191, 210–217.

Davis, D. R. (2009) Declining fruit and vegetable nutrient composition: what is the evidence? *HortScience*, 44, 15–19.

Deppe, C. (384) (2000) *Breed your own vegetable varieties*, 2nd edn. Chelsea Green Publishing, White River Junction, VT.

Doebley, J. F., Gaut, B. S. & Smith, B. D. (2006) The molecular genetics of crop domestication. *Cell*, 127, 1309–1321.

Duggan, T. (2014) New "Girl" is a Monsanto-free tomato. *San Francisco Chronicle*, 2014 November 24, available at: http://www.sfgate.com/homeandgarden/article/New-Girl-is-a-Monsanto-free-tomato-5295723.php (accessed Oct. 11, 2018).

Dyer, G. A., Serratos-Hernández, J. A., Perales, H. R., Gepts, P., Piñeyro-Nelson, A., Chávez, A., Salinas-Arreortua, N., Yúnez-Naude, A., Taylor, E. & Alvarez-Buylla, E. R. (2009) Dispersal of transgenes through maize seed systems in Mexico. *PLoS ONE* 4, e5734, DOI:10.1371/journal.pone.0005734.

Ellstrand, N. C. (2014) Is gene flow the most important evolutionary force in plants? *American Journal of Botany*, 101, 737–753, DOI: 10.3732/ajb.1400024.

Ellstrand, N. C. & Elam, D. R. (1993) Population genetic consequences of small population size: Implications for plant conservation. *Annual Review of Ecology and Systematics*, 24, 217–242.

Endler, J. A. (1992) Natural selection: Current usages. In Keller, E. F. & Lloyd, E. A. (eds) *Key words in evolutionary biology*, 20–224. Harvard University Press, Cambridge, MA.

Esteva, G. & Marielle, C. (eds) (2003) *Sin maíz no hay país*. Consejo Nacional para la Cultura y las Artes, Dirección General de Culturas Populares e Indígenas, México, DF.

Foley, J. A., Ramankutty, N., Brauman, K. A., Cassidy, E. S., Gerber, J. S., Johnston, M., Mueller, N. D., O'Connell, C., Ray, D. K., West, P. C., et al. (2011)

Solutions for a cultivated planet. *Nature*, 478, 337–342, DOI: 10.1038/nature10452.

Fowler, C. & Mooney, P. (1990) *Shattering: Food, politics, and the loss of genetic diversity*. University of Arizona Press, Tucson, AZ.

Frankel, O. H., Brown, A. H. D. & Burdon, J. J. (1995) *The conservation of plant biodiversity*. Cambridge University Press, Cambridge, UK.

Gepts, P. (2001) Origins of plant agriculture and major crop plants. In *Our fragile world: Challenges and opportunities for sustainable development*, 629–637t. EOLSS Publishers, Oxford, UK.

Grünwald, N. J., Cadena Hinojosa, M. A., Covarrubias, O. R., Peña, A. R., Niederhauser, J. S. & Fry, W. E. (2002) Potato cultivars from the Mexican National Program: sources and durability of resistance against late blight. *Phytopathology*, 92, 688–693, DOI: 10.1094/PHYTO.2002.92.7.688.

Harrington, J. F. (1960) Thumb rules of drying seeds. *Crops and Soils*, 13, 16–17.

Hartmann, H. T., Kester, D. E., Davies, F. T., Jr. & Geneve, R. L. (2002) *Plant propagation: Principles and practices*, 7th edn. Prentice Hall, Upper Saddle River, NJ.

Heald, P. J. & Chapman, S. (2009) Crop diversity report card for the twentieth century: diversity bust or diversity boom? *SSRN*, available at: https://papers.ssrn.com/sol3/papers.cfm?abstract_id=1462917 (accessed July 31, 2018), DOI: 10.2139/ssrn.1462917.

Howard, P. H. (2009) Visualizing food system concentration and consolidation. *Southern Rural Sociology*, 24, 87–110.

Howard, P. H. (2015) Intellectual property and consolidation in the seed industry. *Crop Science*, 55, 2489–2495.

Ibarra-Perez, F. J., Ehdaie, B. & Waines, J. G. (1997) Estimation of outcrossing rate in common bean. *Crop Science*, 37, 60–65.

Jaganathan, D., Ramasamy, K., Sellamuthu, G., Jayabalan, S. & Venkataraman, G. (2018) CRISPR for crop improvement: An update review. *Frontiers in Plant Science*, 9, 985, DOI: 10.3389/fpls.2018.00985.

Janick, J. (2005) The origins of fruit, fruit growing, and fruit breeding. *Plant Breeding Reviews*, 25, 255–321.

Jarvis, A., Ramirez-Villegas, J., Nelson, V., Lamboll, R., Nathaniels, N., Radeny, M., Mungai, C., Bonilla-Findji, O., Arango, D. & Peterson, C. (2014) *Farms of the future: An innovative approach for strengthening adaptive capacity*, 119–124. AISA Workshop on Agricultural Innovation Systems in Africa (AISA), 29–31 May 2013, Nairobi, Kenya, available at: http://www.farmaf.org/images/documents/related_materials/AISA_workshop_proceedings_final__March_2014.pdf (accessed Oct. 11, 2018).

Jarvis, D. I., Brown, A. H. D., Cuong, P. H., Collado-Panduro, L., Latournerie-Moreno, L., Gyawali, S., Tanto, T., Sawadogo, M., Mar, I., Sadiki, M., et al. (2008) A global perspective of the richness and evenness of traditional crop-variety diversity maintained by farming communities. *Proceedings of the National Academy of Sciences of the United States of America*, 105, 5326–5331.

Justice, O. L. & Bass, L. N. (1978) *Principles and practices of seed storage*. USDA, Washington, DC, available at: http://naldc.nal.usda.gov/download/CAT87208646/PDF (accessed Oct. 11, 2018)

Keane, P. (2012) Horizontal or generalized resistance to pathogens in plants. In Cumagun, C. J. (ed.) *Plant pathology*, 327–362. InTech, DOI: 10.5772/30763.

Kislev, M. E., Hartmann, A. & Bar-Yosef, O. (2006) Early domesticated fig in the Jordan Valley. *Science*, 312, 1372–1374, DOI: 10.1126/science.1125910.

Kloppenburg, J. (2005) *First the seed: The political economy of plant biotechnology*, 2nd edn. University of Wisconsin Press, WI.

Kloppenburg, J. (2014) Re-purposing the master's tools: The open source seed initiative and the struggle for seed sovereignty. *Journal of Peasant Studies*, 1–22.

Knispel, A. L., McLachlan, S. M., Van Acker, R. C. & Friesen, L. F. (2008) Gene flow and multiple herbicide resistance in escaped canola populations, *Weed Science*, 56, 72–80.

Lacy, S., Cleveland, D. A. & Soleri, D. (2006) Farmer choice of sorghum varieties in southern Mali. *Human Ecology*, 34, 331–353, DOI: 10.1007/s10745-006-9021-5.

Larson, G., Piperno, D. R., Allaby, R. G., Purugganan, M. D., Andersson, L., Arroyo-Kalin, M., Barton, L., Climer Vigueira, C., Denham, T., Dobney, K., et al. (2014) Current perspectives and the future of domestication studies. *Proceedings of the National Academy of Sciences*, 111, 6139–6146. DOI: 10.1073/pnas.1323964111.

Luedeling, E., Zhang, M. & Girvetz, E. H. (2009) Climatic changes lead to declining winter chill for fruit and nut trees in California during 1950–2099. *PLoS One*, 4, e6166.

McCormack, J. H. (2010) Seed storage and processing. Saving Our Seeds, Charlottesville, VA, available at: http://www.savingourseeds.org/pubs/seed_processing_storage_ver_1pt6.pdf (accessed June 21, 2018).

McCouch, S., Baute, G. J., Bradeen, J., Bramel, P., Bretting, P. K., Buckler, E., Burke, J. M., Charest, D., Cloutier, S., Cole, G., et al. (2013) Agriculture: Feeding the future. *Nature*, 499, 23–24, DOI: 10.1038/499023a.

McGregor, S. E. (1976) *Insect pollination of cultivated crop plants, agricultural handbook No. 496*. Agricultural Research Service, US Department of Agriculture, Washington, DC.

Ménard, L., McKey, D., Mühlen, G. S., Clair, B. & Rowe, N. P. (2013) The evolutionary fate of phenotypic plasticity and functional traits under domestication in manioc: Changes in stem biomechanics and the appearance of stem brittleness. *PLoS ONE*, 8, e74727, DOI: 10.1371/journal.pone.0074727.

Meyer, R. S., DuVal, A. E. & Jensen, H. R. (2012) Patterns and processes in crop domestication: An historical review and quantitative analysis of 203 global food crops. *New Phytologist*, 196, 29–48, DOI: 10.1111/j.1469-8137.2012.04253.x.

Monsanto (2015) *2015 Annual Report*, available at: https://monsanto.com/app/uploads/2017/05/2015_Annual_Report_FullWeb.pdf (accessed June 23, 2018).

Nabhan, G. P. & Rea, A. (1987) Plant domestication and folk-biological change: The Upper Piman/devil's claw example. *American Anthropologist*, 89, 57–73.

NRC (National Research Council of the National Academies) (1992) *Neem: A tree for solving global problems*. NRC, Washington, DC.

Pardey, P., Koo, B., Drew, J., Horwich, J. & Nottenburg, C. (2013) The evolving landscape of plant varietal rights in the United States, 1930–2008. *Nature Biotechnology*, 31, 25–29, DOI: http://www.nature.com/nbt/journal/v31/n1/abs/nbt.2467.html - supplementary-information.

Pope, K. O., Pohl, M. E. D., Jones, J. G., Lentz, D. L., von Nagy, C., Vega, F. J. & Quitmyer, I. R. (2001) Origin and environmental setting of ancient agriculture in the lowlands of Mesoamerica. *Science*, 292, 370–374.

Rao, V. R. (1998) Strategies for collecting of tropical fruit species germplasm. In Arora, R. K. & Rao, V. R. (eds) *Tropical fruits in Asia: Diversity, maintenance, conservation and use*, 18–31. Proceedings of the IPGRI-ICAR-UTFANET Regional Training Course on the Conservation and Use of Germplasm of Tropical Fruits in Asia held at Indian Institute of Horticultural Research, May 18–31, 1997, IPGRI.

Robinson, R. A. (2007) *Return to resistance: Breeding crops to reduce pesticide dependence*, 3rd edn. Sharebooks Publishing, Fergus, Ontario, CA.

Robinson, R. A. (2009) Breeding for quantitative variables. Part 2: Breeding for durable resistance to crop pests and diseases. In Ceccarelli, S., Weltzien, E. & Guimares, E. (eds) *Plant breeding and farmer participation*, 367–390. FAO, in collaboration with ICARDA and ICRISAT Rome, Italy.

Sin maíz no hay país (2018) Campaña nacional en defensa de la soberanía alimentaria y la reactivación del campo mexicano, available at: http://www.sinmaiznohaypais.org/ (accessed June 26, 2018).

Smith, S. E. & Halbrook, K. (2004) A plant genetics primer. *Native Plants Journal*, 5, 105–111.

Soleri, D. (2017) Civic seeds: New institutions for seed systems and communities—a 2016 survey of California seed libraries. *Agriculture and Human Values*, DOI: 10.1007/s10460-017-9826-4.

Soleri, D. & Smith, S. E. (1995) Morphological and phenological comparisons of two Hopi maize varieties conserved in situ and ex situ. *Economic Botany*, 49, 56–77, DOI: 10.1007/BF02862278.

Soleri, D., Smith, S. E. & Cleveland, D. A. (2000) Evaluating the potential for farmer and plant breeder collaboration: A case study of farmer maize selection in Oaxaca, Mexico. *Euphytica*, 116, 41–57, DOI: 10.1023/A:1004093916939.

Sperling, L., Loevinsohn, M. E. & Ntabomvura, B. (1993) Rethinking the farmer's role in plant breeding: Local bean experts and on-station selection in Rwanda. *Experimental Agriculture*, 29, 509–519.

Tin, H., Berg, T. & Bjørnstad, Å. (2001) Diversity and adaptation in rice varieties under static (*ex situ*) and dynamic (*in situ*) management. *Euphytica*, 122, 491–502.

Van Brunt, J. (1985) Ex parte Hibberd: Another landmark decision. *Nature Bio/Technology*, 3, 1059. DOI: 10.1038/nbt1285-1059.

Van Der Plank, J. E. (1966) Horizontal (polygenic) and vertical (oligogenic) resistance against blight. *American Potato Journal*, 43, 43–52.

Van Der Plank, J. E. (1984) *Disease resistance in plants*, 2nd edn. Academic Press, Orlando, FL.

Wehner, T. C. (2008) Watermelon. In Prohens, J. & Nuez, F. (eds) *Vegetables I: Asteraceae, Brassicaceae, Chenopodicaceae, and Cucurbitaceae*, 381–418. Springer, New York, NY, DOI: 10.1007/978-0-387-30443-4_12.

White, H. & Connolly, B. (2011) *Breeding organic vegetables. A step-by-step guide for growers*, Northeast Organic Farming Association of New York, NY, available at: https://eorganic.info/sites/eorganic.info/files/u2/BreedingOrganicVegetables-2011.pdf (accessed June 24, 2018).

Index

Page numbers in **bold** type refer to figures, tables and boxed text.

controls, experimental **88**, 92, **94**
Cooperative Extension Services 170, 194, 195
coping strategies 67, 71–72, 226
correlation coefficients **47–48**
cortical sloughing, roots 248, **248**
cottony-cushion scale *(Icerya purchasi)* 239, 264
cotyledons 126, 150, 163, 290
CRISPR gene editing 285
Croptime calculator (Oregon) 142, 144
cross-pollinated plants **288–289**, 290, **293**
cultivation of soils 178–179, 186, 189
cultural identity and rights 25, **60**, 113
curative responses 227–228
cuticle
    of insects 246, 297
    of plants 138, 235
cuttings 159, **167**, 243
cutworm collars, for seedlings 205, 243, **244**

damage threshold 235
damping off, seedlings 241, 250, **251**
dark green leafy vegetables 15
Darwin, Charles 5
data collection 77, 82, 92–93
data interpretation
    analysis of formal experiments 93–97, **94**, **95**, **96**
    trendlines ('best-fit') **47–48**
daylength requirements 139–140
deciduous perennials 128, 142, 164, 206
decision making
    choice of plants to grow 149, **150**, 155, 234–236
    community participation 31, **80**
    criteria in corporate systems 285, 287
    crop choice influenced by likely conditions 207,
        209, 236
    organic waste management 190, 192, 194
    rainfall probability, estimate using data 211,
        212–213
deficit irrigation 202, 204–205, **205**, 207
demographic trends, US population 55–56, **60**
dependent variables 86, 88, 90
depth of planting **158**, **160**, 165
descriptive observations 76, **77**, **94**
determinate growth 128
dew point 111
diagnosis
    goals, in food gardening 226, 228–229
    observational skills and methods 247–249
    specific symptoms and causes 253–263, **254**
diet quality, health impacts 13–22
dietary fiber 16, 19
dioecious plants **129**, 130, **130**, 166
diseases 225, 249–250, 264
    checking transplants for 164
    human, from fungal pathogens **242**
    pathogen identification 247

seed-borne, control methods 150
spread prevention in vegetative propagation
    167, 243
diversity
    crops, types and varieties 138–139, 155–157, 227,
        273–276, **274**
    effects on natural biological control 236, 240
    microbial, in soils 193, 240–241
    strategic management 238–239
    see also genetic diversity
domestication, crops 149, 234, 272, **273**, **274**
dormancy
    in plant growth (buds), *see* vernalization
    seeds 153–154, **154**, **273**
drainage
    cold air 111–112
    properties of soils 132, 177–178
    water 112, 200, **201**, 220
drip irrigation systems 159, **209**, 216–217, 221
drip line, roots 248, **249**
drought
    increased probability with climate change 52, 206
    severe events, IPCC predictions **49**
drought responses **137**
    drought-adapted plant types 110, 137–138,
        207, **207**
    drought-hardened transplants 164
    evolutionary adaptations of stomata 136–137
    leaf loss (drought deciduous plants) 128, 138, 165
    in urban communities 118–119
    water stress 126, 137, 206
drying, seeds **295**, 296

Earth
    atmospheric characteristics **105**, 111, **133**
    axial tilt **102**, 102–103, 106, 108
    evolution of life on 132, 179
ecological thinking
    as approach to food gardening 3–5, **4**,
        27, 185, 190
    drought/heat adaptation management 138
    pest and disease problem strategies 226–229, **227**
    responses to resource scarcity 48, **162**
    source/sink environments 53, 55
    urban contexts 189–190
economics
    consequences of short-term focus 48–49
    economic benefits of food gardens 27–30, 56
    neoliberal policies 30
    structural inequity and poverty **54, 59**
ecosystem services (ES) 26–27
edge effects 92, **92**
emergence
    heterotrophic growth period 163
    problems, and improvement 157, 159, **159**, **160**
    seedling quality and vigor 150, 155

CABI – who we are and what we do

This book is published by **CABI**, an international not-for-profit organisation that improves people's lives worldwide by providing information and applying scientific expertise to solve problems in agriculture and the environment.

CABI is also a global publisher producing key scientific publications, including world renowned databases, as well as compendia, books, ebooks and full text electronic resources. We publish content in a wide range of subject areas including: agriculture and crop science / animal and veterinary sciences / ecology and conservation / environmental science / horticulture and plant sciences / human health, food science and nutrition / international development / leisure and tourism.

The profits from CABI's publishing activities enable us to work with farming communities around the world, supporting them as they battle with poor soil, invasive species and pests and diseases, to improve their livelihoods and help provide food for an ever growing population.

CABI is an international intergovernmental organisation, and we gratefully acknowledge the core financial support from our member countries (and lead agencies) including:

Discover more

To read more about CABI's work, please visit: **www.cabi.org**

Browse our books at: **www.cabi.org/bookshop**, or explore our online products at: **www.cabi.org/publishing-products**

Interested in writing for CABI? Find our author guidelines here: **www.cabi.org/publishing-products/information-for-authors/**